T0178223

Lecture Notes in Computer Science 13987

The series Lecture Notes in Computer Science (LNCS), including its subseries Lecture Notes in Artificial Intelligence (LNAI) and Lecture Notes in Bioinformatics (LNBI), has established itself as a medium for the publication of new developments in computer science and information technology research, teaching, and education.

LNCS enjoys close cooperation with the computer science R & D community, the series counts many renowned academics among its volume editors and paper authors, and collaborates with prestigious societies. Its mission is to serve this international community by providing an invaluable service, mainly focused on the publication of conference and workshop proceedings and postproceedings. LNCS commenced publication in 1973.

Leslie Pérez Cáceres · Thomas Stützle

Editors

Evolutionary Computation in Combinatorial Optimization

23rd European Conference, EvoCOP 2023
Held as Part of EvoStar 2023
Brno, Czech Republic, April 12–14, 2023
Proceedings

 Springer

Editors
Leslie Pérez Cáceres 🄳
Pontificia Universidad Católica de Valparaíso
Valparaíso, Chile

Thomas Stützle 🄳
Université libre de Bruxelles
Bruxelles, Belgium

ISSN 0302-9743   ISSN 1611-3349 (electronic)
Lecture Notes in Computer Science
ISBN 978-3-031-30034-9   ISBN 978-3-031-30035-6 (eBook)
https://doi.org/10.1007/978-3-031-30035-6

This Springer imprint is published by the registered company Springer Nature Switzerland AG
The registered company address is: Gewerbestrasse 11, 6330 Cham, Switzerland

Preface

Metaheuristics and, in particular, evolutionary computation have a history of half a century in which they have provided effective tools for solving complex computational problems. These techniques are essential and fundamental tools to approach problems for which traditional optimization methods face difficulties providing good solutions within reasonable time. These techniques are general-purpose and highly adaptable tools which are commonly inspired by processes observed in the natural world, like natural selection and evolution. The active research in the field is constantly broadening its applications, improving search processes, and extending the techniques' power by combining them with other techniques like machine learning. The research work in these fields is crucial to approach increasingly complex optimization problems that our industrial and scientific development constantly generates. Such complex problems are often combinatorial in nature and describe massive solution spaces, for which finding good solutions challenges the limits of our technology. Combinatorial optimization problems are central to improving systems and processes; consequently, they have and will continue to play a relevant role in industrial and technological development. Metaheuristics and evolutionary computation techniques are currently state of the art techniques when dealing with the most challenging combinatorial optimization problems, and thus the research carried out in these fields has a wide spectrum of applications. The articles in this volume showcase recent theoretical and experimental advances in combinatorial optimization, evolutionary algorithms, metaheuristics, and related research fields.

This volume contains the proceedings of EvoCOP 2023, the 23rd European Conference on Evolutionary Computation in Combinatorial Optimisation. The conference was held in the lovely city of Brno, Czech Republic during April 12–14, 2023. The EvoCOP conference series started in 2001, with the first workshop specifically devoted to evolutionary computation in combinatorial optimization, and it became an annual conference in 2004. EvoCOP 2023 was organized together with EuroGP (the 26th European Conference on Genetic Programming), EvoMUSART (the 12th International Conference on Artificial Intelligence in Music, Sound, Art and Design), and EvoApplications (the 26th European Conference on the Applications of Evolutionary Computation, formerly known as EvoWorkshops), in a joint event collectively known as EvoStar 2023. Previous EvoCOP proceedings were published by Springer in the *Lecture Notes in Computer Science* series (LNCS volumes 2037, 2279, 2611, 3004, 3448, 3906, 4446, 4972, 5482, 6022, 6622, 7245, 7832, 8600, 9026, 9595, 10197, 10782, 11452, 12102, 12692 and 13222). The table on the next page reports the statistics for each of the previous conferences.

This year, 15 out of 32 papers were accepted after a rigorous double-blind process, resulting in a 46% acceptance rate. We would like to acknowledge the quality and timeliness of our high-quality and diverse Program Committee members' work. Each year the members give freely of their time and expertise, in order to maintain the high standards in EvoCOP and provide constructive feedback to help authors improve their papers. Decisions considered both the reviewers' report and the evaluation of the program chairs. The 15 accepted papers cover a variety of topics, ranging from the foundations of evolutionary computation algorithms and other search heuristics, to their accurate design and application to both single- and multi-objective combinatorial optimization problems. Fundamental and methodological aspects deal with runtime analysis, the structural properties of fitness landscapes, the study of metaheuristics core components, the clever design of their search principles, and their careful selection and configuration by means of hyper-heuristics. Applications cover problem domains such as routing, permutation problems and general graph problems. We believe that the range of topics covered in this volume reflects the current state of research in the fields of metaheuristics and combinatorial optimization.

EvoCOP	LNCS vol.	Submitted	Accepted	Acceptance (%)
2023	13987	32	15	46.8
2022	13222	28	13	46.4
2021	12692	42	14	33.3
2020	12102	37	14	37.8
2019	11452	37	14	37.8
2018	10782	37	12	32.4
2017	10197	39	16	41.0
2016	9595	44	17	38.6
2015	9026	46	19	41.3
2014	8600	42	20	47.6
2013	7832	50	23	46.0
2012	7245	48	22	45.8
2011	6622	42	22	52.4
2010	6022	69	24	34.8
2009	5482	53	21	39.6
2008	4972	69	24	34.8
2007	4446	81	21	25.9
2006	3906	77	24	31.2
2005	3448	66	24	36.4
2004	3004	86	23	26.7
2003	2611	39	19	48.7
2002	2279	32	18	56.3
2001	2037	31	23	74.2

We would like to express our appreciation to the various persons and institutions who made EvoCOP 2023 a successful event. Firstly, we thank the local organization team, led by Jiri Jaros and Lukas Sekanina from the Brno University of Technology in the Czech Republic. Our acknowledgments also go to SPECIES, the Society for the Promotion of Evolutionary Computation in Europe and its Surroundings. We extend our acknowledgments to Nuno Lourenço from the University of Coimbra, Portugal, for his dedicated work with the submission and registration system, to João Correia from the University of Coimbra, Portugal, for the EvoStar publicity and social media service, to Francisco Chicano from the University of Málaga, Spain, for managing the EvoStar website, and to Sérgio Rebelo and Tiago Martins from the University of Coimbra, Portugal, for their important graphic design work. We wish to thank our prominent keynote speakers, Marek Vácha and Evelyne Lutton. Finally, we express our appreciation to Anna I. Esparcia-Alcázar from SPECIES, Europe, whose considerable efforts in managing and coordinating *EvoStar* helped towards building a unique, vibrant and friendly atmosphere.

Special thanks also to Christian Blum, Francisco Chicano, Carlos Cotta, Peter Cowling, Jens Gottlieb, Jin-Kao Hao, Jano van Hemert, Bin Hu, Arnaud Liefooghe, Manuel Lopéz-Ibáñez, Peter Merz, Martin Middendorf, Gabriela Ochoa, Luís Paquete, Günther R. Raidl, Sébestien Verel and Christine Zarges for their hard work and dedication at past editions of EvoCOP, making this one of the reference international events in evolutionary computation and metaheuristics.

April 2023

Leslie Pérez Cáceres
Thomas Stützle

Organization

EvoCOP 2023 was organized as a part of EvoStar 2023, jointly with EuroGP 2023, EvoMUSART 2023, and EvoApplications 2023.

Organizing Committee

Conference Chairs

Leslie Pérez Cáceres   Pontificia Universidad Católica de Valparaíso, Valparaíso, Chile

Thomas Stützle   Université libre de Bruxelles, Brussels, Belgium

Local Organization

Jiri Jaros   Brno University of Technology, Czech Republic

Lukas Sekanina   Brno University of Technology, Czech Republic

Publicity Chair

João Correia   University of Coimbra, Portugal

EvoStar Coordinator

Anna Esparcia-Alcázar   Universitat Politècnica de València, Spain

EvoCOP Steering Committee

Christian Blum   Artificial Intelligence Research Institute (IIIA-CSIC), Spain

Francisco Chicano   University of Málaga, Spain

Peter Cowling   Queen Mary University of London, UK

Jens Gottlieb   SAP AG, Germany

Jin-Kao Hao   University of Angers, France

Bin Hu   AIT Austrian Institute of Technology, Austria

Arnaud Liefooghe   University of Lille, France

Manuel Lopéz-Ibáñez   University of Manchester, UK

Martin Middendorf	University of Leipzig, Germany
Gabriela Ochoa	University of Stirling, UK
Luís Paquete	University of Coimbra, Portugal
Günther Raidl	Vienna University of Technology, Austria
Jano van Hemert	Optos, UK
Sébastien Verel	Université du Littoral Cote d'Opale, France
Christine Zarges	Aberystwyth University, UK

Society for the Promotion of Evolutionary Computation in Europe and Its Surroundings (SPECIES)

Penousal Machado (President)
Mario Giacobini (Secretary)
Francisco Chicano (Treasurer)

Program Committee

Richard Allmendinger	University of Manchester, UK
Matthieu Basseur	Université du Littoral Côte d'Opale, France
Christian Blum	Artificial Intelligence Research Institute (IIIA-CSIC), Spain
Alexander Brownlee	University of Stirling, UK
Maxim Buzdalov	ITMO University, Russia
Arina Buzdalova	ITMO University, Russia
Christian Camacho-Villalón	Université libre de Bruxelles, Belgium
Josu Ceberio	University of the Basque Country, Spain
Marco Chiarandini	University of Southern Denmark, Denmark
Francisco Chicano	University of Málaga, Spain
Carlos Coello Coello	CINVESTAV-IPN, Mexico
Carlos Cotta	Universidad de Málaga, Spain
Nguyen Dang	University of St Andrews, UK
Bilel Derbel	University of Lille, France
Marcos Diez García	Fujitsu Research of Europe Limited, UK
Karl Doerner	University of Vienna, Austria
Benjamin Doerr	Ecole Polytechnique, France
Carola Doerr	CNRS and Sorbonne University, France
Talbi El-Ghazali	University of Lille, France
Jonathan Fieldsend	University of Exeter, UK
Carlos M. Fonseca	University of Coimbra, Portugal
Alberto Franzin	Université libre de Bruxelles, Belgium

Bernd Freisleben	Philipps-Universität Marburg, Germany
Carlos Garcia-Martinez	University of Córdoba, Spain
Adrien Goeffon	University of Angers, France
Andreia Guerreiro	University of Lisbon, Portugal
Jin-Kao Hao	University of Angers, France
Geir Hasle	SINTEF Digital, Norway
Mario Inostroza-Ponta	Universidad de Santiago de Chile, Chile
Ekhine Irurozki	Telecom Paris, France
Thomas Jansen	Aberystwyth University, UK
Andrzej Jaszkiewicz	Poznan University of Technology, Poland
Marie-Eleonore Kessaci	Université de Lille, France
Ahmed Kheiri	Lancaster University, UK
Frederic Lardeux	University of Angers, France
Rhydian Lewis	Cardiff University, UK
Arnaud Liefooghe	University of Lille, France
Manuel López-Ibáñez	University of Manchester, UK
Jose A. Lozano	University of the Basque Country, Spain
Gabriel Luque	University of Málaga, Spain
Krzysztof Michalak	Wroclaw University of Economics and Business, Poland
Elizabeth Montero	Universidad Nacional Andrés Bello, Chile
Nysret Musliu	Vienna University of Technology, Austria
Gabriela Ochoa	University of Stirling, UK
Pietro Oliveto	University of Sheffield, UK
Beatrice Ombuki-Berman	Brock University, Canada
Luís Paquete	University of Coimbra, Portugal
Mario Pavone	University of Catania, Italy
Paola Pellegrini	French Institute of Science and Technology for Transport, France
Francisco B. Pereira	Polytechnic Institute of Coimbra, Portugal
Pedro Pinacho	Universidad de Concepción, Chile
Daniel Porumbel	Conservatoire National des Arts et Métiers, France
Jakob Puchinger	EM Normandie Business School, France
Abraham Punnen	Simon Fraser University, Canada
Günther Raidl	Vienna University of Technology, Austria
María Cristina Riff	Universidad Técnica Federico Santa María, Chile
Marcus Ritt	Universidade Federal do Rio Grande do Sul, Brazil
Eduardo Rodriguez-Tello	CINVESTAV – Tamaulipas, Mexico
Andrea Roli	Universitá di Bologna, Italy
Hana Rudová	Masaryk University, Czech Republic

Contents

Fairer Comparisons for Travelling Salesman Problem Solutions Using Hash Functions

Mehdi El Krari[1]([✉])[iD], Rym Nesrine Guibadj[2][iD], John Woodward[3][iD], and Denis Robilliard[2][iD]

[1] Computational Optimisation and Learning Lab, School of Computer Science, University of Nottingham, Nottingham, UK
mehdi@elkrari.com
[2] Université du Littoral Côte d'Opale, EA 4491 - LISIC, Calais, France
{rym.guibadj,denis.robilliard}@univ-littoral.fr
[3] Operational Research Group, School of Electronic Engineering and Computer Science, Queen Mary University of London, Mile End Road, London E1 4NS, UK
j.woodward@qmul.ac.uk

Abstract. Fitness functions fail to differentiate between different solutions with the same fitness, and this lack of ability to distinguish between solutions can have a detrimental effect on the search process. We investigate, for the Travelling Salesman Problem (TSP), the impact of using a hash function to differentiate solutions during the search process. Whereas this work is not intended to improve the state-of-the-art of the TSP solvers, it nevertheless reveals a positive effect when the hash function is used.

Keywords: Hash functions · Combinatorial Problems · Travelling Salesman Problem · Local Search · Genetic Algorithms · Memetic Algorithms

1 Introduction

The way a solution of a Combinatorial Optimisation Problem (COP) can be represented is a key issue to design an efficient search algorithm to solve it. A representation associates an encoding, that can be easily evaluated during the search algorithm. For example, if we consider the Travelling Salesman Problem (TSP), a solution to this problem is a tour in which all the cities are listed in the order they are visited, and each city is visited only once. This solution can be represented using different encodings [8,16]: binary, graphs, permutations etc. In the permutation representation of the TSP, this is interpreted as a sequence of cities in which the first and the last elements are connected. The cost of a sequence depends on the order of the cities in the permutation.

The permutation is used as a solution encoding in many other COPs. They can be found in various application areas such as assignment problems [11,14,17], scheduling problems [2], routing problems [17], etc.

© The Author(s), under exclusive license to Springer Nature Switzerland AG 2023
L. Pérez Cáceres and T. Stützle (Eds.): EvoCOP 2023, LNCS 13987, pp. 1–15, 2023.
https://doi.org/10.1007/978-3-031-30035-6_1

An evaluation function, that associates a fitness measure to each solution, should be defined in order to (i) distinguish two solutions based on their quality and (ii) guide the search process. Often these two purposes are expected to be met by a single function [1].

The mapping from solutions space to fitness values may belong to one of the following cases:

- 1-to-1 mapping: each fitness value is associated to only one solution
- n-to-1 mapping: several different solutions have the same fitness value

While a canonical form was proposed in genetic programming by Woodward [21], it still not evident to find it for many COPs. It is common for a fitness function to map different solutions to the same fitness value. This means the metaheuristic cannot distinguish solutions based solely on their fitness values, and this loss of information may impede the search ability of the metaheuritics.

When the search space, i.e. the set of all the feasible solutions for a given instance, has many solutions with the same fitness value, this often results in large regions containing *plateaus*. A metaheuristic may repeatedly return to recently visited solutions as it wanders around the plateau as the fitness function does not provide any helpful information. We say a cycle occurs when the search process returns to an already visited solution again. This term is mentioned in the literature [3,9] to describe the same phenomena. The problem of *cycling* may lead the metaheuristic to be confined to a particular area of the search space.

Another issue arising with population-based search techniques, such as Genetic Algorithms, is the premature convergence of the metaheuristic when different solutions have the same fitness in the last generations. Indeed, convergence measures are mainly based on population diversity to terminate the evolution. Usually, the diversity of the population is measured by assessing the similarity among solutions based on their fitness. One of the definitions of convergence in an evolutionary process is when a certain percentage of the population has the same fitness, thus indicating that the evolutionary process has stagnated [12].

Differentiating solutions by the mean of their respective fitness values is motivated by the low complexity induced by the comparison. It can even be constant ($\mathcal{O}(1)$) for some COPs, such as the TSP. On the other hand, differentiating solutions by their respective encoding (permutations, binary strings, etc.) is entirely accurate but more expensive. Comparing two permutations, for example, is linear ($\mathcal{O}(n)$), which can likely increase the complexity of the whole metaheuristic from $\mathcal{O}(n^k)$ to $\mathcal{O}(n^{k+1})$.

We introduced in a previous short paper [5] a new hash function for the TSP. In this study, we show its positive effect to provide relevant information during the search process. Experiments are conducted on the TSP but point to possible use on other COPs. Three metaheuristics are analysed: Iterated Local Search (ILS), Genetic Algorithms (GAs) and Memetic Algorithms (MAs). In this paper, we refer to solutions comparison as the differentiation mechanism to distinguish between two solutions.

The remainder of this paper is organised as follows. A formal definition of TSP is provided in the next section, with an analysis of the fitness values distribution of some TSP instances over the search space. Section 3 introduces a new hash function designed for TSP permutations and gives a comparative study based on the number of collisions. Computational results are then presented in Sect. 4 to show the effect of the hash function on three metaheuristics. Finally, Sect. 5 presents our conclusions and our plans for future work.

2 Collision Analysis of the Fitness Function on the TSP

The TSP is frequently used as a test-bed for designing effective methods to solve general sequencing permutation problems. The problem is modelled with a graph $G = (V, A)$ where $V = \{v_1, ..., v_n\}$ is the vertex set, and $A = \{(v_i, v_j)|v_i, v_j \in V, i \neq j\}$ is the edge set. A non-negative cost (or distance) matrix $C = (c_{ij})$ is associated with A. This paper focuses on the most widely studied form of the problem in which costs are assumed to be symmetric $c_{ij} = c_{ji}$ and satisfy the triangle inequality ($c_{ij} + c_{jk} > c_{ik}$). A feasible TSP solution is a sequence of nodes/cities arranged in a permutation π of size n. Its cost is the sum of the distances of each couple of adjacent cities in the permutation. We define a fitness function to evaluate a TSP permutation π as follows:

$$f_{\text{fit}}(\pi) = \sum_{i=1}^{n-1} c_{\pi_i \pi_{i+1}} + c_{\pi_n \pi_1} \tag{1}$$

It is common practice in evolutionary computation to use the fitness function, f_{fit}, to compare solutions. It is well-known that many solutions may map to the same fitness value, but, to the best of our knowledge, no prior work sheds light on how much the fitness values are repeated over a search space nor how they are distributed over its solutions. To do so, we chose 39 instances from the TSP benchmark TSPLIB [18] (sizes n range from 51 to 575). We then explore the search space in two different ways. Firstly, a set S_{LO} is composed of n^2 local optima obtained with an ILS framework to get as many neighbouring solutions as possible. Secondly, a set S_{rand} is built, containing $10 \times n^2$ random solutions. These samples were generated in such a way that all the solutions are distinct. This means that they do not contain two identical permutations. We compute in the first part of this section the number of collisions occurring in each sample for each instance.

We say that we have a collision between two solution s_1, s_2 if $f(s_1) = f(s_2)$, where the function f outputs a numeric value. We then examine how these collisions are distributed over the fitness values in Sect. 3.2. As an example, if 4 solutions s_1, s_2, s_3, s_4 map to the same value, we count 4 *repetitions* ($f(s_1) = f(s_2) = f(s_3) = f(s_4)$), and 6 *collisions* (($s_1, s_2), (s_1, s_3), (s_1, s_4)$,$(s_2, s_3), (s_2, s_4), (s_3, s_4)$). Thus the number of collisions may exceed the sample size.

2.1 Too Many Collisions for the Fitness Function

To determine if the fitness function as a comparison tool can affect a metaheuristic, we measure the collisions over the above-mentioned samples and list them in Table 1. For each instance, we expose the sample size, $|S_{LO}|$ and $|S_{rand}|$, and the number of collisions, C_{LO} and C_{rand}, computed by comparing all the solution pairs (these values are rounded at $1E3$—precise values are displayed in Table 2); then the number of the different fitness values, Fit_{LO} and Fit_{rand}, retrieved in each sample.

Getting collisions from large samples of solutions is not surprising, especially when it comes to local optima that share common edges between them. But the number of collisions we have in Table 1 exceeds our expectations. Indeed, a very high number of collisions is observed in almost all samples, with up to millions of collisions for the smallest ones. Moreover, according to Fit_{LO} and Fit_{rand}, we notice very small sets of fitness values to whom the solutions of S_{LO} and S_{rand} are mapping. In other words, the large set of solutions is distributed over a small set of fitness values, making some fitness values very repetitive.

Table 1. Collision analysis of the fitness function on the TSP

Instance	$|S_{LO}|$ 1E3	C_{LO} 1E3	Fit_{LO}	$|S_{rand}|$ 1E3	C_{rand} 1E3	Fit_{rand}	Instance	$|S_{LO}|$ 1E3	C_{LO} 1E3	Fit_{LO}	$|S_{rand}|$ 1E3	C_{rand} 1E3	Fit_{rand}
eil51	3	127	53	26	1,069	577	pr144	21	59	5,889	207	200	98,182
berlin52	3	5	799	27	65	6,833	ch150	23	478	893	225	3,921	10,792
st70	5	230	99	49	1,877	1,170	kroA150	23	129	2,807	225	707	46,224
eil76	6	456	72	58	4,140	766	kroB150	23	136	2,708	225	680	47,674
pr76	6	3	3,788	58	18	43,331	pr152	23	73	4,750	231	197	117,041
gr96	9	11	3,872	92	71	49,124	u159	25	71	5,647	253	530	70,958
rat99	10	457	189	98	3,317	2,613	rat195	38	4,736	310	380	25,839	5,329
kroA100	10	31	2,163	100	170	32,619	d198	39	1,338	1,047	392	3,001	38,231
kroB100	10	33	2,028	100	173	32,317	gr202	41	405	3,256	408	3,439	36,831
kroC100	10	30	2,168	100	169	32,881	ts225	51	128	12,854	506	825	170,884
kroD100	10	31	2,034	100	186	30,723	gr229	52	178	10,401	524	922	167,730
kroE100	10	34	1,970	100	168	33,161	gil262	69	16,165	308	686	96,690	4,930
rd100	10	71	1,025	100	599	12,060	a280	79	14,369	414	784	90,807	6,746
eil101	10	1,376	80	102	10,622	975	lin318	101	2,219	3,951	1,011	10,188	78,223
lin105	11	59	1,443	110	289	26,055	rd400	160	18,439	1,398	1,600	81,554	29,261
pr107	11	41	2,549	114	58	73,334	fl417	174	28,258	1,424	1,739	31,432	80,618
pr124	15	37	4,028	154	116	82,802	gr431	186	2,027	15,233	1,954	34,979	91,416
bier127	16	18	7,199	161	187	68,143	pcb442	196	7,548	4,649	1,858	7,753	303,216
ch130	17	351	734	169	2,419	9,609	rat575	331	234,671	592	3,306	676,190	16,898
gr137	19	37	5,593	188	179	89,176							

2.2 Distribution of Collisions over Fitness Values

We count for each fitness value in each sample S_{LO} how many times it appears. We observed from the different gathered data a high similarity of distribution of repetitions between the different instances. For a better observation of these distributions, we draw for each instance a scatter plot with ascendant and linear scale axes, where each dot depicts the number of repetitions (Y-axis) of one fitness value (X-axis). The latter is illustrated by its gap (δ) from the optimal solution's fitness, calculated with the formula 2.

$$\delta = \frac{f_{\text{fit}}(solution) - f_{\text{fit}}(optimal)}{f_{\text{fit}}(optimal)} \times 100(\%) \tag{2}$$

As expected, the similarity of distributions induces similarly shaped plots. It allowed us to classify them into two main distributions, represented in Fig. 1. Below each exposed plot in the figure, we provide (i) the instance name; (ii) the maximal value of the abscissa axis, i.e. the gap between the highest fitness value existing in S_{LO} and the optimal solution (x_{max}); (iii) the maximal value of the ordinate axis, which is the most important repetition observed in S_{LO} (y_{max}).

In 19 instances (from the 39 studied), we observed the distribution of repetitions with a bell curve. The first row of plots in Fig. 1 shows 3 examples of these instances where we can notice a distribution close to normality. While the different x_{max} values are in a fairly narrow range and don't depend on the collisions caused by f_{fit}, there is a strong correlation between the density of the plots and their respective y_{max} value: the larger the value of y_{max}, the smaller the thickness of the curve will be and vice versa. To illustrate this correlation better, we put in our examples a pair of instances with approximately equal sizes (rat195 and gr202), where the scatter plot becomes more sparse when y_{max} decreases.

For instances with a low y_{max} (which doesn't mean a low number of collisions), the repetitions become more sparse on the plot until we move away from the normal distribution. The second row of the figure exposes 3 examples of the 17 instances where the repetitions form an area with a shape close to a bell. Unlike the first class of instances, the number of distinct fitness values is more important but each one appears in the sample less frequently.

Inspecting the collisions occurring in large samples of solutions reveals a high number of repetitions of the fitness values. This can mislead metaheuristics when the fitness function is used for comparing solutions. The analysis of the distribution of fitness repetitions unveils two major classes of distribution. A first one where instances have a few (distinct) fitness values but with high repetition. Then a second one with a reversed tendency. Such information can be exploited to predict how a trajectory-based metaheuristic can be influenced. For example, in a tabu search context, a high number of repetitions of fitness values may lead to short cycles when the algorithm considers each visited solution by its fitness value. These hypotheses are verified and validated in Sect. 4.

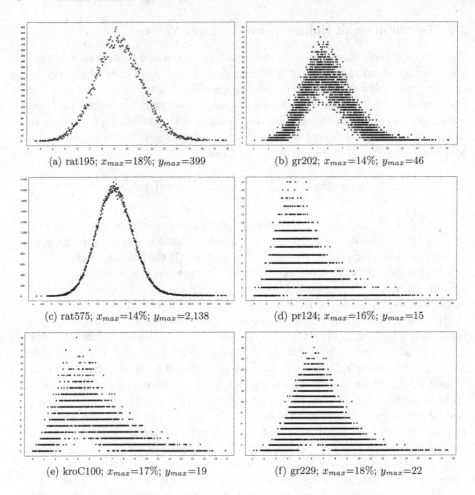

(a) rat195; x_{max}=18%; y_{max}=399   (b) gr202; x_{max}=14%; y_{max}=46

(c) rat575; x_{max}=14%; y_{max}=2,138   (d) pr124; x_{max}=16%; y_{max}=15

(e) kroC100; x_{max}=17%; y_{max}=19   (f) gr229; x_{max}=18%; y_{max}=22

Fig. 1. Distribution of fitness repetitions and their gaps from the optimal solution

3   Hash Functions for a Reliable Comparison

The analysis shown in the previous section is a strong motivation to search for an efficient alternative to the objective-based fitness function to compare solutions. Hash functions for COPs (specifically permutation-based ones) have been proposed to yield a lower number of collisions than the given fitness function.

3.1   Existing Hash Functions

Woodruff and Zemel introduced three hash functions in [20] . The first function, h_1, is based on multiplying pseudo-random integers ρ_i with each element of the solution vector π_i. The maximum integer value $MAXINT$ is used to avoid overflow. The second function, h_2, makes use of a matrix P of a pre-computed random weight, while the last function, h_3, replaces an entry $P(i,j)$ with $P(i) \times P(j)$ which is equivalent to replacing the matrix P with a long vector of pre-computed random weights. Since the authors claimed h_3 is better than h_2, the latter will not appear in our comparative study.

$$h_1 = (\sum_{i=1}^{n} \rho_i \pi_i)\%(MAXINT + 1) \tag{3}$$

$$h_2 = \sum_{i=1}^{n} P(\pi_i, \pi_{i+1}) \tag{4}$$

$$h_3 = (\sum_{i=1}^{n} P(\pi_i) P(\pi_{i+1}))\%(MAXINT + 1) \tag{5}$$

The three hash functions were designed taking into account the following properties:

1. Computation and update of the hash value should be as fast as possible, and in any case as fast as the fitness function. The hash value update after applying a move on a candidate solution should preferably be computed in $\mathcal{O}(1)$ time.
2. The hash values should be in a range that results in reasonable storage requirements and comparison effort.
3. The function should guarantee a low collision probability to minimise the risk of two permutations having identical hashes.

Toffolo et al. [19] employed the two hash functions defined in Eq. 6 and Eq. 7 to rapidly evaluate a newly explored route of the CVRP. The function h_p is a multiplicative hash which depends on the visited permutation. The second function, h_s, is an additive hash that depends on the set of visited customers. These hash functions were used with two different values of ρ. It is set to the prime number 31, or to the smallest prime number greater than the number of customers. To prevent overflow during multiplication, the values ρ^i were bounded taking the rest of the integer division by a large number. While a solution in a CVRP instance uses a subset of customers, it is not the case for the TSP. Using all the customers/nodes for any solution of the latter makes h_s have the same value for a given instance and thus cannot be used as a hash function for the TSP.

$$h_p(\pi) = \sum_{i=1}^{n} \rho^i \pi_i \tag{6}$$

$$h_s(\pi) = \sum_{i=1}^{n} \rho^{\pi_i} \tag{7}$$

3.2 The Proposed Function

In addition to the three properties (previously mentioned in Sect. 3.1) that a hash function should acquire, we propose in this paper to implement a hash function with an added characteristic.

While the existing functions are based on vectors of large random values, we want to design our hash function (η) with only the solution and the instance data already stored by the heuristic. This is more challenging since random values help to reduce the number of collisions considerably. Conversely, solution data can increase collisions due to the correlation and similarities we may observe between a pair of solutions. The fitness function, based on the distance matrix, is a concrete example.

To make sure η obeys the property n° 1, we define a (sub-)hash function η_e for one edge of the permutation. η can be formulated then as described in formula 8. It also ensures to get the same hash value when a permutation is shifted since the solution stays the same in the TSP case.

$$\eta = \eta_e(1, n) + \sum_{i=1}^{n-1} \eta_e(i, i+1) \tag{8}$$

The operands we chose for our hash function are the distance matrix and the set of node identifiers which are n distinct integers in the range $[1; n]$. In addition to the mathematical operators, we define a new operator mod in Formula 9. This definition is an adjustment of the classical modulo operator to ensure having the same hash value when the permutation is symmetrical and prevent the η_e function from returning a zero value ($(a < b) \Rightarrow (a\%b = 0)$).

$$mod(a, b) = max(a, b)\%min(a, b) \tag{9}$$

To lower the number of collisions, we favoured multiplication over addition since it gives more diverse values. The division is dismissed to avoid dealing with precision issues. We decided to involve more the node identifiers rather than the distance matrix. Values of the latter are larger and will quickly lead to memory overflows.

Following all the guidelines mentioned above, we designed the function η_e as shown in formula 10. π_i is the identifier of the ith node in the permutation π. $C = (c_{ij})$ is the distance matrix of the studied instance. Formula 11 defines our hash function η.

$$\eta_e(i, j) = mod(\pi_i, \pi_j) \times (\pi_i + \pi_j) \times (\pi_i \times \pi_j) \times c_{\pi_i, \pi_j} \tag{10}$$

$$\eta = \eta_e(1, n) + \sum_{i=1}^{n-1} \eta_e(i, i+1)$$

$$= mod(\pi_1, \pi_n) \times (\pi_1 + \pi_n) \times (\pi_1 \times \pi_n) \times c_{\pi_1, \pi_n} + \tag{11}$$

$$\sum_{i=1}^{n-1} mod(\pi_i, \pi_{i+1}) \times (\pi_i + \pi_{i+1}) \times (\pi_i \times \pi_{i+1}) \times c_{\pi_i, \pi_{i+1}}$$

3.3 Comparative Study

Table 2 below compares the fitness function f_{fit} and the hash functions η and h_3 (which was chosen as the best of the functions in Sect. 3.1 after a preliminary comparison [1]).

For each instance sample (S_{LO} and S_{rand}), the number of collisions resulting from each function is printed on the table.

The first remark is the significant reduction of collisions made between f_{fit} and the hash functions, which can't be with no effect on the search process. The second observation is the excellent results obtained by η compared with h_3, especially for large instances. Our designed hash function succeeded in getting zero collisions on 36 (resp. 32) instances for S_{LO} (resp. S_{rand}), against only 29 (resp. 10) instances for h_3. The overall average for η is less than one collision in each set of samples, while it is much higher for h_3 in S_{rand}.

This shows that it is possible to design a hash function with fewer collisions than those proposed in the literature.

η can then be embedded in a metaheuristic, with a constant time cost, as a reasonable alternative to fitness evaluation in order to compare solutions. In addition to the three properties listed in Sect. 3.1, this hash function only uses solution data and does not need large vectors of random values, thus reducing its memory footprint. Note that we did not encounter any overflow with our proposed hash function on our set of instances and samples. Nonetheless, in the event of overflow, one could use standard strategies, such as clipping values with a modulo operator (see also [19]).

The following section shows the multiple effects of using the hash functions.

4 Revisiting Some Metaheuristics with Hash Functions

Let's consider f_{comp} the comparison function to check the equality between a pair of solutions. The results of each test/run performed in this section are obtained with $f_{\text{comp}} = f_{\text{fit}}$, then $f_{\text{comp}} = \eta$. We provide the same input in each case. Fifty runs are assigned to each instance/test.

---

[1] The comparison between all the functions is available at https://elkrari.com/hashfunctions/.

Table 2. A comparison of the number of collisions between f_{fit}, h_3 and η obtaines on the samples S_{rand} and S_{LO}

Instance	S_{LO}			S_{rand}			Instance	S_{LO}			S_{rand}		
	f_{fit}	η	h_3	f_{fit}	η	h_3		f_{fit}	η	h_3	f_{fit}	η	h_3
eil51	126,875	0	0	1,068,752	2	0	pr144	59,098	0	0	199,699	0	0
berlin52	5,495	0	0	64,662	0	0	ch150	477,971	0	0	3,921,003	0	4
st70	229,647	0	0	1,876,506	0	0	kroA150	128,766	0	0	707,170	0	4
eil76	456,422	1	0	4,140,456	2	1	kroB150	135,785	0	1	680,086	0	3
pr76	2,838	0	0	18,107	0	0	pr152	72,564	0	0	196,547	0	1
gr96	10,633	0	0	71,134	1	0	u159	70,763	0	0	529,831	0	2
rat99	457,303	0	0	3,317,137	2	0	rat195	4,736,380	0	0	25,838,895	1	5
kroA100	30,902	0	0	170,251	1	0	d198	1,337,616	0	0	3,001,233	0	10
kroB100	32,978	0	1	172,655	0	1	gr202	404,695	0	0	3,439,300	0	13
kroC100	30,277	0	0	169,127	0	1	ts225	128,149	0	0	824,742	0	19
kroD100	30,805	0	0	185,716	0	2	gr229	177,592	0	1	921,695	0	15
kroE100	33,600	0	0	167,732	0	1	gil262	16,165,285	0	0	96,690,038	0	26
rd100	71,443	0	0	598,845	0	1	a280	14,368,564	0	2	90,806,864	0	31
eil101	1,376,150	0	0	10,621,646	1	0	lin318	2,219,290	0	1	10,188,376	0	45
lin105	59,275	0	0	288,911	0	1	rd400	18,439,021	0	2	81,553,529	0	99
pr107	41,307	0	0	58,422	0	1	fl417	28,258,436	1	3	31,432,278	0	97
pr124	36,773	0	0	116,389	0	0	gr431	2,026,980	0	2	34,978,940	0	127
bier127	17,641	7	0	187,417	0	5	pcb442	7,548,302	0	3	34,978,940	0	138
ch130	350,821	0	1	2,419,326	0	4	rat575	234,671,318	0	4	676,190,119	0	382
gr137	37,181	0	0	179,311	0	3	average	8,586,280.54	0.23	0.54	28,794,148.38	0.26	26.72

This section doesn't aim to improve the state-of-the-art. The objective is to provide a comparison of the two scenarios of f_{comp} in the same environment. For each metaheuristic, we implement a basic version with known operators and strategies.

4.1 Cycling Analysis

One of the limitations of ILS is cycling. A cycle occurs when the search returns to an already visited local optimum, which means the algorithm is stuck in a limited region of the search space. To inspect the effect of using hash functions to identify cyclings, we ran an ILS with a stochastic local search [10] by the 2-Opt neighbourhood function [4,7] and a perturbation of n 2-Opt random moves. Each run stops when the first cycle occurs or when the algorithm reaches $50 \times n$ iterations.

Table 3 shows the average number of visited solutions before a cycle arises. For each instance, we note the maximum number of iterations (Max_{iter}), which is also the maximum number of local optima we can visit in each run. For each case of f_{comp}, we record how many times a cycle happened (C), and the average number of visited local optima before ILS stops (or the history size $|H|$). We note zero when no cycle appears during the 50 runs. The last column of the table compares the two cases of f_{comp} with the Mann Whitney U Test (also called Wilcoxon Rank Sum Test) [6,15] The test measures the separation level between the number of iterations made in each case with a p−value.

Table 3. Cycling analysis of the ILS framework when using η then f_{fit} to differentiate solutions.

Instance	Max_{iter}	$f_{comp} = \eta$		$f_{comp} = f_{fit}$		p−value
		C	|H|	C	|H|	
eil51	2550	22	1463.18	50	7.78	6.86E-18
berlin52	2600	50	401.8	50	39.96	6.86E-18
st70	3500	10	1946.5	50	11.98	6.86E-18
eil76	3800	1	3733	50	8.32	6.86E-18
pr76	3800	41	1990.22	50	110.32	6.86E-18
gr96	4800	0		50	97.72	6.86E-18
rat99	4950	2	3054.5	50	11.56	6.86E-18
kroA100	5000	14	3115.64	50	58.24	6.86E-18
kroB100	5000	0		50	57.1	6.86E-18
kroC100	5000	12	2866	50	65.74	6.86E-18
kroD100	5000	0		50	66.94	6.86E-18
kroE100	5000	0		50	60.4	6.86E-18
rd100	5000	0		50	39.88	6.86E-18
eil101	5050	1	4093	50	7.64	6.86E-18
lin105	5250	20	3209.9	50	50.9	6.86E-18
pr107	5350	0		50	85.18	6.86E-18
pr124	6200	50	521.26	50	95.34	1.15E-14
bier127	6350	0		50	141	6.86E-18
ch130	6500	1	1624	50	33.3	6.86E-18
gr137	6850	0		50	91.02	6.86E-18
pr144	7200	50	608.22	50	100.1	2.77E-16
ch150	7500	0		50	35.54	6.86E-18
kroA150	7500	0		50	56.74	6.86E-18
kroB150	7500	0		50	58.72	6.86E-18
pr152	7600	4	6042.25	50	106.12	6.86E-18
u159	7950	0		50	95.2	6.86E-18
rat195	9750	0		50	15.6	6.86E-18
d198	9900	0		50	35.74	6.86E-18
gr202	10100	0		50	63.26	6.86E-18
ts225	11250	0		50	136.72	6.86E-18
gr229	11450	0		50	143.68	6.86E-18
gil262	13100	0		50	16.98	6.86E-18
a280	14000	0		50	19.8	6.86E-18
lin318	15900	0		50	69.88	6.86E-18
rd400	20000	0		50	41.54	6.86E-18
fl417	20850	0		50	38.26	6.86E-18
gr431	21550	0		50	152.64	6.86E-18
pcb442	22100	0		50	70.88	6.86E-18
rat575	28750	0		28	20.54	6.86E-18

At first sight of Table 3, the difference between the two scenarios seems to be broad. ILS cycles prematurely when comparing solutions with their fitness values, with only a few visited local optima. On the other hand, the hash function enables the algorithm to explore extensively, having to make a decision (restart, strong perturbation,...) when the cycle arises. With a p−value almost equal to zero, the Mann Whitney U Test confirms that the ILS algorithm always detects cycles earlier with the fitness function.

4.2 Convergence Speed

Population-based algorithms can also be exposed to misleading information provided by the fitness function. One of the stopping criteria in these metaheuristics is the convergence rate [12], i.e. the similarity of solutions within a population. We now analyse the convergence speed of two population-based algorithms.

The first one is a GA [13] implemented with a tournament selection, one-point crossover and an elitist replacement strategy. The second one is a memetic algorithm with the same genetic operators and the steepest descent applied to each new individual with 2-Opt.

For each scenario of f_{comp}, Fig. 2a (resp. 2b) displays the average (of the 39 studied instances) evolution of the convergence rate in the $5 \times n$ first generations for our genetic (resp. memetic) algorithm. The evolution is represented with dashed curves for $f_{comp} = f_{fit}$ and solid ones for $f_{comp} = \eta$.

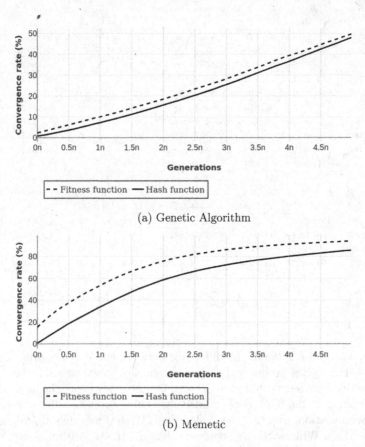

(a) Genetic Algorithm

(b) Memetic

Fig. 2. Evolution of the convergence rate for population-based metaheuristics during $5 \times n$ generations

The two figures reveal the incorrect information given by the fitness function regarding the convergence rates. The difference is tight between the two cases of f_{comp} for the GA. In contrast, the memetic shows a wider gap between the two curves. This means population-based metaheuristics can run for more generations to explore new solutions and regions of the search space.

4.3   Applying Hash Function on Metaheuristics

The previous results of this section warn us of the misguided analysis caused by the fitness function as a comparison function. A hash function can then be more effective in providing the algorithm with better results. Table 4 lists the results of three metaheuristics using η and f_{fit}, respectively, and implemented as follows: (i) An ILS with the steepest descent and perturbations with different strengths depending on the state of the search, run with $10 \times n$ iterations. (ii) A genetic, and (iii) memetic algorithms run with the same operators used earlier, with a stopping criterion of 90% on the convergence rate and a strictly equal number of evaluations for each variant of f_{comp}. The table shows the average gap from the optimal solution (formula 2) to the best solution found in each scenario of f_{comp}.

Table 4. Iterated Local Search, Genetic and Memetic Algorithms run with η, then f_{fit} as a differentiating function.

Instance	ILS		GA		Memetic		Instance	ILS		GA		Memetic	
	η	f_{fit}	η	f_{fit}	η	f_{fit}		η	f_{fit}	η	f_{fit}	η	f_{fit}
eil51	**0.81**	**0.58**	**187.76**	190.95	**0.53**	0.70	pr144	**0.01**	**0.01**	**920.35**	929.78	**0.00**	0.01
berlin52	**0.00**	**0.00**	193.64	**192.26**	**0.00**	**0.00**	ch150	1.88	**1.74**	**522.68**	527.60	**0.37**	0.47
st70	0.29	**0.21**	**296.54**	296.98	**0.09**	0.18	kroA150	1.56	**1.54**	**604.99**	612.51	**0.28**	0.33
eil76	**2.19**	2.23	**246.20**	253.87	**0.38**	0.85	kroB150	1.28	**1.26**	**601.41**	614.83	**0.19**	0.25
pr76	**0.09**	0.13	**291.06**	297.99	**0.04**	0.05	pr152	**0.21**	0.22	**942.39**	949.39	0.11	**0.10**
gr96	0.64	**0.58**	**389.74**	391.14	**0.22**	0.23	u159	**0.99**	1.02	**674.87**	681.97	0.05	**0.01**
rat99	**1.59**	1.76	**393.90**	399.59	**0.27**	0.34	rat195	**3.99**	4.51	**619.70**	622.17	**0.83**	1.09
kroA100	**0.25**	0.26	**473.64**	476.42	**0.04**	0.06	d198	**1.03**	1.07	**729.34**	732.31	**0.16**	0.18
kroB100	**0.56**	0.57	**442.16**	444.14	0.14	**0.11**	gr202	**2.34**	2.43	**421.79**	426.32	**0.29**	0.30
kroC100	**0.32**	**0.32**	**482.51**	490.31	**0.07**	0.08	ts225	**0.35**	**0.35**	**871.65**	873.22	**0.01**	0.03
kroD100	0.77	**0.70**	**444.51**	450.29	**0.15**	0.24	gr229	2.26	**2.31**	**658.88**	673.00	**0.34**	0.39
kroE100	**0.66**	0.75	**461.56**	464.54	0.21	**0.19**	gil262	**3.01**	3.22	**765.70**	775.37	**0.25**	0.64
rd100	0.65	**0.62**	**423.15**	424.12	**0.20**	0.25	a280	**2.94**	3.69	**890.03**	897.99	**0.26**	0.71
eil101	**2.91**	2.98	**303.01**	311.78	**0.48**	1.34	lin318	**2.53**	2.65	**979.80**	991.38	**0.36**	0.41
lin105	0.18	**0.13**	**508.88**	516.17	**0.02**	0.03	rd400	**3.91**	4.37	**983.71**	994.41	**0.57**	0.66
pr107	0.34	**0.29**	**761.09**	772.48	**0.09**	0.11	fl417	**0.88**	1.05	**2,897**	2,927	**0.22**	0.45
pr124	0.03	**0.01**	**745.45**	749.41	0.02	**0.01**	gr431	**3.02**	3.12	**965.36**	971.80	**0.54**	0.60
bier127	**0.86**	0.92	**313.17**	320.58	0.20	**0.19**	pcb442	**3.59**	3.99	**1,095**	1,102	**0.64**	0.71
ch130	1.31	**1.30**	**471.45**	478.66	**0.31**	0.40	rat575	**5.44**	6.52	**1,204**	1,205	**0.89**	1.92
gr137	1.18	**1.04**	**559.23**	564.66	**0.13**	0.16	average	**1.46**	1.55	**659.97**	666.57	**0.25**	0.38

The exposed results confirm the positive effect of using a hash function in different metaheuristic classes, especially for large instances. These improvements are achieved by non-biased runs of the above-mentioned algorithms. While the fitness function led to wrong cycles and premature convergence, the hash function allows the algorithm to make fairer differentiations and then make the right decisions at the right moments.

5   Discussion and Conclusion

Using the fitness function to compare solutions can be harmful to many metaheuristics. This is due to the high number of collisions caused by the fitness function and the significant repetitions in its values. This paper proposes a new effective hash function with respect to the existing ones in the literature. The number of collisions caused by our function η is zero on almost all the generated samples and can be improved by comparing the pair (f_{fit}, η) of values. An analysis of different state-of-the-art heuristics unveiled the positive effect of using a hash function as a comparison tool. While the fitness function misleads the search process to short cycles, we observed longer explorations when using a hash function. A similar effect was noticed on population-based algorithms where the convergence rate increases more slowly with hash values. These improvements were reflected on their respective metaheuristics by reaching better solutions.

While this paper tackled the TSP as one of the most used COP, others can also take advantage of the proposed hash function or by designing new ones. We envisage then for our future works to explore new problems with different solution representations (permutations or binary strings). Genetic programming can be used to produce unbiased, and possibly problem-independent, hash functions.

References

1. Brownlee, A.E., Woodward, J.R., Swan, J.: Metaheuristic design pattern: surrogate fitness functions. In: Proceedings of the Companion Publication of the 2015 Annual Conference on Genetic and Evolutionary Computation, pp. 1261–1264 (2015)
2. Brucker, P.: Scheduling Algorithms. Springer, Heidelberg (2007). https://doi.org/10.1007/978-3-540-69516-5
3. Cai, S., Su, K., Sattar, A.: Local search with edge weighting and configuration checking heuristics for minimum vertex cover. Artif. Intell. **175**(9–10), 1672–1696 (2011)
4. Croes, G.A.: A method for solving traveling-salesman problems. Oper. Res. **6**(6), 791–812 (1958)
5. El Krari, M., Guibadj, R.N., Woodward, J., Robilliard, D.: Introducing a hash function for the travelling salesman problem for differentiating solutions. In: Proceedings of the Genetic and Evolutionary Computation Conference Companion, pp. 123–124. GECCO 2021, Association for Computing Machinery, New York, NY, USA (2021). https://doi.org/10.1145/3449726.3459580
6. Fay, M.P., Proschan, M.A.: Wilcoxon-Mann-Whitney or t-test? on assumptions for hypothesis tests and multiple interpretations of decision rules. Statist. Surv. **4**, 1 (2010)
7. Flood, M.M.: The traveling-salesman problem. Oper. Res. **4**(1), 61–75 (1956)
8. Hartung, E., Hoang, H.P., Mütze, T., Williams, A.: Combinatorial generation via permutation languages. In: Proceedings of the Fourteenth Annual ACM-SIAM Symposium on Discrete Algorithms, pp. 1214–1225. SIAM (2020)
9. Hifi, M., Michrafy, M., Sbihi, A.: A reactive local search-based algorithm for the multiple-choice multi-dimensional knapsack problem. Comput. Optim. Appl. **33**(2–3), 271–285 (2006)

10. Hoos, H.H., Stützle, T.: Stochastic local search: Foundations and applications. Elsevier (2004)
11. Koopmans, T.C., Beckmann, M.: Assignment problems and the location of economic activities. Econometrica **25**(1), 53–76 (1957)
12. Langdon, W.B., Poli, R.: Foundations of Genetic Programming. Springer Science & Business Media, Cham (2013). https://doi.org/10.1007/978-3-662-04726-2
13. Larrañaga, P., Kuijpers, C., Murga, R., Inza, I., Dizdarevic, S.: Genetic algorithms for the travelling salesman problem: A review of representations and operators. Artif. Intell. Rev. Int. Surv. Tutor. J. **13**(2), 129–170 (1999)
14. Loiola, E.M., de Abreu, N.M.M., Boaventura-Netto, P.O., Hahn, P., Querido, T.: A survey for the quadratic assignment problem. Eur. J. Oper. Res. **176**(2), 657–690 (2007)
15. Mann, H.B., Whitney, D.R.: On a test of whether one of two random variables is stochastically larger than the other. Ann. Math. Statist. 50–60 (1947)
16. Michalewicz, Z., Fogel, D.B.: How To Solve It: Modern Heuristics. Springer Science & Business Media, Heidelberg (2013). https://doi.org/10.1007/978-3-662-07807-5
17. Pierskalla, W.: The tri-substitution method for the three-dimensional assignment problem. Can. Oper. Res. Soc. J. **5**(2), 71 (1967)
18. Reinelt, G.: TSPLIB-a traveling salesman problem library. ORSA J. Comput. **3**(4), 376–384 (1991)
19. Túlio A.M., T., Thibaut, V., Tony, W.: Heuristics for vehicle routing problems Sequence or set optimization? Comput. Oper. Res. **105**, 118–131 (2019)
20. Woodruff, D.L., Zemel, E.: Hashing vectors for Tabu search. Ann. Oper. Res. **41**, 123–137 (1993)
21. Woodward, J.R., Bai, R.: Canonical representation genetic programming. In: Proceedings of the First ACM/SIGEVO Summit on Genetic and Evolutionary Computation, pp. 585–592. GEC 2009, Association for Computing Machinery, New York, NY, USA (2009). https://doi.org/10.1145/1543834.1543914

Application of Adapt-CMSA to the Two-Echelon Electric Vehicle Routing Problem with Simultaneous Pickup and Deliveries

Mehmet Anıl Akbay[1]([✉])[iD], Can Berk Kalayci[2][iD], and Christian Blum[1][iD]

[1] Artificial Intelligence Research Institute (IIIA-CSIC), Campus UAB,
Bellaterra, Spain
makbay@iiia.csic.es
[2] Department of Industrial Engineering, Pamukkale University, Denizli, Turkey
cbkalayci@pau.edu.tr

Abstract. This study addresses the two-echelon electric vehicle routing problem with simultaneous pickup and deliveries. In a two-echelon distribution network, large vehicles transport goods from central warehouses to satellites, while smaller and environmentally friendly vehicles distribute goods from these satellites to final customers. The considered problem also includes simultaneous pickup and delivery constraints that usually arise as a reverse logistics practice. A MILP model is developed and solved for small-sized problem instances using CPLEX. Since the tackled problem becomes rather complex because of the multi-tier structure and constraints, solving even small-sized instances using CPLEX requires very long computation times. Therefore, the application of a self-adaptive variant of the hybrid metaheuristic *Construct, Merge, Solve & Adapt* is proposed. In the context of problem instances too large for the application of CPLEX, our algorithm is compared to probabilistic versions of two well-known constructive heuristics. The numerical results show that our algorithm outperforms CPLEX in the context of rather small problem instances. Moreover, it is shown to outperform the heuristic algorithms when larger problem instances are concerned.

Keywords: two-echelon · EVRP · simultaneous pickup and delivery

1 Introduction

Route planning of vehicles used in freight distribution has been one of the primary objectives of logistics for many years. Researchers and practitioners have been developing models and solution methods to find the best routes for a fleet of vehicles transporting products from supply points to demand points. Danzig et al. [13] addressed this problem as a truck dispatching problem for the first time in the literature. After this pioneering work, many variations of the addressed problem have been introduced under the title of vehicle routing problems [16].

Rising concerns about the environment and issues such as noise, traffic congestion, and population growth, especially in large cities, require objectives

L. Pérez Cáceres and T. Stützle (Eds.): EvoCOP 2023, LNCS 13987, pp. 16–33, 2023.
https://doi.org/10.1007/978-3-031-30035-6_2

considering not only the economy but also social and environmental issues in the design of logistics systems. In this line, one of the more recent approaches in logistics is to operate freight distribution using multi-echelon distribution structures [11]. In such a framework, goods are transported from central warehouses to final customers through transshipment facilities, also called satellites. In this way, vehicles with large loading capacities and, generally, higher exhaust emissions are kept out of the city centres. Moreover, goods are distributed from the satellites to customers with vehicles more suitable for their operation in densely populated areas, i.e., vehicles with low carbon emissions and smaller loading capacities.

An additional approach for reducing the environmental impact of logistics activities and, therefore, for preserving the environment and increasing the quality of city life is to utilize environmentally-friendly vehicles for freight distribution. Lately, there has been a remarkable increase in the number of logistics and e-commerce companies using electric vehicles for distribution in urban areas [20]. However, despite the advantages of using electric vehicles in logistics, deriving optimal routing plans for these vehicles is rather complex because of a limited driving range and an en-route charging necessity.

Our Contribution. This study is the first one to address the two-echelon electric vehicle routing problem with simultaneous pickup and delivery (2E-EVRP-SPD). In this problem, we consider that large trucks with internal combustion engines transport products from central warehouses to satellites in the surroundings of cities. Subsequently, smaller electric vehicles distribute goods from these satellites to customers located in the cities. Moreover, as the name of the problem already indicates, it also considers simultaneous pickup and delivery (SPD) constraints for the delivery of goods to customers. When SPD constraints are considered, each customer may have two different demands: (1) the goods to be delivered to the demand point (*delivery demand*), and (2) the goods to be collected from the demand point (*pickup demand*). So, once a vehicle visits a certain customer, both demands must be met simultaneously. This approach usually arises as a reverse logistics practice [14]. However, despite the importance of reverse logistics in terms of sustainability, the number of publications on EVRP-SPD variants is rather limited. Only [1,28] considered SPD constraints within the scope of EVRPs. However, to the best of our knowledge, SPD constraints have yet to be considered within the scope of two-echelon electric vehicle routing problems [18]. Our contributions are as follows. First, the addressed problem is formulated as a mixed integer linear program (MILP). Any general-purpose MILP solver, such as CPLEX[1] or Gurobi[2], may be used to solve this model. However, due to the multi-tier structure of the distribution network, the limited driving range of electric vehicles, and the SPD constraints, the 2E-EVRP-SPD problem is rather complex. In fact, our computational experiments show that CPLEX struggled to solve even small-sized problems to optimality. In fact, in most cases, CPLEX was only able to derive valid solutions with large optimality gaps. Therefore, a recent self-adaptive variant of the hybrid metaheuristic "Construct Merge Solve & Adapt" (CMSA) [6]—called Adapt-CMSA—is developed for being able to solve large-sized

---

[1] https://www.ibm.com/analytics/cplex-optimizer.
[2] http://www.gurobi.com/.

Table 1. Sets and notations

n_d, n_s, n_r, n_c	: Number of central warehouses, satellites, charging stations and customers, respectively
N_D	: Set of central warehouses, $N_D = \{n_{d_1}, ..., n_{d_{n_d}}\}$
N'_D	: Set of dummy central warehouses corresponding to N_D, $N'_D = \{n'_{d_1}, ..., n'_{d_{n_d}}\}$
N_S	: Set of satellites, $N_S = \{n_{s_1}, ..., n_{s_{n_s}}\}$
N'_S	: Set of dummy satellites corresponding to N_S, $N'_S = \{n'_{s_1}, ..., s'_{s_{n_s}}\}$
N_R	: Set of charging stations, $N_S = \{n_{r_1}, ..., n_{r_{n_r}}\}$
N_C	: Set of customers, $N_S = \{n_{c_1}, ..., n_{c_{n_c}}\}$
N_{DS}	: Set of central warehouses and satellites, $N_{DS} = N_D \cup N_S$
N_{SD}	: Set of satellites and dummy central warehouses, $N_{DS} = N_S \cup N'_D$
N_{DSD}	: Set of complete nodes in the first echelon, $N_{DSD} = N_D \cup N_S \cup N'_D$
N_{RC}	: Set of charging stations and customers, $N_{RC} = N_R \cup N_C$
N_{SRC}	: Set of satellites, charging stations and customers, $N_{SRC} = N_S \cup N_R \cup N_C$
N_{RCS}	: Set charging stations, customers and dummy satellites, $N_{RCS} = N_R \cup N_C \cup N'_S$
N_{SRCS}	: Set of complete nodes in the second echelon, $N_{SRCS} = N_S \cup N_R \cup N_C \cup N'_S$

problem instances. CMSA algorithms are based on applying an exact solver iteratively to sub-instances of the original problem instance. In other words, the search space is reduced before the exact solver is applied. Search space reduction is achieved in a bottom-up manner by, first, probabilistically generating valid solutions to the tackled problem and, second, by merging these solutions in order to obtain a sub-instance. Adapt-CMSA [3], on the other hand, was developed to reduce the parameter sensitivity of the standard version of CMSA. Examples of applications of CMSA can be found in [1,4,15].

Related Literature. The two-echelon electric vehicle problem is regarded as a combination of two initially independent research lines: the one on electric vehicle routing problems (EVRPs) and the one on two-echelon vehicle routing problems (2E-VRPs); see [10] and [12]. Works on EVRPS include [5,19], while [24] is a recent example for work on 2E-VRPs. On the other hand, there is still a dearth of literature on two-echelon electric vehicle routing problems (2E-EVRPs). Jie et al. [17] were among the first to propose a 2E-EVRP with battery-swapping stations (BSS). A hybrid algorithm that combines column generation and LNS is proposed to solve the addressed problem. Breunig et al. [7] proposed a metaheuristic approach based on LNS and an exact mathematical programming algorithm that employs decomposition and pricing techniques to solve 2E-EVRP. Moreover, Cao et al. [8] investigated the design of a two-echelon reverse logistics network to collect recyclable waste utilizing a mixed fleet of electric and conventional vehicles. Instead of a single integrated mathematical model, the authors formulated the addressed problem as two distinct models, one for each echelon. Furthermore, Wu and Zhang [26] developed a branch and price algorithm to solve a 2E-EVRP. They tested the proposed solution approach on small and medium-sized instances containing up to 20 customers and two charging stations. Recently, Wang and Zhou [25] introduced a 2E-EVRP with time windows and battery-swapping stations. They developed a MILP model that minimizes transportation, handling, and fixed costs for the vehicles used in the first and second echelons, in addition to battery-swapping costs. However, the time spent on battery swapping is not considered. A variable neighbourhood search (VNS) algorithm was proposed to solve large-sized problem instances.

Finally, Akbay et al. [2] developed a three-index node-based MILP model for
2E-EVRP with time windows and proposed a VNS algorithm to solve large-
sized problem instances. In addition to classical VNS neighbourhood operators,
they employed destroy and repair operators in the shaking step of the algorithm.

2   Problem Description

In the following, we provide a technical description together with a MILP model
of the 2E-EVRP-SPD. For this purpose, the sets and notations from the Table 1
are required. The 2E-EVRP-SPD can be defined on a complete, directed graph
$G(N, A)$ that is formed by the following subsets of nodes: the set of central
warehouses (also called depots) (N_D), the set of satellites (N_S), the set of charging stations (N_R), and the set of customers (N_C). Note that N_S and N_R also
include multiple copies of each satellite and charging station to allow multiple visits to any of the satellites and charging stations. The set of arcs on the
other hand (A) includes (1) arcs that connect central warehouses and satellites
$A^1 = \{(i, j) \mid i \neq j \text{ and } i, j \in N_{DSD}\}$ and (2) arcs that connect satellites, customers and charging stations $A^2 = \{(l, m) \mid l \neq m \text{ and } l, m \in N_{SRCS}\}$. Each
arc $(i, j) \in A^1$ is associated with a distance d^1_{ij} and each arc $(l, m) \in A^2$ is
associated with a distance d^2_{lm}.

Two different fleets of vehicles, each one homogeneous in itself, serve in the first
and second echelons in order to meet customer demands. A fleet of large trucks
with internal combustion engines are located in a central warehouse and transfer
products between the central warehouses and the satellites, while a fleet of electric
vehicles is present at the satellites and transfer products between satellites and
customers (demand points). In the first echelon, a truck with a loading capacity
of Q^1 starts its tour from a central warehouse, visits one or more satellites, and
returns to the central warehouse from which the tour started. Not all satellites have
to be visited by large trucks unless there is a demand for pickup and/or delivery.
Furthermore, a satellite can be visited by multiple large vehicles if the delivery or
pickup demand of the satellite exceeds the vehicle capacity. In the second echelon,
on the other hand, each customer with a delivery demand $D^2_i > 0$ or a pickup
demand $P^2_i > 0$ (or both) must be served by an electric vehicle with a loading
capacity of Q^2. An electric vehicle starts its tour with a fully charged battery (B)
and the vehicle's battery is consumed in proportion to the distance travelled. The
constant h represents the battery consumption rate of an electric vehicle per unit
distance travelled. If a charging station is visited, the electric vehicle's battery is
fully charged up to level B with a constant charging rate of $g > 0$.

Our MILP model contains the following binary decision variables. A decision variable x_{ij} takes value 1 if arc $(i, j) \in A^1$ is traversed, and 0 otherwise.
Moreover, a decision variable y_{ij} takes value 1 if arc $(i, j) \in A^2$ is traversed,
and 0 otherwise. Next, decision variables $BSCa_i$ and $BSCd_i$ record the battery
state of charge on arrival, respectively departure, at (from) vertex $i \in N_{SRCS}$.
Furthermore, for each arc $(i, j) \in A^1$, variable u^1_{ij} denotes the remaining cargo
to be delivered to satellites of the route, while v^1_{ij} denotes the amount of cargo
already collected (picked up) at already visited satellites. Similarly, for each arc

$(i, j) \in A^2$, variable u_{ij}^2 denotes the remaining cargo for the route, while v_{ij}^2 denotes the amount of cargo already collected at visited customers. Since the demand of each satellite depends on the customers serviced through it, decision variables D_i^1 and P_i^1 are introduced to calculate, respectively, the delivery and pickup demands of satellites. Finally, variable z_{ij} takes value 1 if customer (i) is serviced from satellite (j), and 0 otherwise. The MILP model can then be stated as follows.

$$\textbf{Min} \sum_{i \in N_{DS}} \sum_{j \in N_{DSD}} d_{ij}^1 * x_{ij} + \sum_{l \in N_{SRC}} \sum_{m \in N_{SRCS}} d_{lm}^2 * y_{lm} + \sum_{j \in N_{DSD}} x_{0j} * c^{lv}$$

$$+ \sum_{i \in N_S} \sum_{j \in N_{SRCS}} y_{ij} * c^{ev} \tag{1}$$

$$\sum_{j \in N_{SD}} x_{ij} \leq 1 \qquad\qquad \forall i \in N_S \tag{2}$$

$$\sum_{i \in N_{DS}, i \neq j} x_{ij} - \sum_{i \in N_{SD}, i \neq j} x_{ji} = 0 \qquad\qquad \forall j \in N_S \tag{3}$$

$$\sum_{i \in N_{DS}, i \neq j} u_{ij}^1 - \sum_{i \in N_{SD}, i \neq j} u_{ji}^1 = D_j^1 \qquad\qquad \forall j \in N_S \tag{4}$$

$$\sum_{i \in N_{DS}, i \neq j} v_{ij}^1 - \sum_{i \in N_{SD}, i \neq j} v_{ji}^1 = P_j^1 \qquad\qquad \forall j \in N_S \tag{5}$$

$$0 \leq u_{ij}^1 \leq Q^1 \qquad\qquad \forall i \in N_D, j \in N_{DS} \tag{6}$$

$$v_{ij}^1 = 0 \qquad\qquad \forall i \in N_D, j \in N_{SD} \tag{7}$$

$$u_{ij}^1 + v_{ij}^1 \leq Q^1 * x_{ij} \qquad\qquad \forall i \in N_D, j \in N_{DS}, i \neq j \tag{8}$$

$$\sum_{l \in N_C} z_{li} * D_l^2 = D_i^1 \qquad\qquad \forall i \in N_S \tag{9}$$

$$\sum_{l \in N_C} z_{li} * P_l^2 = P_i^1 \qquad\qquad \forall i \in N_S \tag{10}$$

$$\sum_{j \in N_{RCS}, i \neq j} y_{ij} = 1 \qquad\qquad \forall i \in N_C \tag{11}$$

$$\sum_{j \in N_{RCS}, i \neq j} y_{ij} \leq 1 \qquad\qquad \forall i \in N_R \tag{12}$$

$$\sum_{i \in N_{SRC}, i \neq j} y_{ij} - \sum_{i \in N_{RCS}, i \neq j} y_{ji} = 0 \qquad\qquad \forall j \in N_{RC} \tag{13}$$

$$\sum_{i \in N_S} z_{li} = 1 \qquad\qquad \forall l \in N_{RC} \tag{14}$$

$$y_{li} \leq z_{li} \qquad\qquad \forall i \in N_S, l \in N_{RC} \tag{15}$$

$$y_{il} \leq z_{li} \qquad\qquad \forall i \in N_S, l \in N_{RC} \tag{16}$$

$$y_{lm} + z_{li} + \sum_{s \in N_S, i \neq s} z_{ms} \leq 2 \qquad\qquad \forall l, m \in N_{RC}, l \neq m, \forall i \in N_S \tag{17}$$

$$\sum_{i \in N_{SRC}, i \neq j} u_{ij}^2 - \sum_{i \in N_{RCS}, i \neq j} u_{ji}^2 = D_j^2 \qquad\qquad \forall j \in N_{RC} \tag{18}$$

$$\sum_{i \in N_{SRC}, i \neq j} v_{ji}^2 - \sum_{i \in N_{RCS}, i \neq j}' v_{ij}^2 = D_j^2 \qquad \forall j \in N_{RC}$$
(19)

$$0 \leq u_{ij}^2 \leq Q^2 \qquad \forall i \in N_S, j \in N_{RCS}$$
(20)

$$v_{ij}^2 = 0 \qquad \forall i \in N_S, j \in N_{RCS}$$
(21)

$$u_{ij}^2 + v_{ij}^2 \leq Q^2 * y_{ij} \qquad \forall i \in N_{RCS}, j \in N_{SRC}$$
(22)

$$0 \leq BSCa_j \leq BSCa_i - (hd_{ij})y_{ij} + B(1 - y_{ij}) \qquad \forall i \in N_C, \forall j \in N_{RCS}, i \neq j$$
(23)

$$0 \leq BSCa_j \leq BSCd_i - (hd_{ij})y_{ij} + B(1 - y_{ij}) \qquad \forall i \in N_{SR}, \forall j \in N_{RCS}, i \neq j$$
(24)

$$BSCa_i \leq BSCd_i \leq B \qquad \forall i \in N_{SR}$$
(25)

$$x_{ij} \in 0, 1 \qquad \forall i \in N_{SRC}, j \in N_{RCS}, l \neq m$$
(26)

$$y_{lm} \in 0, 1 \qquad \forall l \in N_{SRC}, m \in N_{RCS}, l \neq m$$
(27)

In this study, solutions using fewer vehicles—that is, with fewer routes—are preferred over others, even if the total distance travelled is higher than in other routes. Therefore, the objective function does not only consider the travelled distance but adds also an extra cost c^{lv} for each large vehicle used in the first echelon and c^{ev} for each electric vehicle used in the second echelon. Note, in this context, that the number of large vehicles used in a solution is equal to the number of variables on outgoing arcs of a central warehouse with a value of 1. Moreover, the number of electric vehicles used in a solution is equal to the number of variables on outgoing arcs of satellites that have a value of 1. In this way, the objective function (1) minimizes the sum of the total distance travelled and the vehicle costs. Constraints (2) control the connectivity of satellites and constraints (3) guarantee the balance of flow in the first echelon nodes. Constraints (4)-(8) guarantee that the delivery and pickup demands of satellites are satisfied simultaneously by the large vehicles serving in the first echelon. Constraints (9) and (10) determine each satellite's delivery and pickup demand to be the total delivery and pickup demands of those customers served by the relevant satellite. Constraints (11) and (12) control the connectivity of customers and charging stations. Constraints (13) ensure the balance of flow for the second echelon nodes. Constraints (14) guarantee that a customer receives service from only one satellite. Constraints (15)-(17) ensure that a tour started from a satellite ends at the same satellite. Constraints (18)-(22) guarantee that the delivery and pickup demands of customers are satisfied simultaneously by the electric vehicles serving in the second echelon. Finally, constraints (23)-(25) are battery state constraints.

3   Adapt-CMSA for the 2E-EVRP-SPD

In this section, we will describe the Adapt-CMSA algorithm that we designed for the application to the 2E-EVRP-SPD. However, before describing the algorithm we first explain the solution representation.

Solution Representation. Any solution $S = (R^1, R^2)$ produced by the algorithm consists of two sets of routes, R^1 and R^2, where R^1 contains the routes of large vehicles in the first echelon and R^2 contains the routes of the electric vehicles in the second echelon. Each route $t^1 \in R^1$ starts from a central warehouse, visits one or more satellites and returns to the same central warehouse. Each route $t^2 \in R^2$ starts from a satellite, visits a number of locations $v \in N_{RC}$ and returns to the same satellite. As an example, let vector $\mathbf{I}$ contain the complete set of node indexes of an example problem instance with one central warehouse, two satellites, three charging stations and five customers.

$$\mathbf{I} = (\underbrace{0,}_{\text{central warehouse}} \underbrace{1, 2,}_{\text{satellites}} \underbrace{3, 4, 5,}_{\text{charging stations}} \underbrace{6, 7, 8, 9, 10,}_{\text{customers}})$$

A solution S that contains one route in the first echelon (t^1_1) and two routes in the second echelon $(t^2_1$ and t^2_2) is represented as follows:

$$S = (R^1, R^2) \text{ where } \begin{array}{l} R^1 = \left\{ \quad t^1_1 = \{0 \to 1 \to 2 \to 0\} \quad \right\} \\[4pt] R^2 = \left\{ \begin{array}{l} t^2_1 = \{1 \to 7 \to 3 \to 6 \to 1\} \\ t^2_2 = \{2 \to 9 \to 8 \to 5 \to 10 \to 2\} \end{array} \right\} \end{array}$$

Set Covering Based Model. As described in Sect. 4.1, any solution produced by the algorithm is kept in the form of two sets of routes. Similarly, a sub-instance $C = (C^1, C^2)$ in the context of our Adapt-CMSA algorithm consist of two sets, C^1 and C^2, containing those routes that were previously generated by the probabilistic application of solution construction heuristics. For solving such a sub-instance, in this study, we make use of the following set-covering-based MILP model. Note that the other option would have been to use the MILP model outlined in the previous section, with those arc variables fixed to zero whose corresponding arcs do not form part of any of the solutions that were merged into the sub-instance.

Given a sub-instance $C = (C^1, C^2)$, each route $r^k \in C^k$ is associated with a distance, resp. cost, value d^k_r. Moreover, ld^2_{sr} and lp^2_{sr} represent the total delivery and pickup loads of the route $r \in C^2$ serving from satellite $s \in N_S$. As described in Sect. 2, D^1_s and P^1_s refer to delivery and pickup demands of the respective satellite. Parameter a^1_{sr} is set to value one if satellite s is traversed by route r, and 0 otherwise. Moreover, parameter a^2_{ir} is set to value one if customer i is visited by route r, and 0 otherwise. The binary decision variable x_r^k takes value one if the route in the k-th echelon is selected, value zero otherwise. Moreover, dp_{sr} and pp_{sr} refer to the amount of goods delivered to the satellite s by the route r. The set-covering-based model can then be stated as follows.

$$\textbf{Min} \quad \sum_{r \in C^1} d_r^1 * x_r^1 + \sum_{r \in C^2} d_r^2 * x_r^2 + \sum_{r \in C^1} c^{lv} * x_r^1 + \sum_{r \in C^2} c^{ev} * x_r^2 \tag{28}$$

$$\textbf{s.t.} \quad \sum_{r \in C^2} ld_{sr}^2 = D_s^1 \qquad\qquad \forall s \in N_s \tag{29}$$

$$\sum_{r \in C^2} lp_{sr}^2 = P_s^1 \qquad\qquad \forall s \in N_s \tag{30}$$

$$\sum_{r \in C^1} pd_{sr} * a_{sr}^1 = D_s^1 \qquad\qquad \forall s \in N_s \tag{31}$$

$$\sum_{r \in C^1} pp_{sr} * a_{sr}^1 = P_s^1 \qquad\qquad \forall s \in N_s \tag{32}$$

$$\sum_{s \in N_S} pd_{sr} \leq Q^1 * x_r^1 \qquad\qquad \forall r \in C^1 \tag{33}$$

$$\sum_{r \in C^1} pp_{sr} \leq Q^1 * x_r^1 \qquad\qquad \forall r \in C^1 \tag{34}$$

$$\sum_{r \in C^2} a_{ir}^2 * x_r^2 >= 1 \qquad\qquad \forall i \in V \tag{35}$$

$$x_{r_1}^1, x_{r_2}^2 \in 0,1 \qquad \forall r_1 \in C^1, r_2 \in C^2, k \in \{1,2\} \tag{36}$$

$$pd_{sr}, pp_{sr} \geq 0 \qquad\qquad \forall s \in N_S, r \in C^1 \tag{37}$$

The objective function minimizes the sum of the total distance travelled and vehicle costs. Constraints (29) and (30) determine each satellite's delivery and pickup demands. Constraints (31) and (32) ensure that partial deliveries are allowed in the first echelon in case the delivery or pickup demand of the relevant satellite exceeds the capacity of a large vehicle. Constraints (33) and (34) guarantee that the total delivery and pickup load in large vehicles can not exceed the vehicle capacity. Constraint (35) ensures that each customer must be visited at least once. Finally, constraints (36) and (37) control variable domains.

The Adapt-CMSA Algorithm. Algorithm 1 shows the pseudo-code of our Adapt-CMSA algorithm for the 2E-EVRP-SPD. First, the best-so-far solution S^{bsf} is initialized as an empty solution. Then, parameters α_{bsf}, n_a and l_{size} are initialized in lines 4 and 5. The handling of these parameters will be described in detail below.

At each iteration, Adapt-CMSA builds a sub-instance C of the original problem instance as follows. First, C is initialized to the best-so-far solution S^{bsf}. Then, a number of n_a solutions are probabilistically constructed in lines 8–12. The function for the construction of a solution, ProbabilisticSolutionConstruction(S^{bsf}, α_{bsf}, l_{size}), receives—apart from the best-so-far-solution S^{bsf}—two

---

Algorithm 1. Adapt-CMSA for the 2E-EVRP-SPD

1: **input 1:** values for CMSA parameters t_{prop}, t_{ILP}
2: **input 2:** values for solution construction parameters α^{LB}, α^{UB}, α_{red}
3: $S^{bsf} := \emptyset$
4: $\alpha_{bsf} := \alpha^{UB}$
5: Initialize(n_a, l_{size})
6: **while** CPU time limit not reached **do**
7:    $C := S^{bsf}$
8:    **for** $i := 1, \ldots, n_a$ **do**
9:       $S := $ ProbabilisticSolutionConstruction(S^{bsf}, α_{bsf}, l_{size})
10:       LocalSearch1(S)
11:       $C := $ Merge(C, S)
12:    **end for**
13:    $(S^{cplex}, t_{solve}) := $ SolveSubinstance(C, t_{ILP}) {This function returns two objects: (1) the obtained solution (S^{cplex}), (2) the required computation time (t_{solve})}
14:    RemoveDuplicates(S^{cplex})
15:    LocalSearch2(S^{cplex})
16:    **if** $t_{solve} < t_{prop} \cdot t_{ILP}$ and $\alpha_{bsf} > \alpha^{LB}$ **then** $\alpha_{bsf} := \alpha_{bsf} - \alpha_{red}$ **end if**
17:    **if** $f(S^{cplex}) < f(S^{bsf})$ **then**
18:       $S^{bsf} := S^{cplex}$
19:       Initialize(n_a, l_{size})
20:    **else**
21:       **if** $f(S^{cplex}) > f(S^{bsf})$ **then**
22:          **if** $n_a = n^{init}$ **then** $\alpha_{bsf} := \min\{\alpha_{bsf} + \frac{\alpha_{red}}{10}, \alpha^{UB}\}$ **else** Initialize(n_a, l_{size}) **end if**
23:       **else**
24:          Increment(n_a, l_{size})
25:       **end if**
26:    **end if**
27: **end while**
28: **output:** S^{bsf}

---

parameters as input. Here, parameter α_{bsf} (where $0 \leq \alpha_{bsf} < 1$) is used to bias the construction of new solutions towards the best-so-far solution S^{bsf}. More specifically, the similarity between the constructed solutions and S^{bsf} will increase with a growing value of α_{bsf}. Parameter l_{size} controls the number of considered options at each solution construction step. A higher value of l_{size} results in more diverse solutions which, in turn, leads to a larger sub-instance. After the construction of a solution S (line 9), a local search is applied to each route $t^2 \in R^2$, see line 10. Well-known intra-route operators such as, *relocation*, *swap* and *two_opt* are sequentially utilized to improve each route. Each operator uses the best-improvement strategy. The *relocation* operator removes each customer from its current position and inserts it into a different position in the same route. The *swap* neighbourhood considers changing the positions of two selected nodes of the same route. Finally, the *two_opt* neighbourhood considers all possibilities of selecting two non-consecutive nodes in the same route and reversing the node sequence between the two selected nodes.

After the application of local search, the so-called *merge step* is performed in function $\mathsf{Merge}(C, S)$. In particular, every route from S^1 is added to C^1 and every route from S^2 is added to C^2. After probabilistically constructing n_a solutions and merging them to form the sub-instance C, the sub-instance is solved with CPLEX, which is precisely done in function $\mathsf{SolveSubinstance}(C, t_{\mathrm{ILP}})$; see line 13. Hereby, t_{ILP} is the CPU time limit for the application of CPLEX, which is applied to the set-covering model from Sect. 3. Note that the output S^{cplex} of function $\mathsf{SolveSubinstance}(C, t_{\mathrm{ILP}})$ is—due to the computation time limit—not necessarily an optimal solution to the sub-instance. Since the set-covering-based model potentially allows customers to be visited more than once, S^{cplex} may contain some of the customers in multiple routes. In that case, redundant customers are removed using function $\mathsf{RemoveDuplicates}(S^{\mathrm{cplex}})$, see line 14. This function first determines all redundant customers and calculates the distance between the respective customer and the two adjacent nodes. Then, it removes all redundant customers starting from the one with the highest distance value until all customers only appear in exactly one route. Subsequently, a local search procedure is applied to S^{cplex} using inter-route neighbourhood operators *exchange (1,1)* and *shift (1,0)*. The *exchange (1,1)* neighbourhood considers all exchanges of two customers not in the same route. The *shift (1,0)* neighbourhood looks at all possibilities of removing a customer from its current route and inserting it at any position in the other routes. Both operators are applied based on the best-improvement strategy.

The self-adaptive aspect of Adapt-CMSA is to be found in the dynamic change of parameters α_{bsf}, n_a and l_{size}. In the first place, we will describe the adaptation of parameter α_{bsf}. First of all, Adapt-CMSA requires a lower bound α^{LB} and an upper bound α^{UB} for the value of α_{bsf} as input. In addition, the step size α_{red} for the reduction of α_{bsf} must also be given as input. Adapt-CMSA starts by setting α_{bsf} to the highest possible value α^{UB}; see line 4.[3] In case the resulting MILP can be solved in a computation time t_{solve} which is below a proportion t_{prop} of the maximally possible computation time t_{ILP}, the value of α_{bsf} is reduced by α_{red}; see line 22. The rationale behind this step is as follows. In case the resulting MILP can easily be solved to optimality, the search space is too small, caused by a rather low number of routes in C^1 and C^2. In order to increase the size of the MILP, the solutions constructed in $\mathsf{ProbabilisticSolutionConstruction}(S^{\mathrm{bsf}}, \alpha_{\mathrm{bsf}}, l_{size}, d_{\mathrm{rate}}, h_{\mathrm{rate}})$ should be more different to S^{bsf}, which can be achieved by reducing the value of α_{bsf}.

The adaptation of parameters n_a and l_{size} is done in a similar way and with a similar purpose. These parameters are set to their initial values, that is, $n_a := n^{\mathrm{init}}$ and $l_{\mathrm{size}} := l_{\mathrm{size}}^{\mathrm{init}}$ in function $\mathsf{Initialize}(n_a, l_{\mathrm{size}})$, which is called at three different occasions: (1) at the start of the algorithm (line 5), (2) whenever solution S^{cplex} is strictly better than S^{bsf} (line 19), and (3) whenever solution S^{cplex} is strictly worse than S^{bsf} and, at the same time, $n_a > n^{\mathrm{init}}$ (line 22). On the other side, in those cases in which S^{cplex} and S^{bsf} are of the same quality,

---

[3] Remember that solutions constructed with a high value of α_{bsf} will be rather similar to S^{bsf}.

the algorithm can afford to generate larger sub-instances and therefore, the values of the two parameters are incremented in function $\mathsf{Increment}(n_a, l_{\mathrm{size}})$. More specifically, n_a is incremented by n^{inc} and l_{size} is incremented by $l_{\mathrm{size}}^{\mathrm{inc}}$.

Solution Construction. When function $\mathsf{ProbabilisticSolutionConstruction}(S^{\mathrm{bsf}}, \alpha_{\mathrm{bsf}}, l_{size})$ is called, one of two heuristics is randomly selected for solution construction. The first one is our version of the C&W savings algorithm [9], while the second one is our insertion algorithm. In the following, both construction algorithms are described in detail.

1. Probabilistic Clark & Wright (C&W) Savings Algorithm. Our probabilistic version of the C&W savings algorithm starts by assigning each customer either to the nearest satellite or to the satellite to which is assigned in S^{bsf}. After the assignment, set $N_C^s \subseteq N_C$ contains all customers assigned to satellite s, for all $s \in N_S$. Then, the following C&W savings procedure is applied concerning each satellite $s \in N_S$. First, a set of direct routes $R^2 = \{(s - i - s) \mid i \in N_C^s\}$ is created. Subsequently, a savings list L that contains all possible pairs (i, j) of nodes (customers and charging stations) together with their respective savings value σ_{ij} is generated. Hereby, σ_{ij} is calculated as follows:

$$\sigma_{ij} := d_{si}^2 + d_{sj}^2 - \lambda d_{ij}^2 + \mu |d_{si}^2 - d_{sj}^2| \tag{38}$$

The so-called *route shape parameter* λ adjusts the selection priority based on the distance between nodes i and j [27], while μ is used to scale the asymmetry between nodes i and j [22]. Note that well-working values for these parameters are obtained by parameter tuning which is presented in Sect. 5. Note also that the savings list L contains, at all times, only those entries (i, j) such that (1) node i and node j belong to different routes, and (2) both i and j are directly connected to the satellite of their route. For executing the C&W savings algorithm, the following list of steps is iterated until the savings list L is empty.

1. First, based on the current savings values of the entries in L, a new value q_{ij} is calculated for each entry $(i, j) \in L$ as follows:

$$q_{ij} := \begin{cases} (\sigma_{ij} + 1) \cdot \alpha_{\mathrm{bsf}} & \text{if } S_{ij}^{\mathrm{bsf}} = 1 \\ (\sigma_{ij} + 1) \cdot (1 - \alpha_{\mathrm{bsf}}) & \text{otherwise} \end{cases} \tag{39}$$

Here, $S_{ij}^{\mathrm{bsf}} = 1$ if node i and node j are successively visited in at least one route of S^{bsf}, and 0 otherwise. The savings list L is then sorted according to non-increasing values of q_{ij}. Finally, a reduced saving list L_r that contains the first (maximally) l_{size} elements of the whole savings list is created.

2. Next, an entry (i, j) is chosen from L_r with respect to the following probabilities:

$$\mathbf{p}(ij) := \frac{q_{ij}}{\sum_{(i',j') \in L_r} q_{i'j'}} \quad \forall \, (i, j) \in L_r \tag{40}$$

Note that, the higher the value of parameter $\alpha_{\mathrm{bsf}} \in [0, 1]$, the stronger is the bias towards choosing arcs—that is, transitions from a customer i to a customer j—that appear in the best-so-far solution S^{bsf}.

3. Then, the two routes corresponding to nodes i and j are merged. The four possible cases for merging two routes are as follows:

Case1: t_1^2 :< s-i-...-s > t_2^2 :< s-j-...-s > rev$(t_1^2) - t_2^2$ t_m^2 :< s-...-i-j...-s >
Case2: t_1^2 :< s-i-...-s > t_2^2 :< s-...-j-s > rev(t_1^2) - rev(t_2^2) t_m :< s-...-i-j...-s >
Case3: t_1^2 :< s-...-i-s > t_2^2 :< s-j-...-s > $t_1^2 - t_2^2$ t_m^2 :< s-...-i-j...-s >
Case4: t_1^2 :< s-...-i-s > t_2^2 :< s-...-j-s > t_1^2 - rev(t_2^2) t_m :< s-...-i-j...-s >

Based on the way in which nodes i and j are directly connected to a satellite, one or both of the routes must be reversed in order to be able to connect nodes i and j. In this context, note that the reversed version of a route t_1^2 is denoted by rev(t_1^2). If the merged route t_m^2 is infeasible in terms of vehicle capacity, the merged route is eliminated and the respective pair of nodes is removed from the savings list. A new candidate is selected following the procedure in the previous step. If the merged route is battery infeasible, a charging station is inserted into the infeasible route. The corresponding procedure determines the first customer in the route at which the vehicle arrives with a negative battery level and inserts the charging station between this customer and the previous customer. For this purpose, the charging station that least increases the route distance is selected and inserted between the respective nodes. If this insertion is not feasible, the previous arcs are considered instead in the same manner. In those cases in which the route is still infeasible after charging station insertion, it is eliminated, and the respective pair of nodes are removed from the savings list. A new candidate is selected following the procedure described in the previous step. This procedure is repeated while the savings list is not empty. After merging, some of the charging stations that were previously added to the routes may become redundant. Those charging stations are removed from the merged route.

4. The savings list L must be updated as described above.

After constructing the routes in the second echelon, the same procedure is applied to construct routes for the large vehicles in the first echelon. The first difference in the procedure for first echelon routes is that all aspects related to batteries and charging stations are not considered. Second, a satellite is allowed to be visited by multiple large vehicles in case the demand exceeds the vehicle's loading capacity.

2. Probabilistic Insertion Algorithm. This heuristic constructs a solution by sequentially inserting each customer into the available routes until no unvisited customer remains. Similar to the C&W savings algorithm, the algorithm first constructs the routes for the second echelon. The first route is initialized by inserting a randomly chosen customer between the satellite that is nearest to this customer. Then, a cost list L formed by all unvisited customers and all possible insertion positions together with their respective cost values is generated. The insertion cost of customer i between nodes j and k is calculated using the following equation: $c(j, i, k) = d_{ji}^2 + d_{ik}^2 - d_{jk}^2$. Then, q_{ij} is calculated for each entry $(j, i, k) \in L$ as follows:

$$
q_{jik} := \begin{cases} (c(j,i,k) + 1) \cdot (1 - \alpha_{\mathrm{bsf}})(1 - \alpha_{\mathrm{bsf}}) & \text{if } S_{ji}^{\mathrm{bsf}} = 1 \text{ and } S_{ik}^{\mathrm{bsf}} = 1 \\ (c(j,i,k) + 1) \cdot \alpha_{\mathrm{bsf}} & \text{if } S_{ji}^{\mathrm{bsf}} = 0 \text{ and } S_{ik}^{\mathrm{bsf}} = 0 \\ (c(j,i,k) + 1) \cdot \alpha_{\mathrm{bsf}}(1 - \alpha_{\mathrm{bsf}}) & \text{otherwise} \end{cases}
$$

$$(41)$$

Next, an entry (j, i, k) is chosen from L_r with respect to the probabilities calculated using Eq. (40). The customer is inserted into the respective position if the vehicle capacity allows for this. Moreover, in case of battery infeasibility, a charging station is inserted into the route as explained above during the description of the C&W savings algorithm. If the insertion leads to infeasibility in terms of vehicle load capacity, a new tour is initialized with the respective customer and the nearest satellite.

4   Experimental Evaluation

All experiments were performed on a cluster of machines with Intel® Xeon® 5670 CPUs with 12 cores of 2.933 GHz and a minimum of 32 GB RAM. CPLEX version 20.1 was used in one-threaded mode within Adapt-CMSA for solving the respective sub-instances and in standalone mode for solving the MILP models representing the complete problem instances.

Problem Instances. A subset of the 2E-EVRP problem instances introduced by [7] were used to test the performance of the proposed algorithm. In particular, from each category (small, medium, large) we chose one set of instances, henceforth called Set1, Set2 and Set3. Instance characteristics are presented in the first three columns of each result table. Since the original problem instances only come with delivery demands, we had to modify them by adding pickup demands. For this purpose, the delivery demand of each customer was separated into delivery and pickup demands using the approach from [23]. The resulting instances are provided at https://github.com/manilakbay/2E-EVRP-SPD.

Parameter Tuning. In order to find well-working parameter values for Adapt-CMSA we utilized the scientific tuning software irace [21]. Instances 100-5-1, 100-5- 2b, 100-10-1, 100-10-2b, 200-10-1, and 200-10-2b were used for the tuning process. Note that in the case of numerical parameters, the precision of irace was fixed to two positions behind the comma. irace was applied with a budget of 2000 algorithm applications. The time limit for each problem instance was set to 900 CPU seconds. A summary of the parameters, their domains, and values selected for the final experiments are provided in Table 2.

Computational Results. In the context of small problem instances, we compare the performance of Adapt-CMSA with the standalone application of CPLEX. As CPLEX is not applicable in a standalone manner to the large-size problem instances, we compare Adapt-CMSA with our probabilistic C&W savings algorithm (pC&W) and our probabilistic sequential insertion algorithm

Table 2. Parameters, their domains, and the chosen values as determined by irace.

Parameter	Domain	Value	Description
λ	$[1,2]$	1.67	Route shape parameter (C&W algorithm)
μ	$[0,1]$	0.32	Asymmetry scaling (C&W algorithm)
l_{size}^{init}	$\{3,5,10,15,20\}$	15	Initial list size value
l_{size}^{inc}	$\{1,3,5,10,20\}$	20	List size increment
n^{init}	$\{100,200,300,500,1000\}$	1000	Initial nr. of constructed solutions
n^{inc}	$\{50,100,200,300,400,500\}$	100	Increment for the nr. of constructed solutions
t_{ILP}	$\{5,10,15,20,25\}$	20	CPLEX time limit (seconds)
α^{LB}	$[0.6,0.99]$	0.77	Lower bound for α_{bsf}
α^{UB}	$[0.6,0.99]$	0.8	Upper bound for α_{bsf}
α_{red}	$[0.01,0.1]$	0.02	Step size reduction for α_{bsf}
t_{prop}	$[0.1,0.8]$	0.79	Control parameter for bias reduction

(pSI). The parameters of both algorithms were set in the same way as for their application within Adapt-CMSA. Moreover, the same computation time limit was used as for Adapt-CMSA, that is, both algorithms were repeatedly applied until a computation time limit of 150 CPU seconds (small problem instances), respectively 900 CPU seconds (large problem instances), was reached. Moreover, Adapt-CMSA, pC&W and pSI were applied 10 times to each problem instance. In order to make a fair comparison, after each application of pC&W and pSI, LocalSearch1() and LocalSearch2() are sequentially applied in order to improve the generated solutions. Finally, note that we fixed the cost of each large vehicle used in a solution to 1500 and the cost of each electric vehicle used in a solution to 1000.

The structure of the result tables is as follows. Instance names are given in the first column. The subsequent block of three columns indicates the number of satellites, the number of charging stations and the number of customers. Columns 'n_{lv}' and 'n_{ev}' provide the number of large, respectively electric, vehicles utilized by the respective solutions. In the case of Adapt-CMSA, pC&W and pSI these numbers refer to the best solution found within 10 independent runs. In the case of Adapt-CMSA, pC&W, and pSI, columns 'best' show the distance values of the best solutions found in 10 runs, while additional columns with the heading 'avg.' provide the average distance values of the best solutions of each of the 10 runs. Next, columns with the heading 'time' show the computation time (in seconds) of CPLEX and the average computation times of Adapt-CMSA to find the best solutions in each run. Note that the time limit for CPLEX was set to 12 h. Finally, columns 'gap(%)' provide the gap (in percentage) between the best-found solutions and the best lower bounds found by CPLEX. Note that, in case the gap value is zero, CPLEX has found an optimal solution.

The following observations can be made. First, CPLEX was only able to provide feasible solutions for nine out of 12 small and medium-sized problem instances, without being able to prove optimality within the computation time of 12 h; see Tables 3 and 4. For the remaining three small and medium-sized instances, CPLEX could not even find a feasible solution. Adapt-CMSA could find the same results provided by CPLEX for seven of these instances. In the case of five instances, Adapt-CMSA was even able to improve the solutions obtained by CPLEX, respectively was able to provide a solution in those cases in which CPLEX failed to provide one. Moreover, the average computation time required for Adapt-CMSA to find the best solution in each run is considerably lower than the time required for CPLEX. More specifically, while CPLEX found its best solutions on average in almost 12 h, Adapt-CMSA was able to do on average in approx. 10 s for the small-sized instances. Concerning the large-sized instances, see Table 5, Adapt-CMSA significantly outperforms both pC&W and pSI, both in terms of best performance and average performance. Note that, even though pC&W provides better distance value than Adapt-CMSA, this could be achieved using more electric vehicles than the solution found by Adapt-CMSA.

Table 3. Computational results for small-sized instances - Set1.

Instances	Characteristics			CPLEX					Adapt-CMSA				
name	n_s	n_r	n_c	n_{lv}	n_{ev}	best	time	gap(%)	n_{lv}	n_{ev}	best	avg.	time
n22-k4-s6-17	2	4	21	2	3	**5174**	43076.7	12.8	2	3	**5174**	5174.0	2.1
n22-k4-s8-14	2	4	21	2	3	**4870**	7465.1	15.4	2	3	**4870**	4870.0	1.9
n22-k4-s9-19	2	4	21	2	3	**4750**	43070.1	8.2	2	3	**4750**	4750.0	3.1
n22-k4-s10-14	2	4	21	2	3	**5442**	43075.3	19.9	2	3	**5442**	5442.0	5.9
n22-k4-s11-12	2	4	21	2	3	5357	43019.9	34.8	2	3	**5290**	5318.8	9.9
n22-k4-s12-16	2	4	21	2	3	**3691**	43074.0	8.4	2	3	**3691**	3695.7	3.4
average	-	-	-	2	3	4880.7	37130.2	16.6	2	3	**4869.5**	**4875.1**	4.4

Table 4. Computational results for medium-sized instances - Set2.

Instances	Characteristics			CPLEX					Adapt-CMSA				
name	n_s	n_r	n_c	n_{lv}	n_{ev}	best	time	gap(%)	n_{lv}	n_{ev}	best	avg.	time
n33-k4-s1-9	2	4	21	2	3	7506	19749.5	10.2	2	3	**7479**	7499.7	76.5
n33-k4-s2-13	2	4	21	2	3	**7358**	43058.1	12.2	2	3	**7358**	7365.9	68.7
n33-k4-s3-17	2	4	21	-	-	-	-	-	2	3	**7538**	7567.8	42.4
n33-k4-s4-5	2	4	21	-	-	-	-	-	2	3	**7947**	8122.5	14.9
n33-k4-s7-25	2	4	21	2	3	**7880**	43054.2	9.3	2	3	**7880**	7887.8	15.2
n33-k4-s14-22	2	4	21	-	-	-	-	-	2	3	**8173**	8173.0	17.7
average	-	-	-	-	-	-	-	-	2	3	**7729.2**	7769.5	39.2

Table 5. Computational results for large-sized instances - Set3.

Instances	Characteristics			pC&W					pIns					Adapt-CMSA				
name	n_s	n_r	n_c	n_{lv}	n_{ev}	best	avg.	time	n_{lv}	n_{ev}	best	avg.	time	n_{lv}	n_{ev}	best	avg.	time
100-5-1	5	10	100	2	15	13402	13742.7	375.2	2	20	13895	14482.2	552.6	2	13	12428	13050.5	446.5
100-5-1b	5	10	100	2	8	10083	10592.4	534.6	2	9	10278	11024.3	472.2	2	6	9419	9727.6	336.8
100-5-2	5	10	100	2	17	8277	8663.4	582.2	2	20	8729	9153.4	426.5	2	16	7982	8161.3	639.1
100-5-2b	5	10	100	2	8	7173	7352.9	612.1	2	9	7076	7425.2	360.4	2	7	6689	7198.1	494.7
100-5-3	5	10	100	3	15	9480	9610.0	601.7	2	19	9422	10227.3	390.1	2	13	8487	9018.3	674.4
100-5-3b	5	10	100	2	15	8901	9607.7	419.3	2	20	9385	9942.1	367.5	2	13	8581	8984.0	512.1
100-10-1	10	11	100	2	20	9936	10038.1	632.6	2	22	10188	10631.0	421.8	2	16	9767	10176.8	364.4
100-10-1b	10	11	100	2	9	9268	9344.9	456.0	2	10	9252	9593.9	547.9	2	7	8745	9013.6	502.9
100-10-2	10	11	100	2	18	8491	8610.5	560.5	2	20	9218	9380.2	461.3	2	15	8036	8450.4	476.0
100-10-2b	10	11	100	2	9	7769	8000.9	587.6	2	9	7907	8365.5	419.4	2	7	7106	7335.7	377.7
100-10-3	10	11	100	2	18	**8620**	**8734.0**	641.3	2	19	9163	9419.2	269.7	2	13	8803	9038.8	354.1
100-10-3b	10	11	100	2	11	7821	7828.6	276.3	2	8	8189	8433.4	661.6	2	7	7501	7713.9	559.0
200-10-1	10	20	200	2	36	13674	14020.8	612.2	2	46	14458	14902.1	396.3	2	31	13622	14013.9	771.8
200-10-1b	10	20	200	2	18	11356	11699.2	354.8	2	21	11661	12086.5	485.1	2	14	10977	11529.5	777.2
200-10-2	10	20	200	2	34	10718	11037.1	625.9	2	48	12151	12296.5	316.8	2	31	10816	11452.6	698.0
200-10-2b	10	20	200	2	17	8773	8922.9	683.6	2	22	9574	9992.5	478.0	2	15	9110	9501.6	745.9
200-10-3	10	20	200	2	33	14494	14689.4	441.2	2	46	15357	15836.1	299.7	2	29	13869	14247.3	840.9
200-10-3b	10	20	200	2	18	10922	11170.3	472.1	2	22	11357	11785.8	479.5	2	13	9943	10301.8	770.0
average	-	-	-	2.1	17.7	9953.2	10203.7	526.1	2.0	21.7	10403.3	10832.1	433.7	2.0	14.8	9548.9	9939.8	574.5

5 Conclusion and Outlook

This study described the application of a self-adaptive version of the hybrid metaheuristic Adapt-CMSA to the two-echelon electric vehicle routing problem with simultaneous pickup and deliveries. At each iteration, the algorithm first creates a sub-instance of the considered problem instance by merging the best-so-far solution with a number of solutions probabilistically generated using two different solution construction mechanisms, a C&W savings heuristic, and an insertion heuristic. The resulting sub-instance is then solved by the application of the MILP solver CPLEX. Preliminary computational experiments showed that making use of the classical MILP model for the purpose of solving sub-instances was not feasible. Therefore, a set-covering-based model was used and solved with CPLEX. Computational experiments were performed on 12 small- and medium-sized problem instances and on 18 large-sized instances. The proposed approach was evaluated and compared to CPLEX in the context of the small- and medium-sized problem instances, and to probabilistic versions of the C&W savings heuristic and the insertion heuristic for large-sized instances. Numerical results indicated that Adapt-CMSA exhibits superior performance for problem instances of all size ranges. In the future, we aim to deepen the analysis of the algorithm on a wider set of instances.

Acknowledgements. This paper was supported by grants TED2021-129319B-I00 and PID2019-104156GB-I00 funded by MCIN/AEI/10.13039/501100011033. Moreover, M.A. Akbay and C.B. Kalayci received support from the Technological Research Council of Turkey (TUBITAK) under grant number 119M236. The corresponding author was funded by the Ministry of National Education, Turkey (Scholarship Prog.: YLYS-2019).

References

1. Akbay, M.A., Kalayci, C.B., Blum, C.: Application of cmsa to the electric vehicle routing problem with time windows, simultaneous pickup and deliveries, and partial vehicle charging. In: Di Gaspero, L., Festa, P., Nakib, A., Pavone, M. (eds.) Metaheuristics. MIC 2022. LNCS, vol. 13838, pp. 1–16. Springer, Cham (2022). https://doi.org/10.1007/978-3-031-26504-4_1
2. Akbay, M.A., Kalayci, C.B., Blum, C., Polat, O.: Variable neighborhood search for the two-echelon electric vehicle routing problem with time windows. Appl. Sci. **12**(3), 1014 (2022)
3. Akbay, M.A., López Serrano, A., Blum, C.: A self-adaptive variant of CMSA: application to the minimum positive influence dominating set problem. Int. J. Comput. Intell. Syst. **15**(1), 1–13 (2022)
4. Arora, D., Maini, P., Pinacho-Davidson, P., Blum, C.: Route planning for cooperative air-ground robots with fuel constraints: an approach based on CMSA. In: Proceedings of GECCO 2019 - Genetic and Evolutionary Computation Conference, pp. 207–214. Association for Computing Machinery, New York, NY, USA (2019)
5. Asghari, M., Al-e Hashem, S.M.J.M.: Green vehicle routing problem: a state-of-the-art review. Int. J. Prod. Econ. **231**, 107899 (2021)
6. Blum, C., Pinacho Davidson, P., López-Ibáñez, M., Lozano, J.A.: Construct, merge, solve & adapt: a new general algorithm for combinatorial optimization. Comput. Oper. Res. **68**, 75–88 (2016)
7. Breunig, U., Baldacci, R., Hartl, R.F., Vidal, T.: The electric two-echelon vehicle routing problem. Comput. Oper. Res. **103**, 198–210 (2019)
8. Cao, S., Liao, W., Huang, Y.: Heterogeneous fleet recyclables collection routing optimization in a two-echelon collaborative reverse logistics network from circular economic and environmental perspective. Sci. Total Environ. **758**, 144062 (2021)
9. Clarke, G., Wright, J.W.: Scheduling of vehicles from a central depot to a number of delivery points. Oper. Res. **12**(4), 568–581 (1964)
10. Conrad, R.G., Figliozzi, M.A.: The recharging vehicle routing problem. In: Proceedings of the 2011 Industrial Engineering Research Conference, p. 8. IISE Norcross, GA (2011)
11. Crainic, T.G., Ricciardi, N., Storchi, G.: Advanced freight transportation systems for congested urban areas. Transp. Res. Part C Emerg. Technol. **12**(2), 119–137 (2004)
12. Crainic, T.G., Ricciardi, N., Storchi, G.: Models for evaluating and planning city logistics systems. Transp. Sci. **43**(4), 432–454 (2009)
13. Dantzig, G.B., Ramser, J.H.: The truck dispatching problem. Manage. Sci. **6**(1), 80–91 (1959)
14. Dethloff, J.: Vehicle routing and reverse logistics: the vehicle routing problem with simultaneous delivery and pick-up. OR-Spektrum **23**(1), 79–96 (2001)
15. Dupin, N., Talbi, E.G.: Matheuristics to optimize refueling and maintenance planning of nuclear power plants. Journal of Heuristics **27**(1), 63–105 (2021)
16. Eksioglu, B., Vural, A.V., Reisman, A.: The vehicle routing problem: a taxonomic review. Comput. Indus. Eng. **57**(4), 1472–1483 (2009)
17. Jie, W., Yang, J., Zhang, M., Huang, Y.: The two-echelon capacitated electric vehicle routing problem with battery swapping stations: formulation and efficient methodology. Eur. J. Oper. Res. **272**(3), 879–904 (2019)
18. Koç, Ç., Laporte, G., Tükenmez, İ: A review of vehicle routing with simultaneous pickup and delivery. Comput. Oper. Res. **122**, 104987 (2020)

19. Kucukoglu, I., Dewil, R., Cattrysse, D.: The electric vehicle routing problem and its variations: a literature review. Comput. Indus. Eng. **161**, 107650 (2021)
20. Lellis, C.: These 21 companies are switching to electric vehicle fleets (2021). https://www.perillon.com/blog/21-companies-switching-to-electric-vehicle-fleets
21. López-Ibánez, M., et al.: The irace package: iterated racing for automatic algorithm configuration. Oper. Res. Perspect. **3**, 43–58 (2016)
22. Paessens, H.: The savings algorithm for the vehicle routing problem. Eur. J. Oper. Res. **34**(3), 336–344 (1988)
23. Salhi, S., Nagy, G.: A cluster insertion heuristic for single and multiple depot vehicle routing problems with backhauling. J. Oper. Res. Soc **50**(10), 1034–1042 (1999)
24. Sluijk, N., Florio, A.M., Kinable, J., Dellaert, N., Van Woensel, T.: Two-echelon vehicle routing problems: a literature review. Eur. J. Oper. Res. (2022)
25. Wang, D., Zhou, H.: A two-echelon electric vehicle routing problem with time windows and battery swapping stations. Appl. Sci. **11**(22), 10779 (2021)
26. Wu, Z., Zhang, J.: A branch-and-price algorithm for two-echelon electric vehicle routing problem. Complex Intell. Syst. 1–16 (2021)
27. Yellow, P.: A computational modification to the savings method of vehicle scheduling. J. Oper. Res. Soc. **21**(2), 281–283 (1970)
28. Yilmaz, Y., Kalayci, C.B.: Variable neighborhood search algorithms to solve the electric vehicle routing problem with simultaneous pickup and delivery. Mathematics **10**(17), 3108 (2022)

Real-World Vehicle Routing Using Adaptive Large Neighborhood Search

Vojtěch Sassmann[1], Hana Rudová[1]([⊠]) [iD], Michal Gabonnay[2],
and Václav Sobotka[1] [iD]

[1] Faculty of Informatics, Masaryk University, Brno, Czech Republic
hanka@fi.muni.cz
[2] Wereldo.com, Brno, Czech Republic

Abstract. Our work addresses a real-world freight transportation problem with a broad set of characteristics. We build upon the classical work of Ropke and Pisinger [10] and propose an effective realization of the adaptive large neighborhood search (ALNS) with constant time complexity for a large portion of frequent steps in insertion and removal heuristics at the cost of additional pre-calculations. Our minimization process handles different objectives with cost models of heterogeneous vehicles. We demonstrate the generic applicability of the proposed solver on various vehicle routing problems. With the help of the standard Li & Lim benchmarks [6] for pickup and delivery with time windows, we show its capabilities compared to the best-found solutions and the original ALNS. Experiments on real-world delivery routing problems provide a comparison with the original implementation by the company Wereldo in OR-Tools [8], where we achieve significant cost savings, faster runtime, and memory savings by order of magnitude. Performance on large-scale real-world instances with more than 300 vehicles and 1,200 pickup and delivery requests is also presented, achieving less than an hour runtimes.

Keywords: Vehicle routing problem · Pickup and delivery with time windows · Adaptive large neighborhood search · Freight transportation

1 Introduction

With today's scale, freight transportation presents a broad field of study for research [13]. The vehicle routing problem (VRP) and its numerous variants can be used to formalize real-world transportation problems. State-of-the-art classification and taxonomic review of VRP can be found in [1,3,11]. The specific reviews [4,9] classifies VRP and its variants based on the used metaheuristics. Among them, the adaptive large neighborhood search (ALNS) [10] belongs to common metaheuristics applied to routing problems. Even though its adaptiveness was not found to be crucial for the improvement [12], we will demonstrate its generic application on complex and large-scale routing problems, which can have very different characteristics.

The problems we consider have heterogeneous fleets as it is the case for many other works reviewed in [5]. We will work with pickup and delivery with

© The Author(s), under exclusive license to Springer Nature Switzerland AG 2023
L. Pérez Cáceres and T. Stützle (Eds.): EvoCOP 2023, LNCS 13987, pp. 34–49, 2023.
https://doi.org/10.1007/978-3-031-30035-6_3

time windows (PDPTW), where each request includes the service of pickup and delivery locations, which may have multiple time windows. As usual, requests have their capacities, and we need to consider both volume in the number of pallets and weight. Time constraints are also represented by the necessary service time at each physical location and the maximal route duration constraint. The maximal number of physical locations per route is constrained as well. Finally, we have multiple depots available [7]. Altogether, our problems combine multiple constraints, as is common in rich VRPs [2].

Let us now summarize the specific contributions of this work.

- We describe the formal model of our problem as an extension of [10] with a unique combination of real-world constraints and objectives.
- We extend ALNS to handle minimization with real-world objectives and propose an efficient realization of its insertion and removal heuristics. It allows for a constant time complexity for many frequent operations instead of the linear one wrt. to the length of the route. To achieve that, we have identified crucial information to store and pre-calculate.
- We verify our implementation on the standard Li & Lim [6] benchmarks to compare it with the best so-found results and with the original ALNS.
- We compare our solver with implementation in OR-Tools [8] provided by the company Wereldo and demonstrate significant improvement in the solution's costs, faster computation, and memory savings by order of magnitude.
- We demonstrate the efficiency of our solver on a large scale real-world problem with more than 300 vehicles and 1,200 pickup and delivery requests, where results were computed within less than an hour.

Overall, we have proposed and implemented a generic algorithm capable of solving a broad family of routing problems. We are glad to see that the Wereldo company nowadays uses the described solver for their everyday operation and specific case studies.

2 Mathematical Model

In this section, we describe the formal mathematical model of the PDPTW variant as an extension of the model described by Ropke and Pisinger [10].

We consider n requests and m vehicles. We denote the pickup nodes by P, where $P = \{1, ..., n\}$ and the delivery nodes by D, where $D = \{n + 1, ..., 2n\}$. Request i is represented by a pickup node i and a delivery node $i+n$, i.e., there are $2n$ tasks to be serviced. We denote the set of all vehicles by K, where $|K| = m$. We also denote the set of all pickup and delivery nodes by N ($N = P \cup D$). This set represents all locations that have to be visited. For each vehicle k ($k \in K$) we consider its starting terminal τ_k ($\tau_k = 2n + k$) and its ending terminal τ'_k ($\tau'_k = 2n+m+k$). In practice, the ending and starting terminals can represent the same location. In our model, we define them separately to distinguish between the start and end of vehicles. For each pickup or delivery node i ($i \in N$), we consider a service time s_i needed at each location to load or unload the goods.

Each vehicle can have different pricing. For each vehicle k, we define its minimum price C_k^{min}, the price per kilometer when the vehicle is empty C_k^{empty}, and the price per kilometer when the vehicle is fully loaded C_k^{full}. If the vehicle is partially loaded, the price per kilometer is calculated from the C_k^{empty} and C_k^{full} values based on the vehicle's current load. For each vehicle k, we also define its maximum route duration F_k and the maximum number of trips W_k (different physical locations on the route), where $W_k \geq 3$, so each vehicle can carry at least one of the requests. There can be at most three trips for a vehicle with a single request: start-pickup, pickup-delivery, and delivery-end.

Each pickup and delivery can have a different number of possible time windows. We define ρ_i ($\rho_i > 0$) for each node i as the number of its time windows. Then we define the time windows TW_i for each location where $TW_i = \{[a_{ir}, b_{ir}] \mid r \in \{1, ..., \rho_i\}\}$. a_{ir} and b_{ir} are the earliest and latest possible times, respectively.

For each request, we consider its weight and volume. Therefore, for each location i, we define the differences of weight l_i and the volume h_i after serving the location i. These amounts must be positive numbers for all pickup locations and negative for all delivery locations. For each pickup i and its delivery $i + n$ it has to be true that $l_i = -l_{i+n}$ and $h_i = -h_{i+n}$. This ensures that the amount of loaded goods equals the amount of unloaded goods. Each vehicle $k \in K$ has its weight limit Q_k and a volume limit U_k.

We define a graph with all locations by $G = (V, A)$. Where $V = N \cup \{\tau_1, ..., \tau_m\} \cup \{\tau_1', ..., \tau_m'\}$ and $A = V \times V$. This graph represents all locations from our problem, the pickup and delivery locations, and the starting and ending terminals. For each vehicle k, we define a subgraph $G_k = (V_k, A_k)$ where $V_k = N \cup \{\tau_k\} \cup \{\tau_k'\}$ and $A_k = V_k \times V_k$. Each subgraph G_k contains only the depot of the vehicle k (G contains all depots). For each arc $(i, j) \in A$ we consider its distance d_{ij} ($d_{ij} \geq 0$) and its travel time t_{ij} ($t_{ij} \geq 0$).

We use five decision variables. A binary variable x_{ijk} where $i, j \in V$ and $k \in K$. This variable has a value one if the arc (i, j) is used by a vehicle k and zero otherwise. S_{ik} ($i \in V, k \in K$) is a variable that indicates when the vehicle k arrives at the location i. L_{ik} ($i \in V, k \in K$) is a non-negative integer that indicates the total weight of the goods loaded on the vehicle k after servicing the node i. H_{ik} ($i \in V, k \in K$) is a non-negative integer that indicates the total volume of the goods loaded on the vehicle k after servicing the node i. S_{ik}, L_{ik}, and H_{ik} are only well-defined when the vehicle k visits the location i. z_i ($i \in P$) is a binary variable that indicates if request i is placed in the request bank. The variable is one if the request is placed in the bank and zero otherwise. The request bank is used during the solution process to handle requests not yet assigned to any vehicle.

The total price C_k of vehicle k is defined as

$$C_k = \max(C_k^{min}, \sum_{(i,j) \in A} x_{ijk} d_{ij}(C_k^{empty} + (C_k^{full} - C_k^{empty})\frac{L_{ik}}{Q_k}) . \qquad (1)$$

The objective function minimizes the weighted sum of the prices of all used vehicles and the number of requests that are not scheduled, i.e., they are kept in the request bank. Parameters α and β are used to adjust these weights.

$$\alpha \sum_{k \in K} C_k + \beta \sum_{i \in P} z_i \ . \tag{2}$$

To ensure that each request is either delivered by a single vehicle or that it is placed in the request bank, we define a constraint

$$\sum_{k \in K} \sum_{j \in N} x_{ijk} + z_i = 1 \qquad \forall i \in P \ . \tag{3}$$

If a request is not placed in the bank, its pickup i and delivery $n+i$ have to be performed by the same vehicle k. This constraint is defined as

$$\sum_{j \in V_k} x_{ijk} - \sum_{j \in V_k} x_{j,n+i,k} = 0 \qquad \forall k \in K, \forall i \in P \ . \tag{4}$$

Constraints (5)–(7) together ensures that a correct path from τ_k to τ_k' is constructed for each vehicle k. Each vehicle k has to start its route in its starting location τ_k (5). Also, each vehicle k has to end its route in its ending location τ_k' (6). If a vehicle k visits some location j, except for a depot, the vehicle must also leave this location (7).

$$\sum_{j \in P \cup \{\tau_k'\}} x_{\tau_k,j,k} = 1 \qquad \forall k \in K \ , \tag{5}$$

$$\sum_{i \in D \cup \{\tau_k\}} x_{i,\tau_k',k} = 1 \qquad \forall k \in K \ , \tag{6}$$

$$\sum_{i \in V_k} x_{ijk} - \sum_{i \in V_k} x_{jik} = 0 \qquad \forall k \in K, \forall j \in N \ . \tag{7}$$

If a vehicle k arrives at a location i at S_{ik}, it has to have enough time to load or unload the goods and travel to the next location j before S_{jk}. However, we want to consider the service time only if the vehicle had to take a trip from i to j. The physical locations of i and j might be the same. In this case, the service time is not needed between i and j. This is ensured by

$$x_{ijk} = 1 \implies S_{ik} + s_i \cdot \mathrm{sgn}(d_{ij}) + t_{ij} \leq S_{jk} \qquad \forall k \in K, \forall (i,j) \in A_k \tag{8}$$

where sgn function is used to include/exclude values from the sum when the distance is (non)zero.

The time windows must be kept. For each vehicle k and each of its visited locations i, the time of arrival S_{ik} has to be from one of its intervals $[a_{ir}, b_{ir}]$.

$$a_{ir} \leq S_{ik} \leq b_{ir} \qquad \forall k \in K, \forall i \in V_k, \exists r \in \{1, ..., \rho_i\} \ . \tag{9}$$

For each pickup i, its corresponding delivery $n+i$ has to be performed after the pickup. This is ensured by

$$S_{ik} \leq S_{n+i,k} \qquad \forall k \in K, \forall i \in P \ . \tag{10}$$

The volume and weight load variables L_{ik} and H_{ik} are set by constraints (11) and (12). The vehicle's weight and volume limits Q_k and U_k are obeyed using the constraints (13) and (14).

$$x_{ijk} = 1 \implies H_{ik} + h_j \leq H_{jk} \qquad \forall k \in K, \forall (i,j) \in A_k \ , \qquad (11)$$

$$x_{ijk} = 1 \implies L_{ik} + l_j \leq L_{jk} \qquad \forall k \in K, \forall (i,j) \in A_k \ , \qquad (12)$$

$$L_{ik} \leq Q_k \qquad \forall k \in K, \forall i \in V_k \ , \qquad (13)$$

$$H_{ik} \leq U_k \qquad \forall k \in K, \forall i \in V_k \ . \qquad (14)$$

We have to ensure that each vehicle k starts and ends empty.

$$L_{\tau_k k} = L_{\tau'_k k} = 0 \qquad \forall k \in K \ . \qquad (15)$$

We must ensure that all of the maximum route duration constraints F_k and the maximum number of trips W_k are obeyed (sgn function again allows to include/exclude values when the distance is (non)zero).

$$S_{\tau'_k,k} - S_{\tau_k,k} \leq F_k \qquad \forall k \in K \ , \qquad (16)$$

$$\sum_{i \in N \cup \{\tau_k\}} \sum_{j \in N \cup \{\tau'_k\}} x_{ij} \cdot \text{sgn}(d_{ij}) \leq W_k \qquad \forall k \in K \ . \qquad (17)$$

Finally, we have to set the domains of the used decision variables.

$$x_{ijk} \in \{0,1\} \qquad \forall k \in K, \forall (i,j) \in A_k \ , \qquad (18)$$

$$z_i \in \{0,1\} \qquad \forall i \in P \ , \qquad (19)$$

$$S_{ik} \geq 0 \qquad \forall k \in K, \forall i \in V_k \ , \qquad (20)$$

$$Q_k \geq L_{ik} \geq 0 \qquad \forall k \in K, \forall i \in V_k \ , \qquad (21)$$

$$U_k \geq H_{ik} \geq 0 \qquad \forall k \in K, \forall i \in V_k \ . \qquad (22)$$

3   Our Approach

Our approach uses the adaptive large neighborhood search (ALNS) algorithm based on Ropke and Pisinger [10] (see page 6). In the beginning, we generate an initial solution $formerSolution$, or it can be a solution obtained in an earlier run (Line 2). The current solution s' is iteratively modified by rearranging requests. We decide which neighborhood to search by selecting one removal (Line 6) and one insertion heuristics (Line 7) using a standard adaptive heuristic selection mechanism. The worst removal, Shaw removal, and random removal are complemented by the greedy insertion and regret insertions heuristics [10]. In each iteration, a random number of q requests to remove (Line 8) is generated as in [10]. If it is impossible to insert some of the removed requests (Line 9), these requests are stored in a request bank in the new solution s'. In the reinsertion process, requests from the request bank are also reinserted, if possible. At the end of each iteration, we check if we have found a new best solution (Line 10). Acceptance of the new modified solution (Line 12) is decided by the simulated annealing combined with the heuristics with noisy objective function from [10].

```
1  function ALNS(formerSolution):
2      s ← formerSolution
3      s_best ← formerSolution
4      while not reached max iteration do
5          s' ← s
6          h^r ← choose removal heuristic
7          h^i ← choose insertion heuristic
8          remove q requests from s' using h^r
9          reinsert unassigned requests to s' using h^i
10         if f(s') < f(s_best) then
11             | s_best ← s'
12         if accept(s', s) then
13             | s ← s'
14     end
15     return s_best
```

3.1   Two-Stage Minimization

To minimize the total vehicle price, we run the ALNS in two stages. In the first stage, we set all of the vehicle's minimum prices $C_k^{\min}$ to zero. This encourages the ALNS to put some of the requests into the larger, more expensive vehicles. For each vehicle, we also set its $C_k^{\text{full}} = C_k^{\text{empty}}$. This is again to achieve greater diversification during the first stage. We take the best solution from the first stage and use it as a former solution in the second stage. In this stage, we use all prices with their original values. Note that for problems without the minimum price $C_k^{\min}$ component, running the ALNS in one stage is sufficient.

This approach was inspired by a similar two-stage approach from [10]. We keep the same approach when solving problems with the standard lexicographic objective aiming to minimize the number of vehicles first and the distances next. In the first stage, the whole fleet is available for the ALNS initially. The search terminates as soon as it finds a feasible solution. In the next ALNS run, one of the used vehicles is removed, and the search is repeated. We repeat the ALNS until a feasible solution with the given number of vehicles is not found (within the iteration limit for the whole first stage). In the second stage of the algorithm, we run the regular ALNS where the former solution is the best feasible solution with the minimum number of vehicles used.

3.2   Insertion Optimization

To build upon the work [10], we propose various methods which allow processing insertions (this section) and removals (next section) effectively. The most time-consuming operation during insertions is finding the best position for a request in a route. Also, this operation is the most time-consuming for the whole solver.

When trying to find the best request position for each route, we must find the best position for the pickup and delivery in the route's actions. A series of actions represent each route. Each action represents one location $i \in N$. No two actions can represent the same location since each location should be served by exactly one vehicle (however, they can represent the same physical location). Suppose we have a route with y actions. After adding the new request, the number of actions will increase by two. This new request is represented by a new pickup action η^p and a new delivery action η^d. The possible positions κ^p for the pickup action η^p can be from the interval $[0, y]$. With such selected pickup position κ^p, the possible delivery action η^d has a position κ^d from interval $[\kappa^p + 1, y + 1]$. An example of such a new route with positions from 0 to 7 is $(\eta_0, \eta^p, \eta_1, \eta_2, \eta^d, \eta_3, \eta_4, \eta_5)$ where $\kappa^p = 1$ and $\kappa^d = 4$. Actions $\eta_0, ..., \eta_5$ are already present in the route. η^p and η^d are the new pickup and delivery actions.

With these possible combinations of κ^p and κ^d, we have to verify that such route satisfies all of the constraints defined in Sect. 2. We also need to calculate the price of the extended route such that we can compute the insertion cost. This is performed in this given order: (1) validation of capacities (weight + volume), (2) validation of the maximum number of trips, (3) validation of time constraints, and (4) calculation of the cost. If any of these validation fails, the following operations are skipped. This order is essential as the first two validations are the fastest, and the cost should be calculated only for valid routes. The process of validations 1–3 is described next. The cost of the solution is computed by iterating over the whole route since it was not identified as a crucial bottleneck.

Validation of Capacities. In Eqn. 11 and 12, we have defined constraints for vehicles' weight and volume capacities. Both constraints can be handled using the same validation mechanism which we demonstrate on the weight. When we add a new request at positions κ^p and κ^d, we have to check that the vehicle's capacity is not overreached. We have to check this at the κ^p position and all other positions between κ^p and κ^d. Actions before κ^p and after κ^d are not affected in terms of capacity and do not need to be checked.

The number of actions that need to be checked is linear in terms of the route's length. However, with some pre-calculation, it can be done in a constant time. We keep several values for all actions η_e, where $e \in \{0, ..., y - 1\}$ and y is the route's length) for all routes. First, we calculate for each action η_e its weight reserve Δ_e^q. This value represents the difference in the vehicle's weight capacity and the weight of loaded goods after performing the given action. An example of a route with its weight reserves is shown in Table 1 (left).

Then, for each action η_e, we keep a list of values $\nu_{e,j}$ where $j \in \{e, ..., y - 1\}$. This list represents the minimum weight reserve of actions from interval $\langle e, j \rangle$, i.e., $\nu_{e,j} = \min\{\Delta_e^q, ..., \Delta_j^q\}$. Note that for each action η_e, we keep a different number of the $\nu_{e,j}$ values. The action at the start of the route has the maximum number of such values equal to the route's length. The last action has only one value for itself. An example of these calculated values is shown in Table 1 (right) for the route on the left. The best way to calculate these values is to start with j set to its maximum value $j = y - 1$ and decrease it. This way, we can reuse

Table 1. Example of a route with weight reserves for a vehicle capacity 200.

action	η_0	η_1	η_2	η_3	η_4	η_5
weight change	+20	+70	+50	-50	-20	-70
current load	20	90	140	90	70	0
weight reserve Δ_e^q	180	110	60	110	130	200

	$j=0$	$j=1$	$j=2$	$j=3$	$j=4$	$j=5$
$e=0$	180	110	60	110	130	200
$e=1$	110	60	60	110	130	
$e=2$	60	60	60	110		
$e=3$	60	60	60			
$e=4$	60	60				
$e=5$	60					

the previously calculated values. To calculate value $\nu_{e,j}$ we can use

$$\nu_{e,j} = \left\{ \begin{array}{ll} \Delta_j^q, & \text{for } e=0 \\ \min\left(\Delta_j^q, \nu_{e-1,j+1}\right), & \text{for } e>0 \end{array} \right\} . \tag{23}$$

When we try to assign a new request at positions κ^p and κ^d, we only need to check if the request's weight is lower or equal to the value $\nu_{\kappa^p-1,\kappa^d-2}$. The value $\kappa^p - 1$ represents a position of the action right before the pickup position κ^p. The value $\kappa^d - 2$ represents the position of the action right before the delivery position κ^d. For insertion optimization example on page 7, we get action η_0 for $\kappa^p - 1$ and we get action η_2 for $\kappa^d - 2$.

This mechanism does not work when we try to place the new pickup at the beginning of a route ($\kappa^p = 0$) or when the route is empty. In the first case, we compare the request's weight with the value $\nu_{\kappa^p,\kappa^d-2}$. In the second case, when the route is empty, we compare the request's weight with the vehicle's capacity.

If the capacity check fails, we can also perform one more optimization; we can skip the next possible delivery positions, which are higher than the current κ^d. If the capacity check fails for κ^d, it will always fail for any other higher values.

This pre-calculation has a quadratic time complexity relative to the route length. It seems inefficient to do such pre-calculation. The important thing is that this pre-calculation is run only when a route is truly modified. When we calculate cost increases for all of the unassigned requests, we can reuse these values for all of them. When we finally insert one request, we must recalculate only the route where the request was just inserted.

Validation of Maximum Number of Trips. In Eq. 17, we limit each vehicle's maximum number of trips. A trip can happen between two actions. If the distance between the locations of these actions is non-zero, the vehicle must perform a trip. When assigning a new request to a route, the check of the overall number of trips has linear time complexity relative to the route's length. We again propose a constant time validation.

It is insufficient to compare the number of actions in the route since this does not correspond to the number of trips due to the possibly same physical locations. There can be a sequence of actions representing the same real-world location, for example, pickups of multiple requests at the same location. Between such actions, there are no trips. We keep information about the number of trips on each route. This value is recalculated whenever we remove or reinsert some requests (with a linear time complexity relative to the length of the route). When we try to assign a new request at positions κ^p and κ^d, we check if we have added some new trips.

For this new request, we define i^p as the pickup location of this request. The i^p is a location from the set of all pickup locations P. We find the location i^{p-1} where the vehicle was previously. If $\kappa^p = 0$ holds, the previous location is the vehicle's starting location τ_k. If $\kappa^p > 0$, we get the previous location from the action on position $\kappa^p - 1$. We also need to find the next location i^{p+1}, where the vehicle goes from i^p. If $\kappa^p = \kappa^d - 1$, the next location is the new request's delivery location. Otherwise, the next location is taken from the action on position $\kappa^p + 1$.

We define $trips^p$ as the number of new trips caused by the new pickup being added; initially, we set it to zero. If the distance between i^{p-1} and i^p is not equal to zero, we increase the $trips^p$ by one. If the distance between i^p and i^{p+1} is not equal to zero, we increase the $trips^p$ by one. Before adding the new pickup, we also need to check if there was already a trip between i^{p-1} and i^{p+1}. If so, we decrease the $trips^p$ by one. This gives us the number of new trips caused by the new pickup. Similarly, we can calculate $trips^d$, which represents the number of new trips caused by the new request's delivery. We also need to check if we did not add the same trip twice in cases where the new delivery is right after the new pickup ($\kappa^p = \kappa^d - 1$). If so, we decrease the $trips^d$ by one.

We add up $trips^p$ and $trips^d$ with the previous number of trips, which has been pre-calculated, to get the total number of trips $trips^{total}$. Finally, we can check that the $trips^{total} \leq W_k$.

Validation of Time Constraints. Constraints specified in Eqs. 8–9 define valid arrival times for all vehicles. They ensure that all vehicles arrive at their locations in corresponding time windows. To check all these constraints, we iterate over the whole route and calculate arrival times. We assume that each vehicle takes the first time window, which is possible to use.

In Eqn. 16, we have also defined a maximum route time. When checking the time windows, we must calculate arrival times for all actions. This allows us to check the maximum duration time without additional calculation.

Same Locations Optimizations. In our model, we have mentioned that some locations can represent the same physical locations. Among a route's actions, there can be a sequence of actions representing the same physical location. The order in which these actions are processed in the location cannot affect the route's price because we consider the service time only for the last of them, and also, in the company instances, the service times are all equal. This allows us to optimize the best positions of η^p and η^d. If we detect such a sequence of actions, and the action η^p represents the same real-world location, we try to assign the action η^p only at the beginning of this sequence. We also do the same when finding the best position of η^d.

This optimization is critical to solving problems with many same physical locations efficiently. Notably, it allows the efficient application to problems where many or even all requests are delivery-only.

3.3 Removal Optimization

The removal heuristics are not as much time-consuming as the insertion heuristics. Still, their efficient implementation is worth consideration.

In the worst removal heuristic, we try to find requests, which increases the most the cost of the solution. Let $\Delta(s, r)$ denotes the difference in costs of solution s when the request r is scheduled and when it is removed from the solution. When we remove a request r, some of the $\Delta(s, r)$ values must be recomputed.

The time complexity of calculating such values has a linear time complexity relative to the route's length. For very long routes, it makes sense to propose constant time recalculation. Suppose we have a route with a length equal to y. This route has actions ζ_e where $e \in \{0, ..., y - 1\}$. For all these actions, we keep information about the distance the vehicle traveled from its starting node to the location represented by this action. Also, we keep the information about the vehicle's load after performing it for each action. Finally, we keep information about the price of the whole route.

Suppose we have a request for which we want to calculate the value $\Delta(s, r)$. To do so, we only need to calculate the price changes in the route. This request must have a pickup and a delivery action on this route. We denote these actions by ζ_p and ζ_d, where $p, d \in \{0, ..., y - 1\}$ and $d > p$ hold. Let Δ^p and Δ^d denote the price decreases caused by differences in distances caused by removing the pickup and delivery locations from the route. We have $\Delta^p = price(\zeta_{p-1}, \zeta_p) + price(\zeta_p, \zeta_{p+1}) - price(\zeta_{p-1}, \zeta_{p+1})$. The $price()$ is a function that returns the price of traveling from the first action's location to the second. We have to use the vehicle's current load when calculating these prices. When computing the $price(\zeta_p, \zeta_{p+1})$, we have to consider a higher price per kilometer because the vehicle had a bigger load. ζ_{p-1} represents the action previous to ζ_p and ζ_{p+1} represents the following action. We have $\Delta^d = price(\zeta_{d-1}, \zeta_d) + price(\zeta_d, \zeta_{d+1}) - price(\zeta_{d-1}, \zeta_{d+1})$. In this case, the price per kilometer in $price(\zeta_{d-1}, \zeta_d)$ is higher than in the other two because the vehicle was more loaded.

When we remove the request, we also affect the price of the path from action ζ_{p+1} to ζ_{d-1} since the vehicle has a lower load. We can calculate the price decrease caused by this load change from the precalculated information. We denote the price decrease Δ^{load}. We take the distance traveled from the starting node to the action ζ_{d-1} and subtract it from a distance traveled to action ζ_{p+1}. This way, we can get the distance from ζ_{p+1} to ζ_{d-1}. Using this distance and the weight of the removed request, we can calculate the Δ^{load}. To get the total price, we use $newPrice = oldPrice - \Delta^p - \Delta^d - \Delta^{load}$.

If the route remains empty after removing the request r, we do not need to calculate anything. In this case, the new price would equal zero, and we can return the old price as the cost difference. Furthermore, if the pickup and delivery actions are right behind each other ($\zeta_p = \zeta_{d-1}$), we need to make sure we do not calculate the same price differences twice in Δ^p and Δ^d. If the pickup action is the first in the route, we use the vehicle's starting location to calculate Δ^p. Similarly, if the delivery action is the last in the route, we use the vehicle's ending location to calculate Δ^d.

4   Experimental Evaluation

Our solver was implemented in the programming language Go version go1.15.15 linux/amd64. We used the same values for parameters as described in the original work [10]. For the ALNS running in two stages, we set the number of iterations to 25,000 for each stage. For one-stage ALNS (problems without minimum price only), 25,000 iterations were sufficient.

4.1   Li and Lim Benchmark Instances

For initial experiments, we use Li & Lim [6] benchmark instances[1] generated for the capacitated pickup and delivery with time windows. These data sets are for 50 to 500 requests. Instances are divided into three categories based on the placement of locations, and they are clustered (LC), random (LR), and combined random-clustered (LRC). Instances LC1, LR1, and LRC1 have short scheduling horizons, while LC2, LR2, and LRC2 have longer horizons.

The first set of our experiments aimed to demonstrate the influence of the algorithmic optimizations from Sects. 3.2 and 3.3. We benchmarked six instances (one of each category) for data sets with 100, 400, and 1,000 tasks. The average improvement in runtime was 34.4 %, 42.6 %, and 51.1 %, respectively. We have obtained even more significant improvement for the real-world data (see their description in Sect. 4.2) with 84.9 %, 78.3 %, and 84.2% runtime improvement for 40, 90, and 140 tasks, respectively, since the existence of the same physical locations allows for additional improvement.

We have selected three instances from each category for further experiments in this section. Experiments on all instances were run 100 times. We have executed our experiments using a grid service provided by MetaCentrum[2]. Each run was limited to 1 CPU core and 1 GB of RAM. In this grid environment, our processes were sharing CPUs with other processes running at the same time. Because of that, the runtimes could vary depending on the current workload.

Comparison to Best Found Solutions. Table 2 (left) shows an overview of how many best solutions we could find depending on the instance size. We can see that for the smallest instances, we were able to find the best solutions for 16 of 18 instances. With the increasing instance size, we could find fewer best solutions. However, it is worth noting that the best solutions published on the Sintef web page were achieved by many runs of various solvers. Therefore, the quality of these solutions is very high. We expect that with more runs of our solver, we will find some of more best solutions. Furthermore, for the instances in which we did not find the best solution, our quality is very close to the best solutions.

Figure 1 (left) shows the difference in the number of vehicles used in our best-found solutions and the best so far found solutions. We have vertically

[1] https://www.sintef.no/projectweb/top/pdptw/li-lim-benchmark/.
[2] https://metavo.metacentrum.cz/en/.

Table 2. The number of instances where we were able to find the best so far found solutions (left) and comparison of our best solutions to the original ALNS's best solutions (right) for 18 instances.

instances size	100	200	400	600	800	1,000
best found	16	13	8	4	3	3

instances size	100	200	400	600	800	1,000
better best	0	9	13	14	15	13
equal best	17	9	4	3	3	4
worse best	1	0	1	1	0	1

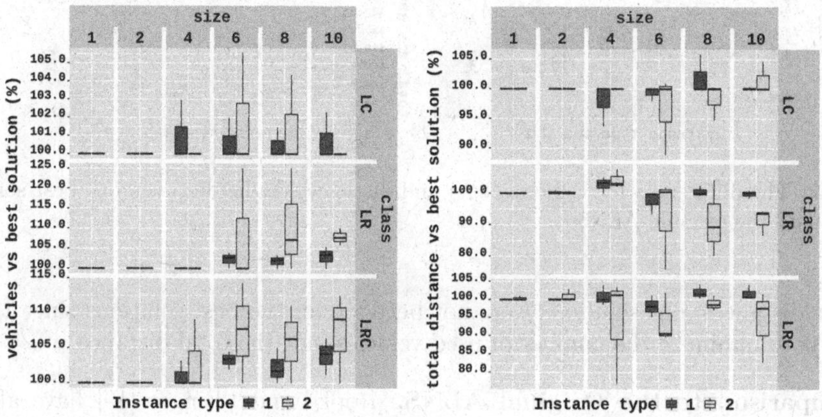

Fig. 1. The differences for vehicles (left) and distances (right) between our best solutions and the best so far found solutions.

categorized the results by the instance category and horizontally by the sizes of the instances (1 to 10 corresponds to 100 to 1,000 tasks). We have also separated them by the length of the schedule horizon (1 represents the shorter horizon, and 2 represents the longer horizon).

We can see in the results that our solver has performed better in the clustered instances than in the random and random clustered. In the clustered instances, the search space is more constrained by the time windows. Also, the number of vehicles used usually corresponds to the number of clusters in these data instances. Therefore, it is easier for the solver to find a solution with the minimum number of vehicles. Not surprisingly, we can see that our solution has performed worse with the increasing size of instances. Our solver has also performed worse for instances with longer schedule horizons (type 2). Because of the longer schedule horizon, it is possible to construct longer routes with fewer vehicles. These solutions are for our solver more challenging to find because they usually require a longer distance.

Figure 1 (right) shows the difference in distances in best solutions obtained by our solver against the best so far found solutions. Our solutions had a lower total distance traveled for the instances with a longer schedule horizon (type 2). This is because our solutions used a higher number of vehicles. Therefore, the solutions could have a lower total distance. It is worth noting that despite the

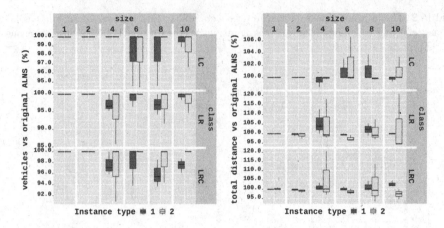

Fig. 2. The differences for vehicles (left) and distances (right) between our best solutions and the original ALNS.

lower distances, our solutions were not better than the best solutions since the goal is to minimize the number of used vehicles, not the total distance.

Comparison to the Original ALNS. Ropke and Pisinger [10] have also performed experiments on the Li & Lim [6] datasets. For instances with 400 tasks and lower, they performed 10 runs. For the other instances, they have performed only 5 runs. They have presented their data about their best solutions, average solutions, and average runtimes.

Table 2 (right) compares our best-found solutions and the best solutions found by the original ALNS implementation. We can see that we have managed to find a better solution in most of the instances. However, this is also because we have performed much more runs than Ropke and Pisinger [10]. Figure 2 compares the quality of our best solutions and the best solutions presented in the original work. We can see that we have very similar results. In many instances, we have managed to find a solution with a lower number of vehicles. Of course, in most cases, this results in a higher total distance.

4.2   Real-World Instances

Comparison with OR-Tools. We have received 12 problem instances with 40–140 requests from company Wereldo. These requests have one or two time windows for pickup and delivery. In each of these instances, all requests have the same real-world pickup location. These instances usually have up to 10 different vehicle models. Each model had different prices per kilometer and capacity. For each of these instances, we have also received results of 10 runs performed by the Wereldo's original solver, which was implemented using OR-Tools [8]. They have a two-stage approach as well. In the first stage, they also ignore the minimum vehicle prices and use the minimum prices per kilometer, ignoring the vehicle's

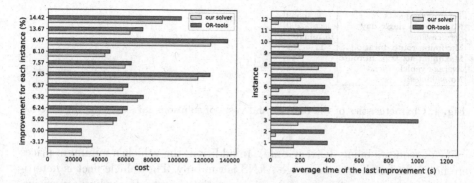

Fig. 3. Comparison of costs (left) and average times (right) of best solutions achieved by our solver and company's solver in OR-tools.

current load. In these results, we received the costs of the best solutions found in both stages and the times when they were found. However, the OR-Tools did not improve the previous solution in the second stage in any of the runs. Therefore, we will use only the results from the first stage for comparison. We have also received information about the peak memory usage obtained during both stages. The company runs these experiments on the Amazon Web Services[3] (AWS). With our solver, we also performed 10 runs for each of these instances on a regular laptop with the Intel Core i7-8550U CPU (1.80GHz).

Figure 3 (left) compares the costs of our best and the company's solutions. We have found a better solution in 10 out of 12 instances with 5–14% improvement. In one instance, the best solutions were equal, and in the last one, the best company's solution was slightly better. In this instance, we have noticed that the company's solver had managed to find a solution with one less vehicle, which led to a solution with a lower cost. Our solver could not find a solution with this lower number of vehicles in any of its runs. In Fig. 3 (right), a comparison of averages of times of the last improvements is provided. The experiments were not executed in the same environment. Still, it is notable that our solver running on a regular laptop was faster. The average memory peak usage provided by the company was 480–570 MB for the first stage and 2.9–4.1 GB in the second stage. Our overall peak memory usage was 100–120 MB in both stages, which is a significant improvement. Note that the high memory usage in the second stage is caused by the fact that the relative vehicle's price per kilometer is not well optimized in the OR-Tools. It is not even recommended to be used, and the developers have practically abandoned this feature.

Large-Scale Problems. These experiments ran on CentOS8, Intel Xeon Skylake 2.3 GHz, 8 CPUs and 16 GB RAM, and each run was limited to 1 CPU. The Wereldo company provided data with characteristics listed in Fig. 4 (left). Data were separated into 5 single-day instances. While the earlier real-world data sets were delivery-only, now we have requests with pickup and deliveries

---

[3] https://aws.amazon.com/.

Total no. requests	6,539
No. requests single day	cca 1,200
Service time	15 minutes
Maximum route duration	11 hours
Maximum no. trips in route	15
Average vehicle speed	19 m/s
No. vehicles	323

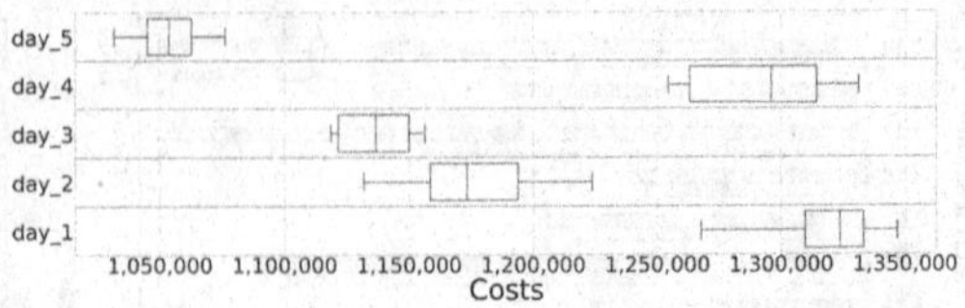

Fig. 4. Characteristics of the dataset and costs of solutions for single day instances.

(85 % of requests being delivery-only). In addition, the vehicle's minimum prices are not considered, and one-stage ALNS is run only. The vehicle fleet is heterogeneous both in capacities as well as in operational costs. Regarding capacities, a significant portion of the fleet (68.1 %) is composed of large trucks capable of accommodating 66 or 72 pallets and up to 24 tons of load. The remaining 6 vehicle types are very different, with capacities from 4 to 25 pallets. There are 8 cost models of vehicles where 71.8 % of them have cost X per kilometer with others ranging between $0.4X$ and $3.1X$. In Fig. 4, we can see the resulting costs for 10 runs of single-day instances. We achieved average runtimes 52.7 ± 1.7 min.

5   Conclusion

This work considers real-world freight transportation problems with many characteristics. We provided a formal model for a unique combination of constraints and objectives which were necessary for the representation of our problems. We proposed a solver based on the adaptive large neighborhood search, which allows us to solve a wide variety of routing problems efficiently. To achieve that, we have concentrated on the complexity of insertion and removal heuristics which constitutes the heart of the search procedure. We have identified frequent validation steps for insertion heuristics, which can be processed in constant instead of linear time relative to the route length. To achieve that, we have identified crucial information to store and precompute. Similarly, constant time recalculation is also proposed to compute differences between the costs of solutions for removal heuristics. The minimization process is enhanced to handle real-world objectives for heterogeneous vehicles where the minimum vehicle's prices necessitate two stages of the ALNS.

In the paper, we demonstrate the generic application of the solver on problems with different characteristics. First, we verify results on the standard Li & Lim benchmarks by comparison with the best-found solutions and the original ALNS solver. These benchmarks represent base pickup and delivery problems with time windows. Further real-world problems from the company with 40–140 requests have the same pickup location, time windows for both pickup and location, and contain all features described in our formal model, including the minimum price. Experimental results show better solution quality, faster computation, and significant memory saving compared with the company's solver in OR-Tools. The last type of problem contains about 85 % delivery-only requests, and its vehicles do not need to consider minimum vehicle prices. Including this data set allows us to demonstrate less than one-hour runs on large-scale problems with 1,200 requests and more than 300 heterogeneous vehicles. To conclude, the solver is used by company Wereldo in everyday operations, and they have also applied it to various case studies for different customers.

Acknowledgements. Computational resources were supplied by the project "e-Infrastruktura CZ" (e-INFRA LM2018140) provided within the program Projects of Large Research, Development and Innovations Infrastructures.

References

1. Braekers, K., Ramaekers, K., Van Nieuwenhuyse, I.: The vehicle routing problem: state of the art classification and review. Comput. Indus. Eng. **99**, 300–313 (2016)
2. Caceres-Cruz, J., Arias, P., Guimarans, D., Riera, D., Juan, A.A.: Rich vehicle routing problem: survey. ACM Comput. Surv. **47**(2), 1–28 (2014)
3. Eksioglu, B., Vural, A.V., Reisman, A.: The vehicle routing problem: a taxonomic review. Comput. Indus. Eng. **57**(4), 1472–1483 (2009)
4. Elshaer, R., Awad, H.: A taxonomic review of metaheuristic algorithms for solving the vehicle routing problem and its variants. Comput. Indus. Eng. **140**, 106242 (2020)
5. Koç, Ç., Bektaş, T., Jabali, O., Laporte, G.: Thirty years of heterogeneous vehicle routing. Eur. J. Oper. Res. **249**(1), 1–21 (2016)
6. Li, H., Lim, A.: A metaheuristic for the pickup and delivery problem with time windows. In: Proceedings 13th IEEE International Conference on Tools with Artificial Intelligence. ICTAI 2001, pp. 160–167 (2001)
7. Montoya-Torres, J.R., Franco, J.L., Isaza, S.N., Jiménez, H.F., Herazo-Padilla, N.: A literature review on the vehicle routing problem with multiple depots. Comput. Indus. Eng. **79**, 115–129 (2015)
8. Perron, L., Furnon, V.: OR-Tools. https://developers.google.com/optimization/
9. Potvin, J.Y.: State-of-the art review – evolutionary algorithms for vehicle routing. INFORMS J. Comput. **21**(4), 518–548 (2009)
10. Ropke, S., Pisinger, D.: An adaptive large neighborhood search heuristic for the pickup and delivery problem with time windows. Transp. Sci. **40**, 455–472 (2006)
11. Toth, P., Vigo, D.: Vehicle routing: Problems, methods, and applications. Society for Industrial and Applied Mathematics (2014)
12. Turkeš, R., Sörensen, K., Hvattum, L.M.: Meta-analysis of metaheuristics: quantifying the effect of adaptiveness in adaptive large neighborhood search. Eur. J. Oper. Res. **292**(2), 423–442 (2021)
13. Vidal, T., Laporte, G., Matl, P.: A concise guide to existing and emerging vehicle routing problem variants. Eur. J. Oper. Res. **286**(2), 401–416 (2020)

A Multilevel Optimization Approach for Large Scale Battery Exchange Station Location Planning

Thomas Jatschka[1]([✉]), Tobias Rodemann[2], and Günther R. Raidl[1]

[1] Institute of Logic and Computation, TU Wien, Vienna, Austria
{tjatschk,raidl}@ac.tuwien.ac.at
[2] Honda Research Institute Europe, Offenbach, Germany
tobias.rodemann@honda-ri.de

Abstract. We propose a multilevel optimization algorithm (MLO) for solving large scale instances of the Multi-Period Battery Swapping Station Location Problem (MBSSLP), i.e., a problem for deciding the placement of battery swapping stations in an urban area. MLO generates a solution to an MBSSLP instance in three steps. First the problem size is iteratively reduced by coarsening. Then, a solution to the coarsest problem instance is determined, and finally the obtained solution is projected to more fine grained problem instances in reverse order until a solution to the original problem instance is obtained. We test our approach on benchmark instances with up to 10000 areas for placing stations and 100000 user trips. We compare MLO to solving a mixed integer linear program (MILP) in a direct way as well as solving the instances with a construction heuristic (CH). Results show that MLO scales substantially better for such large instances than the MILP or the CH.

Keywords: multilevel optimization · mixed integer linear programming · E-mobility

1 Introduction

Electric vehicles (EVs) are becoming an increasingly popular way of transportation for the general public. However, a major inconvenience for the owners of an EV is the long time it takes to recharge a vehicle's battery. For smaller vehicles, such as electric scooters, a promising way to overcome this problem is to exchange a vehicle's battery instead of recharging it. Batteries of electric scooters are compact enough such that users can exchange their depleted batteries for fully charged ones at dedicated battery exchange stations within a short time and without assistance. At such stations batteries are recharged, and can later be provided to customers again.

We consider the Multi-Period Battery Swapping Station Location Problem (MBSSLP) as introduced in [3], where the setup costs for stations should be

T. Jatschka—Acknowledges the financial support from the Honda Research Institute Europe.

L. Pérez Cáceres and T. Stützle (Eds.): EvoCOP 2023, LNCS 13987, pp. 50–65, 2023.
https://doi.org/10.1007/978-3-031-30035-6_4

minimized while a certain amount of customer demand needs to be satisfied. Each of the swapping stations is assumed to have a configurable number of slots at which batteries are charged and can be exchanged. Moreover, it is assumed that customers who want to change batteries specify their trip data (origin, destination, approximate time) online and are automatically assigned by the system to an appropriate station for the exchange (if one exists). As not every customer is willing to travel to a suggested station, e.g., if the detour is too long, the MBSSLP also considers a customer dropout which scales exponentially with the length of the detour induced by traveling to the assigned station.

In [3] a large neighborhood search (LNS) was presented for solving MBSSLP instances with up to roughly 2000 potential locations at which battery swapping stations can be placed and 8000 origin-destination (O/D) pairs that describe the customer trips. However, for real world applications especially the number of O/D-pairs can be magnitudes higher. The LNS proposed in [3] applies a destroy-and-repair scheme and uses a mixed integer linear programming based heuristic for repairing solutions. Unfortunately, this technique does not scale well to much larger instances. Finding good solutions for huge instances is in general a difficult task, even for metaheuristics, and one often resorts to clustering, refinement, or partitioning approaches that reduce the problem size or decompose the problem into smaller subproblems.

In this work we propose a multilevel optimization (MLO) approach for addressing large MBSSLP instances with tens of thousands of potential station areas and up to one hundred thousand user trips at which users need to swap batteries. The presented MLO is based on the algorithmic framework proposed by Walshaw [15], which has already been adapted to various mobility applications such as the traveling salesman problems [14], bike sharing station planning [7], and vehicle routing [10]. The basic idea of this approach is to generate a sequence of coarsened problem instances – referred to as multilevel coarsening – in which the problem sizes become successively smaller. After the coarsening process, a solution to the coarsest problem instance is generated. This solution is then iteratively projected to the problem instances of the multilevel coarsening in reverse order. Hence, after the final projection a solution to the original problem is obtained.

The underlying problem structure of the MBSSLP is given by a bipartite graph with one set representing the areas in which stations can be built and the other set representing the O/D-pairs. An MBSSLP instance is coarsened by first partitioning the nodes of the underlying graph and then deriving a coarsened graph by contracting the nodes in each partition. For solving the coarsest problem instances and for iteratively projecting solutions we make use of mixed integer linear programs (MILPs). Hereby, the MILP projecting a solution to a more detailed problem instance can be decomposed into subproblems, with each subproblem being responsible for projecting a single node of the underlying graph.

Our approach is experimentally evaluated on artificial benchmark scenarios generated as in [3]. To get a grasp of the quality of the solutions found by MLO we compare the solutions to solutions generated by a MILP solver as well as solutions obtained from a construction heuristic for instances with up to tens of thousands potential station areas and up to one hundred thousand O/D-pairs. Results show that MLO scales substantially better than the MILP or the construction heuristic, generating reasonably good results in a short amount of time.

In the next Section we discuss related work. Section 3 provides a formal definition of the MBSSLP. Afterwards, in Sect. 4, we detail our MLO approach. Section 5 describes the benchmark instances and presents and discusses experimental results. Section 6 concludes this work with including outlook on promising future work.

2 Related Work

The MBSSLP, originally proposed in [3], is a capacitated multiple allocation facility location problem [8] and is loosely based on the capacitated deviation-flow refueling location model introduced in [2]. The MBSSLP also takes into account that not every customer is willing to travel to a predestined station when the detour is too long. Such customer satisfaction factors are modeled via a decay function as also done in, e.g., [6,13]. To the best of our knowledge, no multilevel optimization approach has yet been proposed for problems regarding the distribution of battery swapping stations or vehicle charging stations. In [5] a survey of charging station locations problems is provided. This review provides an overview of the size of instances which were considered in related work. From the approaches that also consider capacities of stations, a simulated annealing approach is described [16] to generate solutions for instances with up to 1400 potential station locations and about $15000\,\Omega$/D-pairs. The instances for the MBSSLP in [3] contained up to 2000 potential locations and $8000\,\Omega$ O/D-pairs.

The MLO framework for optimization problems was originally proposed by Walshaw [15] and has been applied to various applications, such as bike sharing [7], vehicle routing [10], or traveling salesman problems [14]. In [11] Valejo et al. give on overview of MLO approaches for complex networks.

3 The Multi-period Battery Swapping Station Location Problem

The Multi-Period Battery Swapping Station Location Problem (MBSSLP) was originally proposed in [3]. We slightly modify the original MBSSLP formulation from [3] in certain aspects. First, we now consider a cyclic time horizon instead of a non-cyclic one as this appears to be more relevant in practice. More specifically, we assume a time horizon of one day that is discretized into equally long consecutive time intervals, for example hours. These intervals are indexed by $\mathcal{T} = \{0,\ldots,t_{\max}-1\}$. As we consider the planning horizon to be cyclic the predecessor of the first interval is assumed to be the last one and the successor of the last one the first interval. Second, we generalize the problem in the sense that instead of speaking of specific potential locations for service stations, we consider areas that may have more than one station. This extension is done in foresight of our MLO approach.

We assume battery swapping stations can be set up in any of n different areas referred to as set the L. Each area $l \in L$ has associated a maximum number of possible stations $r_l \in \mathbb{N}_{>0}$, a maximum number of possible battery charging slots $s_l > 0$ at each of these stations, fixed setup costs $c_l \geq 0$ for setting up one station in this area, and building costs per slot $b_l \geq 0$.

In contrast to some other work [1] that uses detailed multi-agent simulations to optimize system parameters, we model customers in an aggregated way as estimated travel demands in the form of a set of origin-destination (O/D) pairs (i.e., trips) Q and corresponding numbers $d_{qt} > 0$ of how often the need of swapping batteries is expected to arise within each time interval $t \in \mathcal{T}$ for each O/D-pair $q \in Q$. Let $m = |Q|$ be the number of O/D-pairs. As trips in an urban environment, as we consider it, are usually rather short, we assume for simplicity that trips start and end in the same time interval and can be completed with swapping batteries at most once.

Similarly to [6], we consider the satisfaction of users in dependence of detour lengths. Users will tend to avoid swapping batteries at trips for which detours to a swapping station are longer or for some other reason less convenient. To this end we associate each tuple (q, l) with $q \in Q, l \in L$ for each time interval $t \in \mathcal{T}$ with a value $g_{qlt} \in [0,1]$ representing the satisfaction of customers. We make this factor also dependent on time as, e.g., in peak hours users are likely more hesitant to make a certain detour than in hours with not much traffic, respectively.

Let us now define the bipartite undirected graph $G = (Q, L, E)$, where the node sets Q and L correspond to the O/D-pairs and the areas for building swapping stations, respectively, Edge set $E \subseteq Q \times L$ shall include an edge (q, l) for each O/D-pair $q \in Q$ and area $l \in L$ whenever a swapping station with l could potentially satisfy (part of) the demand d_{qt}, i.e., $g_{qlt} > 0$ for at least one $t \in \mathcal{T}$. By $N(q) \subseteq L$, for $q \in Q$, we denote the set of adjacent nodes of node q, which corresponds to the subset of areas that are able to service O/D-pair Q. Vice versa, $N(l) \subset Q$, for any $l \in L$, denotes the adjacent nodes of the area node l, and thus, the O/D-pairs area l may service.

The number of time intervals required for completely recharging a battery is referred to as t^c. We make here the simplifying assumption that charging any battery always takes the same time and only completely recharged batteries are provided to customers again. We denote the set of time intervals in which a battery is not yet fully charged when returned to a station at time $t \in \mathcal{T}$ as $\mathcal{T}^{ch}(t)$ which is defined as $\mathcal{T}^{ch}(t) = \{((t+i) \bmod t_{max}) \mid i = 0, \ldots, t^c\}$.

In the original MBSSLP formulation a solution is feasible if a minimum amount of total customer demand d_{min} is satisfied. In foresight of our MLO approach we relax this condition in cases where d_{min} exceeds the total amount of demand that can be satisfied in any solution to G, referred to as $d_{max}(G)$, and define a solution to be feasible if at least $\min(d_{min}, d_{max}(G))$ demand is satisfied.

For the development of MLO, we also store for each edge $(q, l) \in E(G)$ the maximum demand $\hat{d}_{qlt}$ that can be assigned from q to l in each time interval $t \in \mathcal{T}$, which is calculated by $\hat{d}_{qlt} = \min\left(\frac{\bar{d}_l}{g_{qlt}}, d_{qt}\right)$ where $\bar{d}_l$ refers to the maximal necessary capacity of the stations in an area $l \in L$, i.e., $\bar{d}_l = \min\left(r_l s_l, \max_{t \in \mathcal{T}} \sum_{t' \in \mathcal{T}^{\mathrm{ch}}(t)} \sum_{q \in N(l)} g_{qlt'} d_{qt'}\right)$.

A solution to the MBSSLP is primarily given by a pair of vectors $x = (x_l)_{l \in L}$ with $x_l \in \{0, \ldots, r_l\}$ and $y = (y_l)_{l \in L}$ with $y_l \in \{0, \ldots, \lceil \bar{d}_l \rceil\}$, where x_l indicates the number of swapping stations to be established in area l and y_l represents the respective total number of battery slots at these stations. Moreover, a solution also has to specify which demand is fulfilled where. This is done by variables a_{qlt} that denote the part of d_{qt}, $q \in Q$, which is assigned to an area $l \in N(q)$ in time interval $t \in \mathcal{T}$. Customer satisfaction is considered by multiplying this assigned demand a_{qlt} with the factor g_{qlt} in order to obtain the actually fulfilled demand $\bar{a}_{qlt} = g_{qlt} a_{qlt}$ of O/D-pair q in area l in time interval t.

Based on the variables x, y, a, and $\bar{a}$ the MBSSLP can be expressed as the following MILP:

$$\min \sum_{l \in L}(c_l x_l + b_l y_l) \tag{1}$$

$$x_l \cdot s_l \geq y_l \qquad\qquad l \in L \tag{2}$$

$$\bar{a}_{qlt} = g_{qlt} \cdot a_{qlt} \qquad\qquad t \in \mathcal{T}, \, q \in Q, \, l \in N(q) \tag{3}$$

$$\sum_{l \in N(q)} a_{qlt} \leq d_{qt} \qquad\qquad t \in \mathcal{T}, \, q \in Q \tag{4}$$

$$\sum_{t' \in \mathcal{T}^{\mathrm{ch}}(t)} \sum_{q \in N(l)} \bar{a}_{qlt'} \leq y_l \qquad\qquad t \in \mathcal{T}, \, l \in L \tag{5}$$

$$\sum_{t \in \mathcal{T}} \sum_{q \in Q} \sum_{l \in N(q)} \bar{a}_{qlt} \geq \min(d_{\min}, d_{\max}(G)) \tag{6}$$

$$x_l \in \{0, \ldots, r_l\} \qquad\qquad l \in L \tag{7}$$

$$y_l \in \{0, \ldots, \lceil \bar{d}_l \rceil\} \qquad\qquad l \in L \tag{8}$$

$$0 \leq a_{qlt} \leq \hat{d}_{qlt} \qquad\qquad t \in \mathcal{T}, \, q \in Q, \, l \in N(q) \tag{9}$$

$$0 \leq \bar{a}_{qlt} \leq g_{qlt} \hat{d}_{qlt} \qquad\qquad t \in \mathcal{T}, \, q \in Q, \, l \in N(q) \tag{10}$$

The goal of the objective function (1) is to find a feasible solution that minimizes the setup costs for stations and their battery slots. Inequalities (2) ensure that battery slots can only be allocated to an area $l \in L$ if a sufficient number of stations is opened there. Equalities (3) calculate fulfilled demands $\bar{a}_{qlt}$ by applying the customer satisfaction factors g_{qlt} to the assigned demands a_{qlt}. Constraints (4) enforce that the total demand assigned from an O/D-pair q to areas does not exceed d_{qt} for all $t \in \mathcal{T}$. Inequalities (5) ensure that the capacity y_l is not exceeded at all areas over all time intervals. Note that by using $\bar{a}_{qlt}$ instead of a_{qlt} in (5), we "overbook" areas to consider the expected case, similarly as in [9]. Inequalities (5) also model that swapped batteries cannot be reused for the next t^c time intervals in which they are being charged again. The minimal satisfied demand to be fulfilled over all time intervals is expressed by inequality (6). Finally, the domains of the variables are given in (7)–(10). Note that $d_{\max}(G)$ can be calculated by replacing the objective function (1) with

$$\max \sum_{t \in \mathcal{T}} \sum_{q \in Q^K} \sum_{l \in N(q)} g_{qlt} \cdot a_{qlt} \qquad (11)$$

and removing Constraint (6).

4 Multilevel Refinement Algorithm

Our multilevel optimization approach (MLO) follows the basic scheme proposed by [15] and consists of three steps: iteratively coarsening the problem instance by partitioning and contraction, solving the coarsest instance, and iteratively uncoarsening by projection and possible refinement. During the coarsening step the problem complexity is iteratively reduced by merging areas for setting up stations and O/D-pairs until the size of the problem instance falls below a certain threshold. Then, a solution to the coarsest instance is generated. Afterwards, this solution is successively extended by projecting it to the less coarsened instances and refining it, eventually resulting in a feasible solution to the original instance.

We define a multilevel coarsening for the MBSSLP by the graph sequence $\{G^0, \ldots, G^K\}$ of G with $G^i = (Q^i, L^i, E^i)$, for $i = 0, \ldots, K$. The graph on the lowest level corresponds to the original graph G, i.e., $G^0 = G$. As the original problem graph G, each graph G^i also have respective associated values r_l, s_l, c_l, b_l, and $\bar{d}_l$ for the nodes $l \in L^i$, values d_{qt} for nodes $q \in Q^i$, and values g_{qlt} and $\hat{d}_{qlt}$ for the edges $(q, l) \in E^i$ and $t \in \mathcal{T}$. A graph G^{i+1} with $i \in \{0, \ldots, K-1\}$ is derived from G^i by partitioning Q^i and L^i and merging all nodes within each partition. The vertices $q \in Q^{i+1}$ and $l \in L^{i+1}$ are associated with a non-empty subset of Q^i and L^i, denoted as Q_q^i and L_l^i, referring to the respective partitions of G^i. Hence, it must hold that $Q_q^i \cap Q_{q'}^i = \emptyset$ and $L_l^i \cap L_{l'}^i = \emptyset$ for any $q, q' \in Q^i$, $q \neq q'$ and $l, l' \in L^i$, $l \neq l'$, and $i = 1, \ldots, K$. Finally, (q, l) is an edge in E^{i+1} if there is at least one edge between the nodes Q_q^i and L_l^i in G^i.

Algorithm 1 shows our MLO approach in pseudo-code. Note that in our approach it cannot be guaranteed that a solution obtained for a graph G^{i+1} can be projected to G^i such that the projected solution satisfies at least the same amount of demand as the previous solution. Therefore, after projecting a solution to a graph G^i, we refine it to increase the amount of satisfied demand until the solution becomes feasible w.r.t. G^i.

In the following we describe the concrete steps of our MLO approach in more detail.

Partitioning. A graph G^{i+1} is derived from G^i by first partitioning the node sets of G^i and then contracting the nodes within each partition. For deriving the partitioning of our bipartite graphs we use the same approach as proposed by Valejo et al. [12]. We first generate two unipartite graphs for each vertex set

---

Algorithm 1: MLO

Input : an MBSSLP instance, the number of coarsening steps K
Output: a solution (x, y, a)

1: $i \leftarrow 0$;
2: $G^i \leftarrow G$;
3: **while** $i < K$ **do**
4:      $G^{i+1} \leftarrow \text{coarsen}(G^i)$;
5:      $i \leftarrow i + 1$;
6: **end while**
7: $(x, y, a) \leftarrow$ solve problem w.r.t. G^i;   // coarsest problem instance
8: **while** $i > 0$ **do**
9:      $(x, y, a) \leftarrow$ project solution (x, y, a) for G^i to a solution for G^{i-1};
10:      $(x, y, a) \leftarrow$ refine solution (x, y, a);
11:      $i \leftarrow i - 1$;
12: **end while**
13: **return** (x, y, a);

L^i, Q^i of G^i via one-mode projection, i.e., the vertices of these unipartite graphs are given by the vertices of the corresponding vertex set of G^i with two vertices being adjacent if they have common neighbors in G^i. For calculating weights between two nodes of u, v of L^i or Q^i, respectively, we use the Jaccard similarity measure

$$\chi(u, v) = \frac{|N(u) \cap N(v)|}{|N(u) \cup N(v)|}. \tag{12}$$

Afterwards, each unipartite graph is partitioned independently via *greedy heavy-edge-matching* (GHEM) [11]. GHEM is a variation of *heavy-edge-matching* [4]. The GHEM heuristic iterates over all edges of a graph in descending order and in every iteration partitions the incident nodes of the current edge and removes them from the graph.

Contracting. Recall that we denote a partition of nodes of G^{i-1} as Q_q^{i-1} and L_l^{i-1}, respectively, with $q \in Q^i, l \in L^i$, for $i = 1, \ldots, K$.

When contracting nodes, one has to also aggregate the associated node and edge properties in meaningful ways. To facilitate the later projection of solutions, we split the coarsening of G^{i-1} into two steps, first deriving from G^{i-1} an intermediate graph $\tilde{G}^i$, in which only the partitions on L^{i-1} are merged, and then from $\tilde{G}^i$ the actual G^i in which also the partitions on Q^{i-1} are merged. In the following we denote the node and edge sets of $\tilde{G}^i$ as well as all associated values correspondingly with a tilde, i.e., $\tilde{G}^i = (\tilde{Q}^i, \tilde{L}^i, \tilde{E}^i)$. Moreover, $\tilde{N}^i(\cdot)$ refers to adjacent nodes of nodes in $\tilde{G}^i$, whereas $N^i(\cdot)$ refers only to neighbors of nodes in G^i.

To obtain $\tilde{G}^i$, we thus directly adopt $\tilde{Q}^i = Q^{i-1}$ together with the properties associated with these nodes, i.e., the respective demands. The partitions on L^{i-1} on the other hand are merged to obtain the new area node set $\tilde{L}^i$.

Maximum allowed demand assignments for $(q, l) \in \tilde{E}^i$ pairs are now calculated as

$$\hat{d}_{qlt} = \min \left(\sum_{l' \in L_l^{i-1}} \hat{d}_{ql't}, \ d_{qt} \right), \tag{13}$$

and the customer satisfaction factors are determined as weighted average

$$g_{qlt} = \frac{\sum_{l' \in L_l^{i-1}} \hat{d}_{ql't} \cdot g_{ql't}}{\sum_{l' \in L_l^{i-1}} \hat{d}_{ql't}}. \tag{14}$$

The maximum demand that can be fulfilled by the stations in an area $l \in \tilde{L}^i$ in any time interval is determined by

$$\bar{d}_l = \min \left(\sum_{l' \in L_l^{i-1}} \bar{d}_{l'}, \ \max_{t \in T} \sum_{t' \in T^{ch}(t)} \sum_{q \in \tilde{N}^i(l)} g_{qlt'} \hat{d}_{qlt'} \right). \tag{15}$$

For each $l \in \tilde{L}^i$, s_l, c_l, and b_l are averaged in a weighted manner:

$$s_l = \left\lceil \frac{\sum_{l' \in L_{l'}^{i-1}} \bar{d}_{l'} s_{l'}}{\sum_{l' \in L_l^{i-1}} \bar{d}_{l'}} \right\rceil, \quad c_l = \frac{\sum_{l' \in L_l^{i-1}} \bar{d}_{l'} c_{l'}}{\sum_{l' \in L_l^{i-1}} \bar{d}_{l'}}, \quad b_l = \frac{\sum_{l' \in L_l^{i-1}} \bar{d}_{l'} b_{l'}}{\sum_{l' \in L_l^{i-1}} \bar{d}_{l'}}. \tag{16}$$

Finally, maximum station numbers are derived by

$$r_l = \left\lceil \frac{\bar{d}_l}{s_l} \right\rceil. \tag{17}$$

To finally obtain G^i from the intermediate $\tilde{G}^i$, we directly adopt $L^i = \tilde{L}^i$ together with all the properties associated with these nodes, while merging the partitions on $\tilde{Q}^i$ obtaining the new O/D-pair node set Q^i.

Customer satisfaction factors are aggregated again by taking the weighted average

$$g_{qlt} = \frac{\sum_{q' \in \tilde{Q}_q^{i-1}} \hat{d}_{q'lt} \cdot g_{q'lt}}{\sum_{q' \in \tilde{Q}_q^i} \hat{d}_{q'lt}}. \tag{18}$$

For each edge $(q, l) \in \tilde{E}^i$ and each $t \in T$, the maximum assignable demand is calculated by

$$\hat{d}_{qlt} = \min \left(\sum_{q' \in \tilde{Q}_q^i} \hat{d}_{q'lt}, \ \frac{\bar{d}_l}{g_{qlt}} \right). \tag{19}$$

The demands of these O/D-pairs are aggregated by taking the respective sums for all $q \in \tilde{Q}^i$ and $t \in T$ while also considering the maximal amount of demand d_{qlt} that can be assigned to stations at all adjacent areas $l \in N^i(q)$:

$$d_{qt} = \min \left(\sum_{q' \in \tilde{Q}_q^i} d_{q't}, \ \sum_{l \in N^i(q)} \hat{d}_{qlt} \right). \tag{20}$$

Note that due to the aggregation of the customer satisfaction factors g, it cannot be guaranteed that $d_{\max}(G^i) \geq d_{\min}$. Therefore, as previously discussed we have relaxed the original feasibility criterion of the MBSSLP such that a solution to a graph G^i is feasible if at least $\min(d_{\min}, d_{\max}(G^i))$ demand is satisfied.

Solving the Coarsest Graph. The MILP (1)–(10) is used to generate a solution to the coarsest graph G^K.

Projecting. A solution to a graph G^i is projected to the graph G^{i-1} in two steps. First the solution is projected to $\tilde{G}^i$ and then further projected to G^{i-1}.

Let x, y, and a be defined as described in Sect. 3. Projecting a solution from G^i to $\tilde{G}^i$ is done by solving a linear program (LP) for each $q \in Q^i$:

$$\max \sum_{q' \in Q_q^i} \sum_{l \in \tilde{N}^i(q)} \sum_{t \in \mathcal{T}} \frac{g_{q'lt} a_{q'lt}}{b_l} \tag{21}$$

$$\sum_{l \in \tilde{N}^i(q')} a_{q'lt} \leq d_{q't} \qquad\qquad t \in \mathcal{T},\ q' \in Q_q^{i-1} \tag{22}$$

$$\sum_{q' \in \tilde{N}^i(l) \cap Q_q^i} g_{q'lt} \cdot a_{q'lt} \leq g_{qlt} \cdot a_{qlt} \qquad\qquad t \in \mathcal{T},\ l \in N^i(q) \tag{23}$$

$$0 \leq a_{q'lt} \leq \hat{d}_{q'lt} \qquad\qquad t \in \mathcal{T},\ q' \in Q_q^i,\ l \in \tilde{N}^i(q') \tag{24}$$

Recall that variables a_{qlt} denote the part of d_{qt} w.r.t. $q \in Q$, which is assigned to an area $l \in N(q)$ in time interval $t \in \mathcal{T}$. The objective function of this LP maximizes the ratio of assigned demand to costs for building modules at the respective areas. Constraints (22) ensure that assigned demand does not exceed an O/D-pair's available demand. Constraints (23) ensure that the total demand assigned from all $q' \in Q_q^{i-1}$ to some $l \in \tilde{L}^i$ does not exceed the demand assigned from q to l. Hence, the total number of battery slots required in an area does not increase when projecting the solution to $\tilde{G}^i$. Note that the sub-problems induced by $q \in Q^i$ can be solved independently of each other. However, the total satisfied demand for the obtained solution might be smaller than the satisfied demand in the solution to G^i.

When the solution is projected from $\tilde{G}^i$ to G^{i-1}, we again have one sub-problem for each $l \in \tilde{L}^i$. In this step we also aim to compensate for satisfied demand lost in previous solution projections. Let d_{missing} denote the difference between $d_{\min}$ and the amount of demand satisfied in a solution and let $\delta_{\min}(l) = \sum_{q \in \tilde{N}^i(l)} g_{qlt} \cdot a_{qlt} + d_{\text{missing}}$ for $l \in \tilde{L}^i$. Then, when projecting the solution w.r.t. l, the minimal amount of demand to be satisfied by the areas in L_l^i is $\min(\delta_{\min}(l), d_{\max}(L_l^{i-1}))$ where $d_{\max}(L_l^{i-1})$ is the maximal amount of demand that can be satisfied by the areas in L_l^{i-1}. Moreover, when projecting the solution w.r.t. $\tilde{L}^i$ we not only consider the demand allocated at the areas in the solution but also take into account the so far unassigned demand of all O/D-pairs in $\tilde{Q}^i$. Hence, a sub-problem for $l \in \tilde{L}^i$ induces a sub-instance with the areas L_l^{i-1},

---

Algorithm 2: Project Solution from $\tilde{G}^i$ to G^i

Input : a solution (x, y, a) to $\tilde{G}^i$
Output: a solution to G^{i-1}

1: $\Lambda \leftarrow \{l \in \tilde{L}^i \mid x_l > 0\};$ //areas to be extended

2: $\rho \leftarrow \left(\frac{\sum_{q \in \tilde{N}^i(l)} g_{qlt} \cdot a_{qlt}}{x_l \cdot c_l + y_l \cdot b_l} \right)_{l \in \Lambda};$

3: $\delta_{\min} \leftarrow d_{\min} - \sum_{t \in \mathcal{T}, l \in \Lambda, q \in \tilde{N}^i(l)} a_{qlt};$

4: $\delta \leftarrow \left(d_{qt} - \sum_{l \in \tilde{N}^i(q)} a_{qlt} \right)_{q \in \tilde{Q}^i, t \in \mathcal{T}};$ //unassigned demand

5: **while** $|\Lambda| > 0$ **do**
6:   **if** $\delta_{\min} > 0$ **then**
7:     $l \leftarrow \arg\max_{l \in \Lambda}\{\rho_l\};$
8:   **else**
9:     $l \leftarrow \arg\min_{l \in \Lambda}\{\rho_l\};$
10: **end if**
11: $\Lambda \leftarrow \Lambda \setminus \{l\};$
12: $\delta_{\min} \leftarrow \delta_{\min} + \sum_{q \in \tilde{N}^i(l)} g_{qlt} \cdot a_{qlt};$
13: $\delta_{qt} \leftarrow \delta_{qt} + a_{qlt} \quad \forall q \in \tilde{N}^i(l),\ t \in \mathcal{T};$

14: $(x', y', a') \leftarrow \text{solve}(L_l^{i-1}, \tilde{N}^i(l), \delta_{\min}, \delta);$ //apply MILP (1)–(10)

15: **for** $l' \in L_l^{i-1}$ **do**
16:    $x_{l'} \leftarrow x'_{l'},\ y_{l'} \leftarrow y'_{l'};$
17:    **for** $q \in N^{i-1}(l'),\ t \in \mathcal{T}$ **do**
18:      $a_{ql't} \leftarrow a'_{ql't};$
19:      $\delta_{\min} \leftarrow \delta_{\min} - g_{ql't} \cdot a_{ql't};$
20:      $\delta_{qt} \leftarrow \delta_{qt} - a_{ql't};$
21:    **end for**
22: **end for**
23: **end while**
24: **return** $(x, y, a);$

---

the O/D-pairs $q \in \tilde{N}^i(l)$ with available demands δ_{qt} for $t \in \mathcal{T}$, and a minimal amount of demand to be satisfied $\delta_{\min}(l)$. This sub-instance can then be solved by the MILP (1)–(10). Algorithm 2 gives a detailed description of how δ and $\delta_{\min}(l)$ are calculated for each sub-problem. Note that these sub-problems are no longer independent of each other. Therefore, the order in which they are solved impacts the quality of the projected solution. To keep the setup costs of the projected solution as low as possible, Algorithm 2 chooses the sub-problem induced by the most cost efficient area $l \in \tilde{L}^i$ if d_{missing} is greater than zero as the next one. Otherwise, if the current solution satisfies $d_{\min}$ demand, the sub-problem induced by the least cost efficient area is solved next.

Algorithm 3: Refine Solution

Input : a solution (x, y, a) to G^i
Output: a refined solution to G^i

1: $\Lambda \leftarrow L^i$;
2: $\rho \leftarrow \left((r_l - x_l) \cdot c_l + (\lceil \bar{d}_l \rceil - y_l) \cdot b_l\right)_{l \in \Lambda}$;
3: $\delta_{\min} \leftarrow d_{\min} - \sum_{t \in T, l \in \Lambda, q \in N^i(l)} a_{qlt}$;
4: $\delta \leftarrow \left(d_{qt} - \sum_{l \in N(q)} a_{qlt}\right)_{q \in \tilde{Q}^i, t \in T}$;   //unassigned demand

5: **while** $\delta_{\min} > 0$ **do**
6:     $l \leftarrow \arg\min_{l \in \Lambda}\{\rho_l\}$;
7:     $\Lambda \leftarrow \Lambda \setminus l$;
8:     $\delta_{\min} \leftarrow \delta_{\min} + \sum_{q \in N^i(l)} g_{qlt} \cdot a_{qlt}$;
9:     $\delta_{qt} \leftarrow \delta_{qt} + a_{qlt} \quad \forall q \in N^i(l), \ t \in T$;
10:     $(x', y', a') \leftarrow \text{solve}(l, N^i(l), \delta_{\min}, \delta)$;   //apply MILP (1)–(10)
11:     $x_l \leftarrow x'_l, \ y_l \leftarrow y'_l$;
12:     **for** $q \in N(l'), \ t \in T$ **do**
13:         $a_{qlt} \leftarrow a'_{qlt}$;
14:         $\delta_{\min} \leftarrow \delta_{\min} - g_{qlt} \cdot a_{qlt}$;
15:         $\delta_{qt} \leftarrow \delta_{qt} - a_{qlt}$;
16:     **end for**
17: **end while**
18: **return** (x, y, a);

Refine Solution. After projecting a solution from $\tilde{G}^{i+1}$ to G^i via Algorithm 2, the obtained solution may be infeasible as it cannot be guaranteed that a solution to a sub-problem w.r.t. $l \in \tilde{L}^i$ can actually satisfy $\delta_{\min}(l)$ demand. Therefore, it may be necessary to further refine the obtained solution, i.e., to open additional modules or areas and assign demand to them. Our refinement procedure is shown in Algorithm 3. Similar to Algorithm 2 for each $l \in L^i$ that is not yet fully utilized we define a sub-instance with area l, O/D-pairs $q \in N^i(l)$ with available demands δ_{qt} for $t \in T$, and a minimal amount of demand to be satisfied $\delta_{\min}(l)$. Again, this sub-instance can be solved by the MILP (1)–(10). As long as the solution is infeasible, the sub-instance induced by the cheapest area is chosen and solved. Note that we use Algorithm 3 also as standalone construction heuristic (CH) and will compare the performance of MLO to the performance of CH.

5   Computational Results

We test our approach on artificial instances generated as described in [3]. A total of eight groups of instances identified by their number of station areas n and number of O/D pairs m as (n, m) is created. Each group contains 30 instances. Note that the instances contain some station areas that have no O/D pairs in the vicinity and vice versa. These unconnected nodes are deleted during pre-processing. Table 1 gives an overview over all instance groups. Columns n_{pp} and

Table 1. Test instance groups and the average numbers of nodes after preprocessing.

n	m	n_{pp}	m_{pp}	n	m	n_{pp}	m_{pp}
5000	5000	2354	4936	10000	10000	4658	9457
	12500	3086	12410		25000	6130	24845
	25000	3663	24821		50000	7300	49714
	50000	4194	49634		100000	8394	99407

Table 2. Results obtained by solving the MILP with Gurobi (runtime limit two hours).

n	m	$\gamma_{LB}[\%]$	$n_{feasible}$	$\tau[s]$
5000	5000	3.47	30	7200
	12500	12.01	30	7200
	25000	6.40	25	7200
	50000	4.41	7	7202
10000	10000	21.79	30	7200
	25000	13.77	24	7201
	50000	–	0	–
	100000	–	0	–

m_{pp} list the average numbers of station areas and O/D pairs, respectively after preprocessing. Note that the number of actually usable station areas strongly depends on the number of O/D pairs, i.e., the more O/D pairs a graph with a fixed number of station areas has, the more station areas are adjacent to an O/D pair.

MLO was implemented in Julia[1] 1.8.1 using Gurobi[2] 9.1 as the underlying MILP solver. All test runs have been executed on an AMD EPYC 7402, 2.80GHz machine in single-threaded mode with a global memory limit of 100GB. When solving MILPs during MLO we have set a time limit of ten minutes and terminated the solving earlier when an optimality gap of $\leq 0.5\%$ was reached.

First, Table 2 shows the results for generating solutions to the benchmark instances by solving the MILP (1)–(10) with Gurobi with a runtime limit of two hours. Column γ_{LB} shows average gaps between the best found solutions and the respective lower bounds. Additionally, column $n_{feasible}$ shows for how many instances in a group the solver was able to find a feasible solution. Finally, column τ shows median computation times. We can see that the MILP solver was not able to consistently find a feasible solution within the given time limit for a large part of the instances. In general, the larger the instances, the fewer solutions were found by the MILP solver. However, there are three groups for which the MILP solver was able to find a feasible solution to all instances: $(5000, 5000), (5000, 12500)$, and $(10000, 10000)$. Looking at these groups we can

[1] https://julialang.org/.
[2] https://www.gurobi.com/.

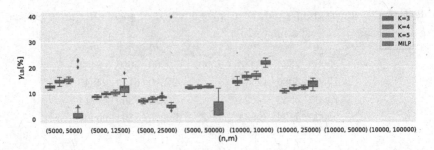

Fig. 1. Comparison of results obtained by a MILP solver to results obtained by MLO with different values for K. Each boxplot covers a single instance group and each instance was evaluated once for each configuration.

also see that the gaps to the best known lower bounds are strongly increasing as the instances become larger. Moreover, for the two largest instance groups the MILP solver was not able to find any feasible solution at all within the time limit.

Next, in Fig. 1 we compare different variants of MLO to the pure MILP app-roach. MLO was able to find feasible solutions for all instances, however, for this comparison we only consider instances to which MLO as well as the MILP solver were able to find solutions. To compare the two approaches we use the best found lowest bound for each instance reported by Gurobi to calculate gaps for the MLO results. We test MLO with different numbers of coarsening levels $K \in \{3, 4, 5\}$. The figure shows that for the reported instances, MLO achieves gaps between 6% to 18%. For the instance groups $(5000, 12500)$, and $(10000, 10000)$, for which the MILP solver found solutions to all instances and the group $(10000, 25000)$, MLO is able to achieve better results than the MILP. Otherwise, the MILP solutions are usually better than the MLO solutions. However, one has to consider, that the MILP solver was not able find solutions to a large part of the instances at all. Additionally, the MILP solver terminated after two hours, while MLO was much faster.

To get a better impression of how well MLO performs, Tables 3, 4, 5 give more detailed results for MLO. As previously mentioned the construction heuristic used for refining solutions after projection can also be used as a standalone construction heuristic (CH). We compare the results obtained by MLO to the results achieved with this construction heuristic as MLO itself is comparable to a construction heuristic in the sense that a complete solution is only obtained at the end of the algorithm. Moreover, as CH generates solutions to all instances within the time limit, we are able to get a better impression of how well MLO performs w.r.t. different values of K and different instance sizes. Columns γ_{CH} show the average gap between the MLO results and the CH results for each instance group. More specifically, let f_{MLO} refer to the objective value of a

Table 3. MLO results for $K = 3$.

n	m	$\gamma_{CH}[\%]$	n_{proj}	$\tau_c[s]$	$\tau_{cp}[s]$	$\tau[s]$
5000	5000	−37.38	5279	6	36	103
	12500	−48.54	13032	12	123	265
	25000	−36.84	24688	28	96	385
	50000	−22.09	47304	77	81	663
10000	10000	−37.62	10156	10	132	241
	25000	−48.93	26035	24	457	747
	50000	−36.88	49348	58	358	954
	100000	−22.09	94635	173	292	1624

Table 4. MLO results for $K = 4$.

n	m	$\gamma_{CH}[\%]$	n_{proj}	$\tau_c[s]$	$\tau_{cp}[s]$	$\tau[s]$
5000	5000	−34.30	5950	6	20	86
	12500	−46.85	14275	11	14	154
	25000	−35.68	26764	26	12	277
	50000	−21.63	51073	71	17	552
10000	10000	−34.32	11466	9	39	149
	25000	−47.21	28480	22	50	325
	50000	−35.59	53482	54	41	603
	100000	−21.46	102171	160	60	1301

Table 5. MLO results for $K = 5$.

n	m	$\gamma_{CH}[\%]$	n_{proj}	$\tau_c[s]$	$\tau_{cp}[s]$	$\tau[s]$
5000	5000	−33.50	6276	6	3	68
	12500	−46.12	14920	11	2	139
	25000	−34.71	27870	25	3	271
	50000	−21.44	53064	70	5	543
10000	10000	−33.67	12113	9	8	117
	25000	−46.75	29759	21	6	281
	50000	−34.76	55703	51	9	576
	100000	−21.42	106172	155	15	1255

solution obtained with MLO and let f_{CH} refer to the objective value of a solution (to the same instance) generated by CH. Then, $\gamma_{CH} = \frac{f_{CH} - f_{MLO}}{f_{CH}} \cdot 100$. Hence, if γ_{CH} is negative, MLO achieved better results, otherwise the better result was achieved by CH. Additionally, the tables also show the average number of sub-problems n_{proj} solved when projecting the solution, the average time τ_c needed for generating all coarsened graphs, the average time τ_{cp} for generating a solution to the coarsest graph, as well as the average total computation time τ. Results indicated that MLO clearly outperforms CH w.r.t. the obtained solution quality, yielding solutions that are up to $\approx 49\%$ better on average. Note that the gaps generally decrease as m increases. Due to a higher number of O/D pairs, a larger number of opened stations is required. Hence, the difference between the worst solution (the solution in which all stations are opened with a maximum number of modules) and an optimal solution becomes smaller. Therefore, one can also expect to achieve "better" results more easily as m increases. As expected, one can also observe that the number of sub-problems that need to be solved during the projection phase increases proportionally to K. However, in general the number of sub-problems solved is roughly equal to m. As the number of sub-problems increases, the total computation times actually decrease as the coarsest problem instance can be solved significantly faster for larger values of K. In general, it seems that the computation times grow proportional to n and m. For the largest instance group MLO needed up to 27 min on average to generate solutions. Moreover, coarsening of the graphs contributes only a small part of the total computation time. Finally, we can also observe that the solution quality slightly deteriorates as K increases. This behavior is not surprising as the coarsest instance becomes a less accurate representation of the original instance the more often it was coarsened.

6　Conclusion and Future Work

We presented a Multilevel Optimization (MLO) approach for solving huge instances of the Multi-Period Battery Swapping Station Location Problem (MBSSLP) proposed in [3]. MLO first generates series of coarsened graphs with each new graph becoming smaller in size. Afterwards, a solution to the coarsest graph is generated and iteratively projected to the coarsened graphs in reverse order until and refined a solution to the original problem graph is obtained.

The approach was tested on artificial benchmark instances with up to 10000 areas for placing stations and 100000 origin-destination (O/D) pairs describing the trips of users. Evaluating our approach on these benchmark instances shows that MLO is able to generate reasonably good solutions within at most half an hour. On the other hand, when formulating and solving the problem as a mixed integer linear program (MILP), the MILP solver struggles to consistently find even feasible solutions to many of the instances. As the size of the instances increases, MLO clearly outperform the MILP solver w.r.t. the achieved solution quality. When confronted with such huge instances, classical metaheuristics often struggle as well to obtain good results as corresponding neighborhood structures are too large to be searched efficiently. Therefore, we instead compared MLO to the construction heuristic (CH) that was used within MLO for refining solutions. Our results show that MLO significantly outperforms CH w.r.t. all instances.

Still, there are multiple ways in which MLO can possibly be improved, especially w.r.t. deriving the coarsened graphs. Using greedy heavy-edge matching we obtain a graph partitioning that contains at most two nodes in each partition. However, it seems promising to adapt this approach such that partitions of larger size can be generated if, for example, one can identify large similarities between multiple nodes. In order to identify similar nodes more reliably it seems also necessary to derive a more problem specific similarity criterion. A possible way to achieve this is to use machine learning for learning the similarity between two nodes, e.g., by training a machine learning model on a smaller set of instances in which only two nodes are coarsened at a time.

A particular issue for MLO is that the amount of satisfied demand can decrease when projecting a solution. A potential way to alleviate this problem is to derive constraints that restrict the possibilities in how a node can be projected.

In this contribution we have explicitly specified how often a graph should be coarsened. Additionally, in each iteration both the set of station areas as well as the set of O/D-pairs is coarsened. In future work we aim to adapt MLO such that a graph is coarsened until both of its vertex sets have in some sense ideal sizes.

Finally, in the real world a solution obtained by MLO is in general not build at once but gradually over time. Therefore, an interesting further problem is to also optimize the schedule by which the stations should be built when realizing the obtained solution.

References

1. Brulin, S., Bujny, M., Puphal, T., Menzel, S.: Data-driven evolutionary optimization of eVTOL design concepts based on multi-agent simulations. In: Proceedings of the American Institute of Aeronautics and Astronautics SciTech Forum (to appear)
2. Hosseini, M., MirHassani, S., Hooshmand, F.: Deviation-flow refueling location problem with capacitated facilities: model and algorithm. Transp. Res. Part D: Transp. Environ. **54**, 269–281 (2017)
3. Jatschka, T., Oberweger, F.F., Rodemann, T., Raidl, G.R.: Distributing battery swapping stations for electric scooters in an urban area. In: Olenev, N., Evtushenko, Y., Khachay, M., Malkova, V. (eds.) OPTIMA 2020. LNCS, vol. 12422, pp. 150–165. Springer, Cham (2020). https://doi.org/10.1007/978-3-030-62867-3_12
4. Karypis, G., Kumar, V.: A fast and high quality multilevel scheme for partitioning irregular graphs. SIAM J. Sci. Comput. **20**(1), 359–392 (1998)
5. Kchaou-Boujelben, M.: Charging station location problem: a comprehensive review on models and solution approaches. Transp. Res. Part C: Emerg. Technol. **132**, 103376 (2021)
6. Kim, J.G., Kuby, M.: The deviation-flow refueling location model for optimizing a network of refueling stations. Int. J. Hydrogen Energy **37**(6), 5406–5420 (2012)
7. Kloimüllner, C., Raidl, G.R.: Hierarchical clustering and multilevel refinement for the bike-sharing station planning problem. In: Battiti, R., Kvasov, D.E., Sergeyev, Y.D. (eds.) LION 2017. LNCS, vol. 10556, pp. 150–165. Springer, Cham (2017). https://doi.org/10.1007/978-3-319-69404-7_11
8. Laporte, G., Nickel, S., da Gama, F.S.: Location Science. Springer, Cham (2015)
9. Murali, P., Ordóñez, F., Dessouky, M.M.: Facility location under demand uncertainty: response to a large-scale bio-terror attack. Soc.-Econ. Plann. Sci. **46**(1), 78–87 (2012). special Issue: Disaster Planning and Logistics: Part 1
10. Pirkwieser, S., Raidl, G.R.: Multilevel variable neighborhood search for periodic routing problems. In: Cowling, P., Merz, P. (eds.) EvoCOP 2010. LNCS, vol. 6022, pp. 226–238. Springer, Heidelberg (2010). https://doi.org/10.1007/978-3-642-12139-5_20
11. Valejo, A., Ferreira, V., Fabbri, R., Oliveira, M.C.F.D., Lopes, A.D.A.: A critical survey of the multilevel method in complex networks. ACM Comput. Surv. **53**(2), 1–35 (2020)
12. Valejo, A., Ferreira, V., de Oliveira, M.C.F., de Andrade Lopes, A.: Community Detection in bipartite network: a modified coarsening approach. In: Lossio-Ventura, J.A., Alatrista-Salas, H. (eds.) SIMBig 2017. CCIS, vol. 795, pp. 123–136. Springer, Cham (2018). https://doi.org/10.1007/978-3-319-90596-9_9
13. Verter, V., Lapierre, S.D.: Location of preventive health care facilities. Ann. Oper. Res. **110**(1), 123–132 (2002)
14. Walshaw, C.: A multilevel approach to the travelling salesman problem. Oper. Res. **50**(5), 862–877 (2002)
15. Walshaw, C.: Multilevel refinement for combinatorial optimisation problems. Ann. Oper. Res. **131**(1), 325–372 (2004)
16. Zockaie, A., Aashtiani, H.Z., Ghamami, M., Nie, Y.: Solving detour-based fuel stations location problems. Comput.-Aided Civ. Infrastruct. Eng. **31**(2), 132–144 (2016)

A Memetic Algorithm for Deinterleaving Pulse Trains

Jean Pinsolle[1,2], Olivier Goudet[2], Cyrille Enderli[1], and Jin-Kao Hao[2(✉)]

[1] Thales DMS France SAS, 2 avenue Gay Lussac, 78852 Elancourt cedex, France
jean.pinsolle@fr.thalesgroup.com
[2] LERIA, Université d'Angers, 2 Boulevard Lavoisier, 49045 Angers, France
{olivier.goudet,jin-kao.hao}@univ-angers.fr

Abstract. This paper deals with the problem of deinterleaving a sequence of signals received from different emitters at different time steps. It is assumed that this pulse sequence can be modeled by a collection of processes over disjoint finite sub-alphabets, which have been randomly interleaved by a switch process. A known method to solve this problem is to maximize the likelihood of the model which involves a partitioning problem of the whole alphabet. This work presents a new memetic algorithm using a dedicated likelihood-based crossover to efficiently explore the space of possible partitions. The algorithm is first evaluated on synthetic data generated with Markov processes, then its performance is assessed on electronic warfare datasets.

Keywords: Memetic algorithm · Markov process · partitioning problem · deinterleaving pulse trains · electronic warfare

1 Introduction

This paper presents an optimization algorithm for deinterleaving data streams that can be described by interleaved Markov processes. Even though such a method can be applied to many fields, the original motivation of this paper is related to radar warning receivers, which are passive sensors performing among other tasks the deinterleaving of pulse trains received from multiple emitters over a common channel.

In this context, pulses are emitted by different radars present in the environment and are intercepted by a single receiver. Each pulse is described by several characteristics called Pulse Description Words (PDW). Some of these features are called *primary* because they are measured in the early stages of the radio frequency signal reception chain, such as time of arrival (ToA), carrier frequency (CF), pulse duration (PD), signal amplitude or angle of arrival (AoA), while others are called *secondary*, such as the time interval between two consecutive pulses (Pulse Repetition Interval, PRI) because they characterize a pulse train. In the case of conventional radars with simple interpulse modulation, basic PRI clustering methods may be sufficient to solve the problem [10,11,13]. For more

L. Pérez Cáceres and T. Stützle (Eds.): EvoCOP 2023, LNCS 13987, pp. 66–81, 2023.
https://doi.org/10.1007/978-3-031-30035-6_5

complex data, multivariate methods such as [3], leveraging on different pulse features (CF, AoA, and PD), have been proposed.

However, modern radars can create much more complex patterns, which may result in a loss of performance of basic clustering methods, and produce errors in the deinterleaving process such as transmitter track proliferation or radar misses. To overcome these limitations, new methods based on inferring mixtures of Markov chains [1] have been proposed in the radar pulse train deinterleaving literature [5]. These methods aim to address complex scenarios by reducing the surplus of clusters found by classical methods. In these interleaved Markov process (IMP) methods, a clustering algorithm is first applied to group the different pulses into different clusters (or letters). Then, in a second step, a partition of these letters into different groups is performed in order to identify the different emitters which could have generated the observed sequence of symbols. This partition of the different symbols is typically done by maximizing a penalized likelihood score, which has been proven consistent under mild conditions on the switch and component processes [14].

In general, such methods for deinterleaving finite memory processes via penalized maximum likelihood raise a challenging combinatorial problem, because finding the optimal partition may require evaluating all the possible partitions of the observed symbols into different groups. Since this search space of all partitions grows exponentially with the number of symbols, an exhaustive search is in general not feasible in a reasonable amount of time. Therefore, heuristics based on greedy criteria have been proposed in [5,14] to provide an approximate solution to this problem in a limited amount of time. However, such greedy searches are prone to easily get stuck in local optima, especially when the search space becomes huge.

In this paper, we propose a new heuristic to solve this deinterleaving partition problem (DPP), by noticing that this problem can be seen as a particular grouping problem. The main contribution of this work is a new memetic algorithm for alphabet partitioning called MAAP, inspired by the memetic framework HEAD [12], which obtains state-of-the-art results for another grouping problem namely the graph coloring problem. The MAAP algorithm takes into account specific features related to penalized entropy estimates in order to speed up the search in the space of all partitions. In addition, it introduces a new likelihood-based crossover capable of sharing low entropy sub-alphabets that will be transmitted to the next generations.

The rest of the paper is organized as follows: Sect. 2 presents the formal background for the deinterleaving of Markov processes. Section 3 presents the settings of the optimization problem. Section 4 describes the proposed memetic algorithm. Section 5 reports the results on synthetic datasets generated with Markov processes, while Sect. 6 provides illustrative examples of radar interceptions in a realistic context. Section 7 discusses the contribution and presents some perspectives for future work.

2   Deinterleaving Markov Processes: Formal Background

In this section, we summarize the formal background of deinterleaving a set of finite memory processes on disjoint subsets and the penalized maximum likelihood method introduced in [14] to solve this problem.

2.1   Interleaved Markov Generative Process

Let $z^n = z_1, \ldots, z_n$ be an observed sequence of n symbols ordered by their time of arrival. Each symbol is drawn from a finite set $\mathcal{A}$ (alphabet).

The underlying generative model of this sequence is assumed to be an *interleaved Markov process* $P = \mathcal{I}_\Pi(P_1, \ldots, P_m; P_w)$, where $m > 0$ is the number of different emitters, P_i is an independent component random process for emitter i, generating symbols in the sub-alphabet $A_i \subset \mathcal{A}$, P_w is a random switch process over the emitters, and $\Pi = \{A_1, \ldots, A_m\}$ is the partition of $\mathcal{A}$ into the sub-alphabets A_i, for $i = 1, \ldots, m$, which are assumed to be non-empty and disjoint.

It is further assumed that all Markov processes are time-homogeneous, independent, ergodic, and with finite memory. Let k_i be the order of P_i, and $\mathbf{k} = (k_1, \ldots, k_m; k_w)$ denote the vector containing the orders of the corresponding processes $(P_1, \ldots, P_m; P_w)$. All states are assumed to be reachable and recurrent, and it is assumed that all symbols $a \in \mathcal{A}$ occur infinitely and their stationary marginal probabilities are positive. There is no assumption on the initial state of the processes.

According to this IMP P, at each time step, $t = 1, \ldots, n$ and given the prefix $z^{t-1} = z_1, \ldots, z_{t-1}$ of the sequence already generated at time $t - 1$, a process P_i is selected by the switch process P_w, then P_i selects a letter z_t from A_i and adds it to the prefix sequence z^{t-1} to form the sequence z^t.

Formally, this generative process can be written as

$$P(z_t|z^{t-1}) = P_w(i|\sigma_\Pi(z^{t-1}))P_i(z_t|z^{t-1}[A_i]), \tag{1}$$

where $\sigma_\Pi(z^{t-1})$ is the sequence of integers $i \in \{1, \ldots, m\}$ derived from the switch selection of the processes P_i to generate the sequence z^{t-1} and $z^{t-1}[A_i]$ is the sub-string of the sequence z^{t-1} obtained by deleting all symbols not in A_i, note that we do not write a sum on i since $P_i(z_t|z^{t-1}[A_i])$ is null for another alphabet than A_i.

By recursive application of Eq. (1), the probability of occurrence of a sequence z^t is then (with a slight abuse of notation)

$$P(z^t) = P_w(\sigma_\Pi(z^t)) \prod_{i=1}^{m} P_i(z^t[A_i]). \tag{2}$$

2.2   Penalized Maximum Likelihood Score

For a Markov process P of order k which generates a sequence u^t of letters drawn from $\mathcal{A}$, the maximum likelihood (ML) of u^t is given by

$$P_k^{ML}(u^t) = \prod_{a^{k+1} \in u^t} P(a^{k+1}|a^k) = \prod_{a^{k+1} \in u^t} \left(\frac{N_{u^t}(a^{k+1})}{N_{u^{t-1}}(a^k)}\right)^{N_{u^t}(a^{k+1})}, \quad (3)$$

with a^k a pattern of k letters in u^t of length k, $P(a^{k+1}|a^k)$ the transition probability from a^k to a^{k+1} and $N_{u^t}(a^{k+1})$ the number of patterns a^{k+1} in u^t. We denote $\hat{H}_k(u^t) = -\log P_k^{ML}(u^t)$ the corresponding ML entropy.

Knowing that the processes are independent and according to Eq. (2), the global ML entropy $\hat{H}_{\Pi,\mathbf{k}}(z^n)$ of a sequence z^n under an IMP model induced by the partition Π and the vector order $\mathbf{k}$ is given by the addition of the ML entropy of each process:

$$\hat{H}_{\Pi,\mathbf{k}}(z^n) = \sum_{i=1}^{m} \hat{H}_{k_i}(z^n[A_i]) + \hat{H}_{k_w}(\sigma_\Pi(z^n)). \quad (4)$$

A global penalized entropy is further defined by adding a penalty term:

$$C_{(\Pi,\mathbf{k})}(z^n) = \hat{H}_{\Pi,\mathbf{k}}(z^n) + \beta\kappa \log n, \quad (5)$$

with β a constant and κ the number of free parameters in the model, which corresponds to the number of free parameters in the different processes:

$$\kappa = \sum_{i=1}^{m} |A_i|^{k_i}(|A_i| - 1) + m^{k_w}(m - 1). \quad (6)$$

Finally, the IMP estimate, i.e., the deinterleaving scheme, is given by minimizing the previous cost function:

$$(\hat{\Pi}, \hat{\mathbf{k}}) = \underset{(\Pi,\mathbf{k})}{argmin}\ C_{(\Pi,\mathbf{k})}(z^n). \quad (7)$$

It is known that the scheme almost surely converges to an equivalent IMP representation as the sequence n approaches infinity [14].

3   Problem Settings and Motivation for this Work

Given an observed sequence z^n of length n, assumed to have been generated from an IMP P defined in the previous section, with unknown number of emitters m and unknown processes P_w and P_i for $i = 1, \ldots, m$, the problem that we address in this paper is to retrieve the partition $\Pi = \{A_1, ..., A_m\}$. This deinterleaving process problem is denoted as DPP in the following. Note that we do not address the problem of retrieving exactly the processes P_w and P_i, which is a more difficult estimation problem.

We assume in this work that each process has a maximum order k_{max}. Therefore, we search for the couple of order vector $\hat{\mathbf{k}} \in \Omega_{k_{max}}$ and partition $\hat{\Pi} \in \Omega_\Pi$ minimizing the global ML entropy $C_{\hat{\Pi},\hat{\mathbf{k}}}$ given by Eq. (5) with $\Omega_{k_{max}}$ the set of possible order vectors given by

$$\Omega_{k_{max}} = \{(k_1, \ldots, k_m; k_w), 1 \le k_i \le k_{max}, i = 1, \ldots, m, w\}, \tag{8}$$

and Ω_Π the search space of the alphabet partitions given by

$$\Omega_\Pi = \{\{A_1, \ldots, A_m\}, \mathcal{A} = \bigcup_{i=1}^m A_i, A_i \cap A_j = \emptyset, 1 \le i, j \le m, 1 \le m \le |\mathcal{A}|\}. \tag{9}$$

Therefore, solving the DPP is a double problem combining an estimation problem consisting in finding the optimal order vector $\mathbf{k}$ for each evaluated partition Π and a combinatorial optimization problem on the space of all partitions Π of the symbol alphabet $\mathcal{A}$.

3.1 Decomposable Score for Estimating Processes Optimal Order

Given a candidate partition $\hat{\Pi} = \cup_{i=1}^m \hat{A}_i$ and order vector $\hat{\mathbf{k}} = (\hat{k}_1, \ldots, \hat{k}_m; \hat{k}_w)$, we first observe that Eq. (5) can be rewritten as

$$C_{\hat{\Pi},\hat{\mathbf{k}}}(z^n) = \sum_{i=1}^m \hat{H}_{\hat{k}_i}(z^n[\hat{A}_i]) + \hat{H}_{\hat{k}_w}(\sigma_{\hat{\Pi}}(z^n)) \tag{10}$$

$$+ \beta \log n \sum_{i=1}^m |\hat{A}_i|^{\hat{k}_i}(|\hat{A}_i| - 1) + \beta \log n \, m^{\hat{k}_w}(m - 1) \tag{11}$$

$$= \sum_{i=1}^m C_{\hat{A}_i,\hat{k}_i}(z^n) + C_{\sigma_{\hat{\Pi}},\hat{k}_w}(z^n), \tag{12}$$

with $C_{\hat{A}_i,\hat{k}_i}(z^n) = \hat{H}_{\hat{k}_i}(z^n[\hat{A}_i]) + \beta \log n \, |\hat{A}_i|^{\hat{k}_i}(|\hat{A}_i| - 1)$, the penalized entropy of the estimated process $\hat{P}_i$ of order $\hat{k}_i$ generating sub-alphabet $\hat{A}_i$, and $C_{\sigma_{\hat{\Pi}},\hat{k}_w}(z^n) = \hat{H}_{\hat{k}_w}(\sigma_{\hat{\Pi}}(z^n)) + \beta \log n \, m^{\hat{k}_w}(m - 1)$, the penalized entropy of the switch process related to partition $\hat{\Pi}$.

With this decomposition, we observe that given an estimated sub-alphabet $\hat{A}_i$, finding the optimal order k_i of the process of each penalized entropy term $C_{\hat{A}_i,\hat{k}_i}(z^n)$ can be done *independently* of the global partition $\hat{\Pi}$ being evaluated. We denote as $C_{\hat{A}_i,k_i^*}(z^n)$, the optimal penalized entropy obtained for the observed sequence z^n and the optimal order $1 \le k_i^* \le k_{max}$.

3.2   A Combinatorial Problem in the Space of Partitions

Since the space search grows exponentially with the number of letters, an exhaustive search is in general not feasible in a reasonable amount of time for the DPP. Greedy searches have recently been proposed by [5,14] to solve this combinatorial problem. However, such greedy local searches are prone to get stuck in local optima.

In this paper, we propose an improved heuristic to find the best partition $\hat{\Pi}$ in the huge search space Ω_Π, by noticing that the studied problem is a particular grouping problem [15]. Given a set S of elements, a grouping problem involves partitioning set S into a number of disjoint groups S_i optimizing a given objective function and possibly satisfying some given constraints.

In the DPP, the alphabet $\mathcal{A}$ corresponds to the set of elements S, and each sub-alphabet A_i corresponds to a group S_i. The task is to find a partition $\hat{\Pi} \in \Omega_\Pi$ such that the global score $f(\hat{\Pi}) = C_{\hat{\Pi}, \mathbf{k}^*}(z^n)$ is minimized (with $C_{\hat{\Pi}, \mathbf{k}^*}(z^n)$ the penalized entropy evaluated for the observed sequence z^n, the partition $\hat{\Pi}$ and the optimal order vector $\mathbf{k}^* \in \Omega_{k_{max}}$ associated).

4   A Memetic Algorithm for Alphabet Partitioning

In this section, we present a new memetic algorithm for alphabet partitioning called MAAP to solve the DPP seen as a grouping problem. Following the main ideas of the HEAD algorithm [12], the proposed algorithm relies on a reduced population of only two individuals and uses a dedicated crossover operator. For local optimization, it employs a tabu search procedure.

4.1   General Framework

The general algorithm architecture of the proposed MAAP algorithm is described in Algorithm 1.

The population is initialized with two random partitions in Ω_Π (see Sect. 4.2). Then at each generation, the algorithm alternates two steps:

1. an intensification phase, where the two individuals of the population Π_1, Π_2 are improved by a tabu search procedure (called TabuAP) during nb_{iter} iterations (see Sect. 4.3). This step produces two individuals Π'_1, Π'_2.
2. a diversification procedure, where two different children Π_1, Π_2 are generated from the two best partitions Π'_1, Π'_2 obtained from the tabuAP local search procedure. The crossover used to generate the two offspring partitions is a dedicated likelihood score-based crossover for the DPP (called GLPX) inspired by the well-known GPX crossover [6] for the graph coloring problem, it is explained in detail in Sect. 4.4.

The following subsections describe each step of the MAAP algorithm.

Algorithm 1. MAAP - Memetic algorithm for alphabet partitioning

---

1: **Input: Observed sequence z^n of n letters drawn from the alphabet $\mathcal{A}$.**
2: **Output: The best partition Π_{best} found so far**
3: Π_1, Π_2, $\Pi_{best} \leftarrow random\_initialization$ ▷ **Section 4.2**
4: **while stop condition is not met do**
5:      $\Pi'_1 \leftarrow TabuAP(\Pi_1, z^n)$ ▷ **Local tabu searches (see Section 4.3)**
6:      $\Pi'_2 \leftarrow TabuAP(\Pi_2, z^n)$
7:      **if** $f(\Pi'_1) < f(\Pi_{best})$ **then**
8:          $\Pi_{best} \leftarrow \Pi'_1$
9:      **end if**
10:     **if** $f(\Pi'_2) < f(\Pi_{best})$ **then**
11:         $\Pi_{best} \leftarrow \Pi'_2$
12:     **end if**
13:     $\Pi_1 \leftarrow GLPX(\Pi'_1, \Pi'_2, z^n)$ ▷ **Crossover operators (see Section 4.4)**
14:     $\Pi_2 \leftarrow GLPX(\Pi'_2, \Pi'_1, z^n)$
15: **end while**
16: **return** Π_{best}

4.2   Initialisation

During the initialization procedure, the partitions Π_1, Π_2, $\Pi_{best} \in \Omega_\Pi$ are randomly built. In order to build a random partition, the letters in $\mathcal{A}$ are considered in alphabetical order. Then at each step, if the partition being constructed has already m groups, the incoming letter a has a probability equal to $\frac{1}{m+1}$ to be placed in each existing group of letters A_i with $i = 1, \ldots, m$, and a probability $\frac{1}{m+1}$ to be placed in a new group A_{m+1}. This process is repeated until all letters are assigned to a sub-alphabet A_i. This procedure allows the creation of a partition randomly and uniformly in the search space Ω_Π.

In order to ensure that the two individuals Π_1 and Π_2 are different in the population at the beginning, this initialization procedure is repeated until the set-theoretic partition distance between Π_1 and Π_2 is greater than 0. The set-theoretic partition distance between two partitions $\Pi_1 = \cup_{i=1}^m A_i$ and $\Pi_2 = \cup_{j=1}^l B_j$ is defined as the minimum number of one-move steps needed to transform Π_1 into Π_2 (up to a group permutation). This distance can be computed by solving a maximum weight bipartite matching problem if we consider each sub-alphabet A_i of Π_1 and B_j of Π_2 as nodes of a bipartite graph connected by edges $e_{ij} = \{A_i, B_j\}$. Each edge e_{ij} has a weight w_{ij} corresponding to the number of letters shared by the two corresponding sub-alphabets A_i and B_j. This matching problem can be solved by the Hungarian algorithm [8] with a time complexity of $O(p^3)$ with $p = max(m, l)$. It produces a matching of maximum cardinality $0 \leq q \leq |\mathcal{A}|$ and the set-theoretic partition distance $D(\Pi_1, \Pi_2)$ is then defined as $|\mathcal{A}| - q$. Note that this distance will also be useful for the experiments. It is indeed a relevant scoring metric that can be used to evaluate the quality of an alphabet partition with respect to a known ground truth when working with simulated data (see Sects. 5.2 and 5.3).

4.3   Tabu Search Procedure

The tabu search procedure for alphabet partitioning (called TabuAP) used during the intensification phase is inspired from the popular TabuCol algorithm for the graph coloring problem [7]. Some adjustments are made to adapt this tabu search to our partitioning problem.

Neighborhood of a Partition. TabuAP explores the search space Ω_Π of all possible partitions that can be formed with the alphabet $\mathcal{A}$, by making transitions from the current solution to one neighboring solution.

A neighboring solution is generated by using the *one-move* operator. For a partition $\Pi = \cup_{i=1}^{m} A_i \in \Omega_\Pi$, the *one-move* operator displaces a letter $a \in A_i$ to a different sub-alphabet $A_j, j \neq i$. Let $\Pi \oplus < a, A_i, A_j >$ be the resulting neighboring partition. We then define the *one-move* neighborhood by

$$ N(\Pi) = \{\Pi \oplus < a, A_i, A_j >: a \in A_i, 1 \leq i \leq m, \ 1 \leq j \leq m+1, m+1 \leq |\mathcal{A}|\}. \tag{13} $$

Notice that with this neighborhood, a letter $a \in A_i$, $i \neq m$ is allowed to be transferred to an existing group A_j for $j = 1, \ldots, m$, with $j \neq i$, or to be placed in a new group A_{m+1}, which increases the total number of groups by one.

Tabu Search. The tabu search procedure iteratively replaces the current solution Π by a neighboring solution Π' taken from the one-move neighborhood $N(\Pi)$ until it reaches a maximum of nb_{iter} iterations of tabu search or the cutoff time for the MAAP algorithm is reached.

At each iteration, TabuAP examines the neighborhood and selects the best admissible neighboring solution Π' to replace Π. A neighboring solution $\Pi \oplus < a, A_i, A_j >$ built from Π is said to be admissible if the associated one-move $< a, A_i, A_j >$ was not registered in a tabu list. Each time such one-move is performed, it is added to the tabu list and forbidden during the $t = r(3) + \alpha|\mathcal{A}|$ next iterations (tabu tenure) where r is a random number uniformly drawn in $1, \ldots, 3$ and α is a hyperparameter of the algorithm set to the value of 0.6.

In order to compute the best admissible partition in the neighborhood, all the differences of global penalized entropy scores $\Delta_{a,j}$, associated with each admissible one-move $< a, A_i, A_j >$ are computed and the move corresponding to the lowest value of $\Delta_{a,j}$ is applied (because it is a minimization problem).

For a move $< a, A_i, A_j >$ applied to the current partition Π and resulting in a new partition $\Pi' = \Pi \oplus < a, A_i, A_j >$, only the penalized entropy of the changing groups and the switch process need to be reevaluated. Indeed, according to Eq. 10,

$$ \Delta_{a,j} = C_{\Pi', \mathbf{k}^{*}} - C_{\Pi, \mathbf{k}^{*}} \tag{14} $$

$$ = C_{A'_i, \hat{k}'^{*}_i} - C_{A_i, \hat{k}^{*}_i} + C_{A'_j, \hat{k}'^{*}_j} - C_{A_j, \hat{k}^{*}_j} + C_{\sigma_{\Pi'}, \hat{k}'^{*}_w} - C_{\sigma_\Pi, \hat{k}^{*}_w}, \tag{15} $$

where $C_{A'_i,\hat{k}'^*_i}$ and $C_{A'_j,\hat{k}'^*_j}$ are respectively the optimal penalized entropy of the new sub-alphabet $A'_i = A_i \backslash a$ and $A'_j = A_j \cup a$ (after moving the letter a from A_i to A_j) with optimal order $\hat{k}'^*_i$ and $\hat{k}'^*_j$; $C_{\sigma_\Pi,\hat{k}^*_w}$ and $C_{\sigma_{\Pi'},\hat{k}'^*_w}$ are respectively the optimal entropy of the switch process of the partitions Π and Π'.

Since $|\mathcal{A}|$ letters can be displaced to at most $|\mathcal{A}| - 1$ sub-alphabets, the size of this neighborhood is bounded by $O(|\mathcal{A}|^2)$. Evaluating a transition toward a neighbor with the one-move operator required to evaluate new penalized entropy, whose time complexity is in $O(n \times k_{max} \times |\mathcal{A}|^{k_{max}+1})$ (n the length of the sequence). Therefore, the overall complexity of this tabu search procedure is $O(nb_{iter} \times n \times k_{max} \times |\mathcal{A}|^{k_{max}+3})$.

4.4   Greedy Likelihood-Based Crossover Operator

The popular greedy partition crossover (GPX) [6] has proven to be very effective for graph coloring [9,12]. The two main principles of GPX are: 1) a solution is a partition of vertices (letters) into color classes (sub-alphabet) and not an assignment of colors to vertices, and 2) large color classes are transmitted to the offspring.

For the DDP, we introduce a new greedy likelihood-based partition crossover called GLPX. GLPX relies on the main principles of the GPX crossover with specific adaptations to our problem. Instead of only prioritizing large groups of letters, which does not make much sense for our problem, we prioritize groups as large as possible, but with as low entropy as possible, because our problem is to minimize the global entropy of the partition over the whole alphabet. A GLPX score for a group A_i is introduced as

$$\begin{cases} \widehat{C}_{A_i}(z^n) = \frac{C_{A_i,k^*_i}(z^n)}{|A_i|-1} & \text{if } |A_i| > 1, \\ \widehat{C}_{A_i}(z^n) = +\infty & \text{if } |A_i| = 1. \end{cases} \tag{16}$$

Given two parent partitions Π_1 and Π_2, the GLPX procedure alternates two steps. First, it transmits to the child the sub-alphabet $\bar{A}$ with the lowest score $\widehat{C}_{\bar{A}}$. After having withdrawn the letters of this sub-alphabet in both parents and having recomputed all scores, it transmits to the child the sub-alphabet $\bar{B}$ with the lowest score $\widehat{C}_{\bar{B}}$ of the second parent. This procedure is repeated until all the letters of the alphabet $\mathcal{A}$ are assigned to the child. For a given parent, if two or more processes have the same lowest score, one of them is selected at random. Note that singletons have infinite scores, and then are randomly selected at the end of the process, when no more groups of at least two letters remain. The GLPX procedure is described in Algorithm 2.

This crossover is asymmetrical like the GPX crossover. As noticed in [12], starting the crossover with parent 1 or parent 2 can produce different offspring solutions. Therefore when used in the MAAP algorithm to generate two new offspring solutions $\Pi_1 = GLPX(\Pi'_1, \Pi'_2, z^n)$ and $\Pi_2 = GLPX(\Pi'_2, \Pi'_1, z^n)$, the two children Π_1 and Π_2 can be very different (in the sense of the set-theoretic partition distance defined in Sect. 4.2).

Algorithm 2. GLPX crossover procedure

1: **Input: parents partitions** $\Pi_1 = \cup_{i=0}^m A_i$, $\Pi_2 = \cup_{i=0}^q B_i$ **and observed sequence** z^n.
2: **Output: Child partition** Π_c
3: $\Pi_c \leftarrow \emptyset$
4: **while** Π_1 **or** Π_2 **are not empty do**
5:      **for** $i = 1, 2$ **do**
6:          $\bar{A} \leftarrow \underset{A \in \Pi_i}{argmin} \; \widehat{C}_A(z^n)$
7:          $\Pi_c \leftarrow \Pi_c \cup \bar{A}$
8:          **for** $a \in \bar{A}$ **do**
9:              $\Pi_1 \leftarrow \Pi_1 \backslash a$
10:         $\Pi_2 \leftarrow \Pi_2 \backslash a$
11:         **end for**
12:      **end for**
13: **end while**
14: **return** Π_c

---

5 Experiments and Computational Results

This section is dedicated to the computational assessment of the proposed algorithm on both synthetic datasets and realistic datasets. Before showing the computational results, we first present the experimental condition.

5.1 Experimental Condition and Reference Algorithm

Parameter Settings. For the TabuAP procedure, the tabu tenure parameter α is set to the value of 0.6 according to [6,12]. The maximal number of iterations for each TabuAP run is set to 50. The penalization parameter β in Eq. (5) is set to $\frac{1}{2}$, which is a common value used in the literature [2], allowing to retrieve the Bayesian Information Criterion (BIC). The maximum order k_{max} for entropy estimation is set to the value of $k_{max} = 1$. Table 1 summarizes the parameter setting for the MAAP algorithm which can be considered the default and was used for all our experiments.

Table 1. Parameter setting in MAAP

Parameter	Description	Value
nb_{iter}	Number of iterations of the TabuAP local search	50
α	Tabu tenure parameter	0.6
β	Penalization parameter entering in Eq. 5	$\frac{1}{2}$
k_{max}	Maximum order for entropy estimation	1

Reference Algorithm. Our MAAP algorithm is compared to the iterated greedy algorithm (iteratedGreedy) for alphabet deinterleaving pulse trains (see Algorithm 1 in [5]). For this iteratedGreedy algorithm, the radius of jump r is set to the value of 2 and the neighborhood radius is set to 1 like in [5]. The maximum number of jumps N is not limited. For this iteratedGreedy algorithm, the entropy evaluation is done with the same function used in the MAAP algorithm, and with the same parameters ($\beta = \frac{1}{2}$ and $k_{max} = 1$). The only difference between MAAP and iteratedGreedy is thus the search heuristic in the space of partition Ω_Π. Both MAAP and iteratedGreedy are coded in Python with the Numpy library and are launched on a computer equipped with Intel Xeon ES 2630, 2.66 GHz CPU.

Evaluation Metric and Stopping Condition. To assess the quality of the best partition Π_{best} found by an algorithm, we compute the set-theoretic partition distance between Π_{best} and the ground truth partition Π_{truth}. The stopping condition for each experiment (on synthetic data and electronic warfare data) is indicated in the corresponding section.

5.2 Experiments on Synthetic Datasets

This section is dedicated to a first computational assessment of the proposed memetic algorithm for the DPP. The data are simulated with an interleaved Markov process $P = \mathcal{I}_\Pi(P_1, ..., P_m; P_w)$ over disjoint sub-alphabets, in the *ideal* framework presented in Sect. 2: independent time-homogeneous, ergodic and finite memory component processes P_w and P_i for $i = 1, ..., m$.

Synthetic Dataset Generation. The datasets are based on synthetic sequences of size n with different numbers of letters $|\mathcal{A}|$ and maximal order equal to 1 to limit the computation time required for entropy estimation.

The following parameters are randomly set up to generate a sequence z^n with an IMP $P = \mathcal{I}_\Pi(P_1, ..., P_m; P_w)$ according to Eq. (1):

- from an alphabet $\mathcal{A}$ of size $|\mathcal{A}|$, a *ground truth* partition $\Pi_{truth} = \cup_{i=1}^m A_i$ is generated with the random initialization procedure as described in Sect. 4.2. m is the number of emitters (groups) associated with this partition;
- for each emitter i ($i = 1, ..., m$), a probabilistic transition matrix Q_i associated with the stochastic process P_i of size $|A_i| \times |A_i|$ is randomly drawn;
- for the switch process, a probabilistic transition matrix Q_w of size $m \times m$ is drawn;
- initial state (letter) of each process P_i is randomly drawn in its correspondent sub alphabet A_i;
- the first emitter is randomly drawn in the set of m emitters;

We consider 4 different configurations ($|\mathcal{A}|, n$) with $|\mathcal{A}| = \{20, 50\}$ and $n = \{10000, 50000\}$. For each configuration, 10 different datasets (z^n, Π_{truth}) are generated. So a total of 40 datasets are obtained. These datasets will be made publicly available.

Table 2. Comparison of MAAP and iteratedGreedy on synthetic datasets generated with interleaved Markov processes. Dominating results (lower scores) are indicated in boldface. Significantly better values are underlined (t-test with p-value of 0.05).

#	Config		iteratedGreedy			MAAP			Config		iteratedGreedy			MAAP		
	$|\mathcal{A}|$	n	$\bar{D}$	$\bar{C}$	time	$\bar{D}$	$\bar{C}$	time	$|\mathcal{A}|$	n	$\bar{D}$	$\bar{C}$	time	$\bar{D}$	$\bar{C}$	time (s)
1	20	10000	0	24612	110	0	24612	101	50	10000	2.4	34606	2253	**1.4**	34593	1771
2	20	10000	0	**26155**	208	0.06	26161	358	50	10000	7.8	34063	1631	**6.57**	**34051**	2240
3	20	10000	0	24381	92	0	24381	121	50	10000	2.1	34968	2909	**1.26**	**34958**	2173
4	20	10000	0	26240	96	0	26240	99	50	10000	4.23	33398	1132	**3.57**	**33391**	2055
5	20	10000	0	24386	77	0	24386	80	50	10000	0.43	34626	1834	0	**34621**	1901
6	20	10000	0	26843	66	0	26843	58	50	10000	11.46	33881	5339	10	**33859**	4426
7	20	10000	0	24100	449	0	24100	103	50	10000	1.56	37051	3041	**0.13**	**37027**	3119
8	20	10000	0	26763	70	0	26763	69	50	10000	**5.56**	34698	1691	6.1	**34682**	3544
9	20	10000	0	24821	82	0	24821	98	50	10000	2.33	33758	3218	**1.63**	33741	3605
10	20	10000	0	26588	82	0	26588	75	50	10000	5.87	36155	9090	4	**36134**	3250
1	20	50000	0	128191	324	0	128191	324	50	50000	0	175904	5216	0	175904	4690
2	20	50000	0.73	119944	298	0	**119533**	383	50	50000	0	163835	8388	0	163835	8350
3	20	50000	1.3	124721	630	0	**124461**	682	50	50000	0	166655	7287	0	166655	7184
4	20	50000	0.87	127508	507	0	**127165**	413	50	50000	0	177233	5979	0	177233	5669
5	20	50000	1.86	132787	388	**0**	**132272**	449	50	50000	0	167299	7288	0	167299	6534
6	20	50000	0.5	124718	425	0	**124559**	382	50	50000	0	180056	4943	0	180056	5006
7	20	50000	0.37	128701	360	0	128701	366	50	50000	0	166328	7619	0	166328	7238
8	20	50000	0.4	120087	583	0.4	128516	493	50	50000	0	166755	7514	0	166755	7450
9	20	50000	0.87	127186	379	0	**126900**	417	50	50000	0	173295	5828	0	173295	6045
10	20	50000	0	127150	552	0	127150	396	50	50000	0	173497	5980	0	173497	5952

Results on Synthetic Data. For each dataset, given the stochastic nature of both algorithms, 30 independent runs are launched. The time limit in seconds for each run is $T_{limit} = 200 * |\mathcal{A}|$ when $n = 10000$ and $T_{limit} = 500 * |\mathcal{A}|$ when $n = 50000$. Once the algorithm reaches this time limit, it returns the best partition Π_{best} found so far with its associated minimum penalized entropy score $C_{\Pi_{best}}(z^n)$.

Table 2 displays the results obtained by the algorithms MAAP and iteratedGreedy on the 40 different datasets generated with 4 different configurations. Columns $\bar{D}$ indicate the average distance relative to the ground truth partition. Columns $\bar{C}$ show the average lowest penalized global entropy obtained by an algorithm and columns *time* correspond to the average time in seconds required by the algorithm to reach its best result. Values in bold mean the algorithm has a better score than the other one. Underlined values mean that the average score obtained for a given algorithm is significantly better than the average score of the other algorithm according to a t-test with p-value of 0.05.

Table 2 shows that both algorithms work efficiently since the distance to the truth partition is often close to zero which validates the relevance of the likelihood-based method used in this context. The comparison between the two algorithms reveals that MAAP obtains significantly better results for several configurations, due to more effective exploration of the search space of all possible partitions.

5.3   Experiments on Electronic Warfare Datasets

In this section we present results on datasets coming from an Electronic Warfare data generator which simulates realistic situations with mobile radar warning receivers. One configuration corresponds to a random draw in a list of known radars and a draw in their relative phasing. We cannot share the content of the generator. For each simulation, a dataset $\mathcal{D}$ consisting in a sequence of pulses with their corresponding frequency (CF) and time of arrival (ToA) is generated. The *ground truth* Π_{truth} (i.e. the association of each pulse to each emitter) is known. The objective is then to retrieve Π_{truth} from the data.

Preprocessing of the Data. A preprocessing step is first performed to obtain the alphabet $\mathcal{A}$ from the dataset $\mathcal{D}$. It consists of clustering pulses with the DBSCAN algorithm [4] based on their frequency. Then, each obtained cluster is associated with a letter in $\mathcal{A}$. The sequence z^n is then obtained by ordering these letters by increasing order of time of arrival (ToA). Since we only use the frequency, the ϵ-neighborhood parameter of DBSCAN corresponds to our precision parameter and is a fixed number of the order of the MHz not specified here.

Illustrated Example. Figure 1a shows an example of a pulse train measured with the frequency and the time of arrival of the different signals. The scales are hidden on purpose. Pulses regrouped in the same cluster after the first pre-processing phase have the same color and are associated with the same letter (from a to l). Figure 1b corresponds to the known ground truth for this scenario (4 emitters):
$\{\{a,b,c\},\{h\},\{d,e,f,g\},\{i,j,k,l\}\}$. Pulses generated by the same emitter have the same color.

We ran the MAAP algorithm on this dataset with default parameters (see Table 1) and Fig. 2 shows the evolution over time of the distance to the ground truth (blue) and the best penalized entropy (red) reached during the search (average, minimum and maximum over 10 runs). The two curves, distance and entropy, have similar variations, meaning that in this case, minimizing the penalized global entropy allows to get closer to the target partition. We observe that in some experiments, MAAP reaches the best target partition (distance of 0) within a few seconds, while in others, it never reaches it in the allotted time with a distance of 4. This highlights that finding a good partition in a limited amount of time is not always easy for these realistic datasets and may depend a lot on the random initiating solution from which the search starts.

Results on Electronic Warfare Data. 10 different scenarios with 5 emitters are generated by the Eletronic Warfare data simulator. The number of observed pulses varies from 10000 to 100000 for these scenarios (the scenarios were cut if the number of points exceeded 100000 and couldn't contain less than 10000 points). We launched the MAAP and iteratedGreedy algorithm [5] with the

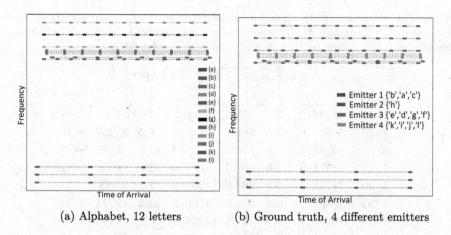

(a) Alphabet, 12 letters   (b) Ground truth, 4 different emitters

Fig. 1. Illustrated example of radar pulses deinterleaving

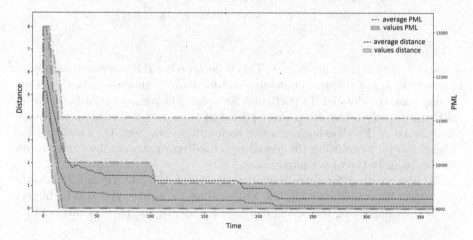

Fig. 2. Evolution over time (seconds) of the distance between the ground truth partition and the current best solution (in blue and left scale), and the corresponding penalized global entropy (in red and right scale), during the search process of the MAAP algorithm. (Color figure online)

same configuration and parameters as presented in Sect. 5.2. Each algorithm is launched 30 times (independent runs) on each dataset with a time limit of three hours.

Table 3 reports the result of these experiments, with the measures of the best distance (D^*) and the average distance ($\bar{D}$) to the known ground truth, the average best penalized global score ($\bar{C}$) obtained over the different runs and the time in seconds required to obtain the best scores.

We first observe in this table that for some scenarios, both algorithms are able to recover or come close to the target partition, but for others, such as

Table 3. Comparison of MAAP and iteratedGreedy on 10 electronic warfare datasets generated with 5 emitters. Dominating results (lower scores) are indicated in boldface. Significant better values are underlined (t-test with p-value 0.05).

sc	Config		iteratedGreedy				MAAP			
	$|\mathcal{A}|$	n	D^*	$\bar{D}$	$\bar{C}$	time	D^*	$\bar{D}$	$\bar{C}$	time
1	13	13302	2	2.0	3354	34	2	2.0	3354	45
2	28	10000	0	7.5	31445	8122	0	**5.5**	**31237**	6047
3	11	69342	3	3.0	18895	78	3	3.0	18895	114
4	11	68571	3	3.0	18872	106	3	3.0	18872	123
5	19	78783	4	7.9	**26443**	265	6	7.6	26548	733
6	24	90891	2	4.9	19550	4711	2	**4.2**	**19271**	3086
7	23	100000	1	7.7	25575	3253	1	**5.6**	**25195**	4470
8	13	100000	5	**5.0**	**88414**	683	4	6.0	88725	871
9	21	100000	8	8.0	107021	2573	8	8.0	107021	3151
10	27	100000	1	6.2	56363	2431	1	**3**	**54717**	4263

scenario 9, they remain far from it. This is because the IMP representation used in this work is not always completely valid for some of these scenarios, as some assumptions are violated. In particular for some scenarios, a certain number of emitters are only active for a short period of time over the whole time frame, which violates the time-homogeneous assumption (see Sect. 1). Therefore, for these datasets, minimizing the global penalized entropy score does not always allow to identify the target partition.

We observe that for three scenarios (2, 7 and 10), MAAP is significantly better than iteratedGreedy [5], but less good on scenario 8; this still highlights the value of improving the search heuristic for solving the DPP in this realistic setting.

6   Conclusions

A memetic algorithm for alphabet partitioning was presented in this work. It is used for the problem of deinterleaving pulse trains generated by multiple emitters and described by interleaved Markov processes.

The results show that the proposed heuristic almost always finds the best partitions for synthetic datasets generated with Markov processes and obtains good results for electronic warfare datasets generated under realistic conditions. For some datasets, it can obtain significantly better results than the recent iterated greedy algorithm [5].

A future work could be to take into account the time delay between different signals to improve the estimation of the different component and switch processes.

Acknowledgments. This work was partially supported by the French Ministry for Research and Education through a CIFRE grant (number N°2022/0062). We are grateful to the reviewers for their valuable comments and suggestions which helped us to improve the paper.

References

1. Batu, T., Guha, S., Kannan, S.: Inferring mixtures of Markov chains. In: Shawe-Taylor, J., Singer, Y. (eds.) COLT 2004. LNCS (LNAI), vol. 3120, pp. 186–199. Springer, Heidelberg (2004). https://doi.org/10.1007/978-3-540-27819-1_13
2. Csiszar, I., Shields, P.: The consistency of the BIC Markov order estimator. In: 2000 IEEE International Symposium on Information Theory (Cat. No.00CH37060), p. 26. IEEE (2000)
3. Davies, C.L., Hollands, P.: Automatic processing for ESM (1982)
4. Ester, M., Kriegel, H., Sander, J., Xu, X.: A density-based algorithm for discovering clusters in large spatial databases with noise. In: Simoudis, E., Han, J., Fayyad, U.M. (eds.) Proceedings of the Second International Conference on Knowledge Discovery and Data Mining (KDD-96), Portland, Oregon, USA, pp. 226–231. AAAI Press (1996)
5. Ford, G., Foster, B.J., Braun, S.A.: Deinterleaving pulse trains via interleaved Markov process estimation. In: 2020 IEEE Radar Conference (RadarConf20), pp. 1–6. IEEE (2020)
6. Galinier, P., Hao, J.K.: Hybrid evolutionary algorithms for graph coloring. J. Comb. Optim. **3**(4), 379–397 (1999)
7. Hertz, A., de Werra, D.: Using tabu search techniques for graph coloring. Computing **39**(4), 345–351 (1987)
8. Kuhn, H.W.: The Hungarian method for the assignment problem. Naval Res. Logistics Q. **2**(1–2), 83–97 (1955)
9. Lü, Z., Hao, J.K.: A memetic algorithm for graph coloring. Eur. J. Oper. Res. **203**(1), 241–250 (2010)
10. Mardia, H.: New techniques for the deinterleaving of repetitive sequences. IEE Proc. F (Radar Signal Process.) **136**(4), 149–154 (1989)
11. Milojević, D., Popović, B.: Improved algorithm for the deinterleaving of radar pulses. IEE Proc. F (Radar Signal Process.) **139**(1), 98–104 (1992)
12. Moalic, L., Gondran, A.: Variations on memetic algorithms for graph coloring problems. J. Heuristics **24**(1), 1–24 (2018)
13. Moore, J., Krishnamurthy, V.: Deinterleaving pulse trains using discrete-time stochastic dynamic-linear models. IEEE Trans. Signal Process. **42**(11), 3092–3103 (1994)
14. Seroussi, G., Szpankowski, W., Weinberger, M.J.: Deinterleaving finite memory processes via penalized maximum likelihood. IEEE Trans. Inf. Theory **58**(12), 7094–7109 (2012)
15. Zhou, Y., Hao, J., Duval, B.: Reinforcement learning based local search for grouping problems: a case study on graph coloring. Exp. Syst. Appl. **64**, 412–422 (2016)

Application of Negative Learning Ant Colony Optimization to the Far from Most String Problem

Christian Blum[1]([✉])[ID] and Pedro Pinacho-Davidson[2][ID]

[1] Artificial Intelligence Research Institute (IIIA-CSIC) Campus of the UAB,
Bellaterra, Spain
`christian.blum@iiia.csic.es`

[2] Department of Computer Science, Faculty of Engineering, Universidad de
Concepción, Concepción, Chile
`ppinacho@udec.cl`

Abstract. We propose the application of a recently introduced version of ant colony optimization—negative learning ant colony optimization—to the far from most string problem. This problem is a notoriously difficult combinatorial optimization problem from the group of string selection problems. The proposed algorithm makes use of negative learning in addition to the standard positive learning mechanism in order to achieve better guidance for the exploration of the search space. In addition, we compare different versions of our algorithm characterized by the use of different objective functions. The obtained results show that our algorithm is especially successful for instances with specific characteristics. Moreover, it becomes clear that none of the existing state-of-the-art methods is best for all problem instances.

Keywords: ant colony optimization · negative learning · combinatorial optimization · far from most string problem

1 Introduction

The family of *sequence (or string) consensus problems* [8] has found important applications, for example, in studying molecular evolution, protein structures, and drug target design. This family of problems includes well known combinatorial optimization problems such as the *closest string problem* (CSP) [10] and the *farthest string problem* (FSP) [19]. In the case of the CSP, the goal is to find a solution—that is, a string—whose total distance to the strings from a given set of input strings is minimal. On the contrary, in the case of the FSP, a solution is sought whose total distance to the input strings is maximal.

The combinatorial optimization problem tackled in this paper—known as the *far from most string problem* (FFMSP) [9]—is also a member of the family of

This paper was supported by grant PID2019-104156GB-I00 funded by MCIN/AEI/10.13039/501100011033.

L. Pérez Cáceres and T. Stützle (Eds.): EvoCOP 2023, LNCS 13987, pp. 82–97, 2023.
https://doi.org/10.1007/978-3-031-30035-6_6

string consensus problems. However, this problem variant has been less studied so far in the literature. In this work, we propose the application of a recent variant of ant colony optimization (ACO), the negative learning ACO variant from [15,16]. This algorithm utilizes negative learning in addition to the standard positive learning mechanism. It was shown to improve significantly over standard ACO variants on a range of hard combinatorial optimization problems such as the multi-dimensional knapsack problem, the minimum dominating set problem, and MaxSAT. One of the main challenges that the FFMSP problem poses to any algorithmic solution is that the objective function has only a rather small range of possible values, which leads to a search space characterized by many large plateaus. In order to deal with this issue, alternative objective functions were designed in [2] and in [14]. In this work, we test a simplified variant of the one from [2], and subsequently, we compare the different alternative objective functions in our negative learning ACO approach. Our results show, first, that none of the proposed objective functions is best for all problem instances. Moreover, the performance of different state-of-the-art methods strongly depends on the problem instance characteristics. In comparison to the state-of-the-art GRASP methods from [4], for example, our algorithm is clearly better for threshold values of 0.8 m (where m is the length of the input strings), while it starts to lose efficiency with growing problem size for threshold values of 0.85 m. Additionally, we are able to show that our algorithm gains an advantage over the GRASP-based memetic algorithm from [7] with growing problem instance size, independently of the threshold value.

1.1 Organization of the Paper

The rest of this paper is organized as follows. The considered optimization problem is technically described—together with related work—in Sect. 2. Next, the developed algorithm is comprehensively described in Sect. 3. Finally, an experimental evaluation and a comparison to competing approaches from the literature is provided in Sect. 5, while conclusions and an outlook to future work can be found in Sect. 6.

2   Far from Most String Problem

A problem instance of the FFMSP is denoted by (Ω, t) where $\Omega = \{s_1, \ldots, s_n\}$ is a set of n input strings over a finite alphabet Σ. Hereby, each of the input strings s_i is of length m, that is, $|s_i| = m$ for all $s_i \in \Omega$. Furthermore, $0 < t < m$ is a fixed threshold value. Henceforth, the j-th character of a string s_i is denoted by $s_i[j]$. Remember also that the *Hamming distance* between two equal-length strings $s_i \neq s_j \in \Omega$ (denoted by $d_H(s_i, s_j)$) is defined as the number of positions at which the corresponding characters in the two strings are different. In technical terms:

$$d_H(s_i, s_j) = |\{k \in \{1, \ldots, m\} \mid s_i[k] \neq s_j[k]\}| \tag{1}$$

A valid solution to the FFMSP problem is any string s of length m over alphabet Σ. The objective function value $f_{\text{orig}}(s)$ of any such string s is defined as follows:

$$f_{\text{orig}}(s) := |\{s_i \in \Omega \mid d_H(s, s_i) \geq t\}| \qquad (2)$$

In other words, the objective function value of a solution/string s is defined as the number of input strings whose Hamming distance with s is greater or equal to the threshold value t.

2.1 Integer Linear Programming Model

As the algorithm proposed in the work makes internal use of a mixed-integer linear programming (MILP) solver for solving sub-instances of the tackled problem instance, a MILP model for the FFMSP is required. In the following, we, therefore, describe the MILP model originally introduced in [2]. There are two sets of binary variables involved in this model. The first set consists of a variable $x_{j,c}$ for each combination of a position $j = 1, \ldots, m$ of a possible solution and for each character $c \in \Sigma$. The second set consists of a binary variable y_i for each input string $s_i \in \Omega$ ($i = 1, \ldots, n$). The MILP model can then be stated as follows.

$$\max \sum_{i=1}^{n} y_i \qquad (3)$$

subject to:

$$\sum_{c \in \Sigma} x_{j,c} = 1 \quad \text{for } j = 1, \ldots, m \qquad (4)$$

$$\sum_{j=1}^{m} x_{j,s_i[j]} \leq (m - t) \cdot y_i \quad \text{for } i = 1, \ldots, n \qquad (5)$$

$$x_{j,c}, y_i \in \{0, 1\}$$

Note that constraints (4) ensure that exactly one character from Σ is chosen for each position j of a solution string. Moreover, constraints (5) ensure that a variable y_i can only have value one if and only if the Hamming distance between input string $s_i \in \Omega$ and a solution string (as defined by the setting of the variables $x_{j,c}$) is greater or equal than the threshold value t.

2.2 Computational Complexity and Previous Work

Among the sequence consensus problems, the FFMSP is one of the computationally hardest. In fact, compared to the other consensus problems, it is much harder to approximate, due to the approximation preserving reduction to the FFMSP from the independent set problem, which is a classical and computationally intractable combinatorial optimization problem. In 2003, Lanctot et al. [9]

proved that for sequences over an alphabet Σ with $|\Sigma| \geq 3$, approximating the FFMSP within a polynomial factor is NP-hard.

Due to the inherent difficulty of solving the FFMSP problem, the research community has mainly focused on heuristic and metaheuristic approaches. The first approach from [12] consists in the construction of a solution by means of a greedy heuristic and the subsequent application of local search. Approaches based on greedy randomized adaptive search procedures (GRASP) were presented in [3–6,14]. The last one from [4] is a hybrid technique that combines GRASP with variable neighbourhood search and with path relinking strategies. Apart from these proposals, the FFMSP was also tackled by means of evolutionary algorithms (EAs). The first one of these was proposed in [6]. The main feature of this algorithm consists in a mechanism for the preservation of diversity in the population. The second EA was presented in [7]. In particular, the proposed EA belongs to the class of memetic algorithms, as it makes use of local search for improving the generated solutions. Moreover, the construction of solutions in this memetic algorithm is based on GRASP. The ACO approach presented in [2] combines a standard ACO algorithm with the subsequent application of CPLEX, warm-started with the best solution found by the ACO approach. Finally, an algorithm based on beam search and the subsequent application of local search was described in [13]. As the GRASP approaches from [4] and the memetic algorithm from [7] were developed independently from each other at approximately the same time, it is currently not clear which one of both is currently state of the art.

As mentioned already before, one of the aspects that causes the FFMSP to be a very challenging problem for metaheuristics—that is, for algorithms based on exploring the search space—is the fact that the set of different objective function values is very small; in particular, the set of possible values for an instance with n input strings is $\{0, \ldots, n\}$. This leads to a search space with many large plateaus, that is, similar solutions will often have exactly the same objective function value. The problem for metaheuristics is that such search spaces do not offer any guidance on where (or in which direction) better solutions than the already found ones might be discovered. As a consequence, metaheuristics do often get stuck on plateaus. In order to alleviate this problem, two previous works [2,14] have made use of an extended objective function developed with the aim of providing better search guidance than the original objective function.

3   The Proposed Algorithm

In this section, we present the application of negative learning ACO to the FFSMP. The main framework of the algorithm is, as in all other existing applications of negative learning ACO [15,16], a $\mathcal{MAX} - \mathcal{MIN}$ Ant System (MMAS) implemented in the Hypercube Framework (HCF) [1].

The pseudo-code of our algorithm, which is henceforth called $\mathrm{Aco}_{\mathrm{neg}}^{+}$, is presented in Algorithm 1. Before we start to describe the algorithm, remember that any string s of length m over alphabet Σ is a valid solution to the problem.

Algorithm 1. Negative learning ACO ($\text{Aco}^+_{\text{neg}}$) for the FFMSP

1: **input:** a problem instance (Ω, t)
2: $cf := 0$, bs_upd := FALSE
3: InitializePheromoneValues($\mathcal{T}, \mathcal{T}^{\text{neg}}$)
4: $s^{bsf} := \text{RunCplex}(t_{\text{init}})$
5: $s^{rb} := s^{bsf}$
6: **while** termination conditions not met **do**
7:     $\mathcal{S}^{\text{iter}} := \emptyset$
8:     **for** $k = 1, \ldots, n_a$ **do**
9:         $s^k := \text{Construct\_Solution}(\mathcal{T}, \mathcal{T}^{\text{neg}})$
10:        $\mathcal{S}^{\text{iter}} := \mathcal{S}^{\text{iter}} \cup \{s^k\}$
11:    **end for**
12:    $s^{ib} := \text{argmax}\{f(s) \mid s \in \mathcal{S}^{\text{iter}}\}$
13:    UpdateSolution(s^{ib}, s^{rb}, s^{bsf})
14:    **if** bs_upd = TRUE **then** $\mathcal{S}^{\text{iter}} := \mathcal{S}^{\text{iter}} \cup \{s^{bsf}\}$ **else** $\mathcal{S}^{\text{iter}} := \mathcal{S}^{\text{iter}} \cup \{s^{rb}\}$ **end if**
15:    $s^{sub} := \text{SolveSubInstance}(\mathcal{S}^{\text{iter}}, \mathcal{T}^{\text{neg}}, t_{\text{sub}})$
16:    UpdateSolution(s^{sub}, s^{rb}, s^{bsf})
17:    ApplyPheromoneUpdate($\mathcal{T}, cf$, bs_upd, s^{ib}, s^{rb}, s^{bsf})
18:    $cf := \text{ComputeConvergenceFactor}(\mathcal{T})$
19:    **if** $cf > 0.999$ **then**
20:        **if** bs_upd = TRUE **then**
21:            $s^{rb} := \text{NULL}$, and bs_upd := FALSE
22:            InitializePheromoneValues($\mathcal{T}, \mathcal{T}^{\text{neg}}$)
23:        **else**
24:            bs_upd := TRUE
25:        **end if**
26:    **end if**
27: **end while**
28: **output:** s^{bsf}, the best solution found by the algorithm

---

The three main actions of our algorithm concern (1) the construction of n_a solutions at each iteration (lines 7–11), (2) solving sub-instances of the tackled problem instances with the ILP solver CPLEX (line 15), and the update of the pheromone values (remaining parts of the algorithm). In the following, we will outline these three aspects in detail. Note also that our algorithm keeps three solutions at all times: (1) the iteration-best solution s^{ib}, the restart-best solution s^{rb}, and the best-so-far solution s^{bsf}. Hereby, s^{ib} is the best solution generated in lines 8–11 at the current iteration, s^{rb} is the best solution found by the algorithm since the last (re-)initialization of the pheromone values, and s^{bsf} is the best solution found by the algorithm since the start.

3.1   Construction of a Solution

A solution s (of length m) is constructed by choosing for each position j ($j = 1, \ldots, m$) exactly one character Σ. This is done on the basis of greedy information and pheromone information. As greedy information, the algorithm uses the inverse of the frequency values of the letters for each position of the input strings. In particular, the frequency value $f_{a,j}$ of a letter $a \in \Sigma$ for a position $1 \leq j \leq m$ is calculated as follows:

$$f_{a,j} := \frac{|\{s_i \in \Omega \mid s_i[j] = a\}|}{n} \quad \forall\, a \in \Sigma \text{ and } 1 \leq j \leq m \tag{6}$$

In addition to the inverse of the frequency values, the algorithm also makes use of a pheromone model $\mathcal{T}$ that contains a standard pheromone value $\tau_{a,j}$ (where $\tau_{\min} \leq \tau_{a,j} \leq \tau_{\max}$) for each $a \in \Sigma$ and $1 \leq j \leq m$. Hereby, $\tau_{\min} := 0.001$ and $\tau_{\max} := 0.999$, as usual for a MMAS algorithm implemented in the HCF. Moreover, the algorithm also utilizes an additional pheromone model $\mathcal{T}^{neg}$ that contains the so-called negative pheromone values $\tau_{\min} \leq \tau_{a,j}^{neg} \leq \tau_{\max}$ for each $a \in \Sigma$ and $1 \leq j \leq m$. In particular, the probability $\mathbf{p}(a \mid j)$ for choosing a character $a \in \Sigma$ for a position $j \in \{1, \ldots, m\}$ is defined as follows:

$$\mathbf{p}(a \mid j) := \frac{f_{a,j}^{-1} \cdot \tau_{a,j} \cdot (1 - \tau_{a,j}^{neg})}{\sum_{b \in \Sigma} f_{b,j}^{-1} \cdot \tau_{b,j} \cdot (1 - \tau_{b,j}^{neg})} \tag{7}$$

In other words, the chance of a letter $a \in \Sigma$ to be selected is rather high if (1) its frequency at position j of the input strings is rather low, (2) its pheromone value for position j is rather high, and (3) its negative pheromone value for position j is rather low. Based on these probabilities a letter is chosen as follows. First, a random value $r \in [0,1]$ is drawn uniformly at random. If $r \leq d_{\text{rate}}$ (where $0 \leq d_{\text{rate}} < 1$ is a parameter of the algorithm), the letter with the highest probability is chosen in a deterministic way. Otherwise, a letter is chosen randomly (roulette wheel selection) based on the letter probabilities. Note that the construction of a solution is performed in function Construct_Solution$(\mathcal{T}, \mathcal{T}^{neg})$ of Algorithm 1 (see line 9).

3.2   Solving Sub-instances

After the construction of n_a solutions—that are then stored in set $\mathcal{S}^{\text{iter}}$—the algorithm first determines the iteration-best solution (line 12) before updating solutions s^{rb} and s^{bsf} (if necessary) with solution s^{ib} (function UpdateSolution$(s^{ib}, s^{rb}, s^{bsf})$ in line 13). Subsequently, either the restart-best solution or the best-so-far solution are added to $\mathcal{S}^{\text{iter}}$, depending on the value of the Boolean control variable bs_upd, whose function will be explained further down.

Then, based on set $\mathcal{S}^{\text{iter}}$, a sub-instance is generated. First, the following sets are defined:

$$\Sigma_j^- := \{a \in \Sigma \mid \nexists\, s \in \mathcal{S}^{\text{iter}} \text{ s.t. } s[j] = a\}, \qquad j = 1, \ldots, m \tag{8}$$

In words, Σ_j^- contains all letters from Σ that do not appear at position j in any of the solutions from $\mathcal{S}^{\text{iter}}$. With these sets, the ILP model from Sect. 2 is restricted by adding the following set of constraints:

$$x_{a,j} = 0 \quad \forall\, j = 1, \ldots, m \text{ and } a \in \Sigma_j^- \tag{9}$$

This restricted ILP model is then solved by the application of the ILP solver CPLEX in function SolveSubInstance($\mathcal{S}^{\text{iter}}$, $\mathcal{T}^{\text{neg}}$, t_{sub}); see line 15. The time limit for CPLEX is set to t_{sub} CPU seconds, which is a parameter of the $\text{Aco}^{+}_{\text{neg}}$ algorithm. Moreover, note that CPLEX is warm-started with the best solution from $\mathcal{S}^{\text{iter}}$ and the value of the CPLEX parameter MIPEmphasis is set to 5, which means that CPLEX will prioritize solution quality over proving optimality. Upon the termination of CPLEX, this function returns the best solution found by CPLEX (s^{sub}) within t_{sub} seconds. Finally, this function also updates the negative pheromone values based on s^{sub} and $\mathcal{S}^{\text{iter}}$. This will be explained further down.

3.3 Update of the Pheromone Values

The update of the standard pheromone values from model $\mathcal{T}$ is governed by the value of a Boolean variable called bs_upd and by the value (cf) of the so-called convergence factor. The way in which the value of bs_upd is initialized (line 2) and changed during the search process of the algorithm (see lines 19–26) is standard for any MMAS algorithm implemented in the HCF. Moreover, the value cf of the convergence factor indicates the current state of convergence of the algorithm. It is calculated by function ComputeConvergenceFactor($\mathcal{T}$), see line 18, in the following way:

$$cf := 2\left(\left(\frac{\sum_{\tau_{a,j}\in\mathcal{T}}\max\{\tau_{\max}-\tau_{a,j},\tau_{a,j}-\tau_{\min}\}}{|\mathcal{T}|\cdot(\tau_{\max}-\tau_{\min})}\right)-0.5\right) \quad (10)$$

The standard pheromone values are then updated in function ApplyPheromone-Update($\mathcal{T}$, cf, bs_upd, s^{ib},s^{rb},s^{bsf}), see line 17, in the following way:

$$\tau_{a,j}:=\tau_{a,j}+\rho\cdot(\psi_{a,j}-\tau_{a,j}) \quad j\in\{1,\ldots,m\},a\in\Sigma \quad (11)$$

where $\psi_{a,j}:=\Delta(s^{ib},a,j)\cdot w^{ib}+\Delta(s^{rb},a,j)\cdot w^{rb}+\Delta(s^{bsf},a,j)\cdot w^{bsf}$ and ρ is a parameter of the algorithm called *learning rate*. Hereby, $\Delta(s,a,j)$ is a function that evaluates to 1 in case $s[j]=a$, and to zero otherwise. Moreover, the values of the weights w^{ib}, w^{rb} and w^{bsf} are determined depending on bs_upd and cf as in any other MMAS algorithm implemented in the HCF as shown in Table 1. Note that it always holds that $w^{ib}+w^{rb}+w^{bsf}=1$. These three weights determine the influence of each of the three solutions (s^{ib}, s^{rb} and s^{bsf}) on the update of the standard pheromone values from $\mathcal{T}$.

Finally, the update of the negative pheromone values from $\mathcal{T}^{\text{neg}}$ is performed as a last action in function SolveSubInstance($\mathcal{S}^{\text{iter}}$, $\mathcal{T}^{\text{neg}}$, t_{sub}), see line 15, as already mentioned before. In particular, only those pheromone values $\tau^{\text{neg}}_{a,j}$ are updated that CPLEX was able to choose in the sub-instance, that is, all $\tau^{\text{neg}}_{a,j}$ with $j\in\{1,\ldots,m\}$ and $a\in\Sigma\setminus\Sigma^{-}_{j}$. The formula for the update is then as follows.

$$\tau^{\text{neg}}_{a,j}:=\tau^{\text{neg}}_{a,j}+\rho^{\text{neg}}\cdot(\psi^{\text{neg}}_{a,j}-\tau^{\text{neg}}_{a,j}) \quad j\in\{1,\ldots,m\},a\in\Sigma\setminus\Sigma^{-}_{j}, \quad (12)$$

Table 1. Values for weights w^{ib}, w^{rb}, and w^{bsf} with respect to the convergence factor value cf and the value of the control variable bs_upd.

bs_upd	FALSE				TRUE
cf	< 0.4	$[0.4, 0.6)$	$[0.6, 0.8)$	≥ 0.8	
w^{ib}	1	2/3	1/3	0	0
w^{rb}	0	1/3	2/3	1	0
w^{bsf}	0	0	0	0	1

where $\psi_{a,j}^{\text{neg}} = 0$ in case $s^{sub}[j] = a$, and $\psi_{a,j}^{\text{neg}} = 1$ otherwise. In other words, all those assignments of a letter a to a position j that formed part of the sub-instance, but which were not chosen for s^{sub}, get their corresponding negative pheromone value increased, while the opposite is the case for the assignments found in s^{sub}.

4 Different Objective Functions

For the reasons outlined before, we tested four different objective functions as a replacement for the original objective function. However, note that these functions can only be used for all comparisons of solutions that arise within the $\text{ACO}_{\text{neg}}^{+}$ algorithm. CPLEX, for solving the sub-instance at each iteration, still makes use of the original objective function.

The first alternative objective function was proposed by Mousavi et al. in [14]. This function, henceforth called $f_{\text{mou}}()$, was designed with the aim of obtaining a search landscape with less plateaus and with a reduced number of local optima. It evaluates a solution taking into account its *likelihood* to lead to better solutions with a rather small number of local search moves. Note that whenever $f_{\text{orig}}(s) > f_{\text{orig}}(s')$ for two valid solutions s and s', it also holds that $f_{\text{mou}}(s) > f_{\text{mou}}(s')$. Therefore, $f_{\text{mou}}()$ can be used in a standalone manner. As $f_{\text{mou}}()$ is difficult to describe in a reduced space, we refer the interested reader to the original publication [14] or to [7], where the authors made use of $f_{\text{mou}}()$ within the memetic algorithm presented in this work.

A second alternative objective function was presented by Blum and Festa in [2]. This function, henceforth called $f_{\text{blu}}()$, is a lexicographic objective function that—as a first criterion—makes use of the original objective function. For the second criterion it uses the following function:

$$h(s) := \sum_{\{s_i \in \Omega | d_H(s, s_i) \geq t\}} d_H(s, s_i) + \max_{\{s_i \in \Omega | d_H(s, s_i) < t\}} \{d_H(s, s_i)\} \tag{13}$$

In words, $h(s)$ takes the sum of the Hamming distances of s with those input strings $s_i \in \Omega$ such that the Hamming distance is at least t. Moreover, to this sum it adds the maximum Hamming distance of s with those input strings $s_i \in \Omega$

such that the Hamming distance is smaller than t. The original objective function and $h()$ are then combined in the following lexicographic way:

$$f_{\text{blu}}(s) > f_{\text{blu}}(s') \text{ iff} f_{\text{orig}}(s) > f_{\text{orig}}(s') \text{ or} \\ (f_{\text{orig}}(s) = f_{\text{orig}}(s') \text{ and } h(s) > h(s')) \tag{14}$$

where s and s' are valid solutions to the problem. The intuition behind $h()$ is the following one. The larger the value of $h(s)$, the lower is the probability that small changes in s lead to a decrease in the original objective function.

Apart from the two functions above, we also test a simplified version of $f_{\text{blu}}()$, which makes use of function $h'()$ instead of $h()$:

$$h'(s) := \max_{\{s_i \in \Omega | d_H(s,s_i) < t\}} \{d_H(s, s_i)\} \tag{15}$$

Note that $h'(s)$, in contrast to $h(s)$, only gives importance to the maximum Hamming distance of s with those input strings $s_i \in \Omega$ such that the Hamming distance is smaller than t. The resulting simplified lexicographic function is henceforth called $f_{\text{sim}}()$.

Finally, we also consider a lexicographic function $f_{\text{com}}()$ that uses $h'()$ as a second criterion and $f_{\text{mou}}()$ as a third criterion:

$$f_{\text{com}}(s) > f_{\text{com}}(s') \textbf{iff} \\ f_{\text{orig}}(s) > f_{\text{orig}}(s') \textbf{or} \\ (f_{\text{orig}}(s) = f_{\text{orig}}(s') \text{and} h'(s) > h'(s')) \textbf{or} \\ (f_{\text{orig}}(s) = f_{\text{orig}}(s') \text{and} h'(s) = h'(s') \text{and} f_{\text{mou}}(s) > f_{\text{mou}}(s')) \tag{16}$$

5   Experimental Evaluation

$\text{ACO}_{\text{neg}}^+$ was implemented in C++ using GCC 10.2.0 for compilation. The experimental evaluation was performed—in single-threaded mode—on a cluster of computers with "Intel® Xeon® CPU 5670" CPUs of 12 nuclei of 2933 MHz MHz and (in total) 32 Gigabytes of RAM. Moreover, all ILPs were solved with IBM ILOG CPLEX V22.1, which is currently the newest version. In the following, we first describe the utilized benchmark sets. Subsequently, the parameter tuning procedure employed for finding well-working parameter values is outlined. Finally, the numerical results are presented and analysed.

5.1   Benchmark Sets

Several related works on the FFMSP introduced sets of benchmark instances. The ones used in this work are the following ones. Ferone et al. [4] made use of a set of instances containing 100 problem instances with random input strings over $\Sigma = \{A, C, T, G\}$ for each combination of $n \in \{100, 200, 300, 400\}$ and $m \in \{200, 600, 800\}$. This set—which contains 1200 problem instances in total—is henceforth called **Ferone**. Note that all instances were solved in previous works with thresholds $t \in \{0.75\,\text{m}, 0.8\,\text{m}, 0.85\,\text{m}\}$. A subset of these instances—that is, those with $n \in \{100, 200\}$—was already used in earlier publications [2,3].

Gallardo and Cotta [7] introduced a set of problem instances with similar specifications as the ones by Ferone et al. [4]. However, instead of 100 random instances per combination of n and m, their set contains only five random instances per combination. Moreover, it is restricted to $n \in \{100, 200\}$. Their set of random instances is henceforth called Gallardo.

Note that in this work we will solve all described problem instances for $t \in \{0.8\,\text{m}, 0.85\,\text{m}\}$. The threshold $t = 0.75\,\text{m}$ was not considered because we noticed (similar to previous works) that the resulting problems are very easily solved to optimality.

5.2 Algorithm Tuning

We used the scientific tuning tool irace [11] for finding well-working parameter values. During preliminary testing, we noticed significant changes in algorithm behaviour and requirements between thresholds $t = 0.8\,\text{m}$ and $t = 0.85\,\text{m}$. Moreover, we also noticed that different algorithm settings were required for instances with $n \in \{100, 200\}$ in comparison to instances with $n \in \{300, 400\}$. Therefore, we applied four different runs of irace:

- Tuning1: for instances with $n \in \{100, 200\}$ solved with $t = 0.8\,\text{m}$.
- Tuning2: for instances with $n \in \{300, 400\}$ solved with $t = 0.8\,\text{m}$.
- Tuning3: for instances with $n \in \{100, 200\}$ solved with $t = 0.85\,\text{m}$.
- Tuning4: for instances with $n \in \{300, 400\}$ solved with $t = 0.85\,\text{m}$.

As tuning instances for each of these runs, we used the first instance (out of 100 instances) for each combination of n and m from the Ferone set, which makes a total of six tuning instances for each of the four tuning runs. The algorithm parameters and the considered domains are shown in Table 2. Moreover, irace was given a budget of 3000 algorithm runs, with a time limit of 600 s per run. The obtained results are shown in Table 3. In particular, for each of the four tuning runs, we provide the four best-ranked parameter settings as delivered by irace.

The following observations can be made. First, there is a clear difference between the parameter settings for $t = 0.8\,\text{m}$ (Tuning1 and Tuning2) and the parameter settings for $t = 0.85\,\text{m}$ (Tuning3 and Tuning4). This concerns, for example, the use of CPLEX. While for $t = 0.8\,\text{m}$ the use of CPLEX is rather reduced (below 10 s for deriving the initial solution, and 0.92 s, respectively 3.69 s, for the application to sub-instances), CPLEX is rather heavily used for $t = 0.85\,\text{m}$. This is already the first indication that the algorithm, most probably, will work better for $t = 0.8\,\text{m}$ than for $t = 0.85\,\text{m}$. Second, the simplified version of f_{blu} (which is used in f_{sim} and f_{com}) seems to be prefered as an objective function when $t = 0.85\,\text{m}$, while f_{blu} seems to be better for $t = 0.8\,\text{m}$. The objective function introduced by Mousavi et al. [14] also contributes as a third criterion in function f_{com} which is best in the context of Tuning4.

Table 2. Algorithm parameters and their domains considered for the tuning process with irace.

Parameter	Considered domain
n_a	$\{3,\ldots,50\}$
d_{rate}	$[0.0, 0.99]$
ρ	$[0.01, 0.5]$
ρ^{neg}	$[0.01, 0.5]$
t_{sub}	$[0.5, 20]$
t_{init}	$\{1,\ldots,50\}$
Obj. func	$\{f_{\text{orig}}, f_{\text{mou}}, f_{\text{blu}}, f_{\text{sim}}, f_{\text{com}}\}$

Table 3. Parameter settings generated by irace

Tuning run	Rank	n_a	d_{rate}	ρ	ρ^{neg}	t_{sub}	t_{init}	Obj. func
Tuning1	1	24	0.86	0.12	0.27	0.92	7	f_{blu}
	2	21	0.89	0.09	0.28	1.04	7	f_{sim}
	3	25	0.88	0.13	0.27	0.54	11	f_{sim}
	4	24	0.93	0.14	0.30	1.21	10	f_{blu}
Tuning2	1	9	0.76	0.27	0.49	3.69	9	f_{blu}
	2	9	0.77	0.28	0.50	3.61	9	f_{blu}
	3	8	0.72	0.22	0.35	2.93	7	f_{blu}
	4	5	0.69	0.20	0.33	3.49	13	f_{blu}
Tuning3	1	45	0.29	0.10	0.11	7.28	37	f_{sim}
	2	29	0.13	0.15	0.16	10.91	27	f_{sim}
	3	49	0.35	0.10	0.36	15.37	14	f_{com}
	4	46	0.29	0.12	0.10	7.24	38	f_{sim}
Tuning4	1	33	0.59	0.21	0.03	18.69	50	f_{com}
	2	32	0.58	0.19	0.04	18.21	50	f_{com}
	3	31	0.57	0.20	0.04	17.71	50	f_{com}
	4	31	0.62	0.23	0.03	17.01	50	f_{com}

However, f_{mou} is never used in any of the best-ranked parameter settings in any of the four tuning runs. We used the first-ranked parameter settings in the respective four cases for the final experimental evaluation.

5.3 Results

$\text{ACO}^+_{\text{neg}}$ was applied exactly once to each of the 1200 problem instances from the Ferone set and 10 times to each of the 60 problem instances from the Gallardo set. A time limit of 600 CPU seconds was used for each run. The results are shown in Tables 4 and 5 in terms of the original objective function values, averaged over all instances of the same combination of n and m, and averaged over all runs in the case of the Gallardo instances. That is, each row in Table 4 (Ferone

Table 4. Numerical results for the **Ferone** instances

n	m	t	CPLEX	HyAco	GRASP	ACO$^+_{neg}$	Impr (%)
	300	0.8 m = 240	70.62	77.84	76.26	84.07	8.00
100	600	0.8 m = 480	71.82	72.97	77.53	88.12	13.66
	800	0.8 m = 640	71.81	70.94	82.17	89.23	8.59
	300	0.8 m = 240	86.71	104.17	94.71	107.41	3.11
200	600	0.8 m = 480	79.56	85.02	80.94	102.42	20.47
	800	0.8 m = 640	75.01	77.95	85.71	97.67	13.95
	300	0.8 m = 240	88.81	n.a.	112.83	120.16	6.50
300	600	0.8 m = 480	67.71	n.a.	83.12	104.95	26.26
	800	0.8 m = 640	65.87	n.a.	90.26	92.88	2.90
	300	0.8 m = 240	92.00	n.a.	119.32	129.29	8.36
400	600	0.8 m = 480	48.94	n.a.	85.99	103.86	20.78
	800	0.8 m = 640	47.18	n.a.	92.86	89.05	−4.10
	300	0.85 m = 255	25.11	28.30	29.54	30.53	3.35
100	600	0.85 m = 510	23.66	22.82	27.47	27.53	0.22
	800	0.85 m = 680	23.93	21.66	26.54	26.61	0.26
	300	0.85 m = 255	22.99	28.59	30.37	32.32	6.42
200	600	0.85 m = 510	22.05	21.90	26.35	27.31	3.64
	800	0.85 m = 680	22.05	20.40	24.42	25.83	5.77
	300	0.85 m = 255	20.81	n.a.	31.83	31.33	−1.57
300	600	0.85 m = 510	21.77	n.a.	24.95	24.68	−1.08
	800	0.85 m = 680	22.41	n.a.	23.53	20.68	−12.11
	300	0.85 m = 255	20.93	n.a.	32.78	31.55	−3.75
400	600	0.85 m = 510	19.46	n.a.	24.56	24.80	0.98
	800	0.85 m = 680	20.38	n.a.	22.82	15.86	−30.50

instances) shows averages over 100 problem instances, while each row in Table 5 presents averages over five problem instances and 10 runs of each algorithm. Note that the final column of each table presents the percentage improvement of the results obtained by ACO$^+_{neg}$ over the current state-of-the-art results. Obviously, in case such a percentage is a negative number, ACO$^+_{neg}$ does not reach the state-of-the-art result in this case.

In the case of the **Ferone** instances, the results of ACO$^+_{neg}$ are compared to the ones of CPLEX (used in standalone mode and with the same computation time limit as ACO$^+_{neg}$), the hybrid ACO approach called HyAco from [2], and the best one of the GRASP approaches from [4]. In other words, each result shown in the column with heading GRASP is the best one obtained from six different GRASP variants. Note that the GRASP variants from [4] were applied with

Table 5. Numerical results, random `Gallardo` instances

n	m	t	CPLEX	MA	GRASP$_{\text{fer}}$	GRASP$_{\text{mou}}$	ACO$_{\text{neg}}^{+}$	Impr (%)
	300	$0.8\,\text{m} = 240$	69.80	84.82	80.78	70.99	83.14	−1.98
100	600	$0.8\,\text{m} = 480$	71.80	87.08	79.12	70.83	86.90	−0.21
	800	$0.8\,\text{m} = 640$	72.20	89.90	79.52	71.08	89.80	−0.11
	300	$0.8\,\text{m} = 240$	87.80	109.58	105.85	83.04	106.00	−3.27
200	600	$0.8\,\text{m} = 480$	78.40	101.23	88.95	80.90	102.48	1.23
	800	$0.8\,\text{m} = 640$	73.60	93.92	80.09	79.77	97.40	3.71
	300	$0.85\,\text{m} = 255$	24.60	32.58	18.41	30.10	30.50	−6.38
100	600	$0.85\,\text{m} = 510$	23.20	28.76	4.89	25.36	27.38	−4.80
	800	$0.85\,\text{m} = 680$	22.80	27.96	2.58	24.33	26.74	−4.36
	300	$0.85\,\text{m} = 255$	24.20	34.49	14.85	32.69	32.32	−6.29
200	600	$0.85\,\text{m} = 510$	22.40	26.17	2.26	25.54	27.10	3.55
	800	$0.85\,\text{m} = 680$	22.20	25.61	0.60	23.71	26.00	1.52

a computation time limit of 30 CPU seconds per run. As ACO$_{\text{neg}}^{+}$ was allowed 600 CPU seconds per run, this might seem unfair. However, the computation time limit of 30 CPU seconds was chosen in [4] because their algorithms did not profit much from longer running times. The results allow us to make the following observations:

- In the case of a threshold value of $t = 0.8\,\text{m}$, ACO$_{\text{neg}}^{+}$ is clearly the new state-of-the-art approach. It outperforms CPLEX, HYACO and the GRASP approaches in all cases, with the exception of the largest case ($n = 400$, $m = 800$) where the best GRASP approach is still better than ACO$_{\text{neg}}^{+}$. The improvement of ACO$_{\text{neg}}^{+}$ over the currently best-known result is maximally 26.26% (in case $n = 300$, $m = 600$). On average the improvement is 10.71%.
- In contrast, for the cases with a threshold of $t = 0.85\,\text{m}$, ACO$_{\text{neg}}^{+}$ improves only slightly over the state of the art in the context of the smaller problem instances ($n \in \{100, 200\}$), while it generally does not reach the state-of-the-art results for the larger problem instances ($n \in \{300, 400\}$). This becomes especially clear in the cases ($n = 300, m = 800$) and ($n = 400, m = 800$).

As already suspected after the tuning experiments, these results show that ACO$_{\text{neg}}^{+}$ seems generally to work better for threshold values of $t = 0.8\,\text{m}$ in contrast to threshold values of $0.85\,\text{m}$. In order to show that the characteristics of the FFMSP strongly depend on the threshold value, we display the evolution of the objective function values of the best-found solutions over time in Fig. 1. Each of the four graphics in Fig. 1 shows the evolution of ACO$_{\text{neg}}^{+}$ for the first 10 (out of 100) problem instances of the chosen case. The cases with threshold value 0.8 m (Figs. 1a and 1b) indicate that the values of the initial solutions provided by CPLEX are much worse than those of the final solutions identified by ACO$_{\text{neg}}^{+}$. Moreover, ACO$_{\text{neg}}^{+}$ is generally able to find improving solutions frequently during the run-time of the algorithm. In contrast, the evolution of ACO$_{\text{neg}}^{+}$ is very

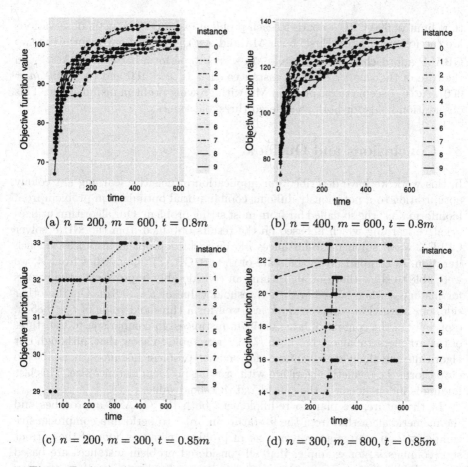

(a) $n = 200$, $m = 600$, $t = 0.8m$

(b) $n = 400$, $m = 600$, $t = 0.8m$

(c) $n = 200$, $m = 300$, $t = 0.85m$

(d) $n = 300$, $m = 800$, $t = 0.85m$

Fig. 1. Evolution of the objective function values (original objective function)

different in the cases with threshold value 0.85 m (see Figs. 1c and 1d). First, the values of the initial solutions provided by CPLEX are much closer to the final values. Moreover, most improving solutions identified by $\text{ACO}^+_{\text{neg}}$ over the run-time concern the secondary function (in f_{sim}, respectively f_{com}). These results, therefore, allow concluding that there is room for improvement of $\text{ACO}^+_{\text{neg}}$ in the case of threshold values 0.85 m.

The comparison to the second state-of-the-art approach—that is, the MA algorithm from [7]—is done in the context of the random instances from the same paper, as already described in Sect. 5.1. Note that this benchmark set only contains rather small problem instances with $n \in \{100, 200\}$ for which $\text{ACO}^+_{\text{neg}}$ worked very well in comparison to the GRASP approaches. Apart from the results of MA and $\text{ACO}^+_{\text{neg}}$, Table 5 also contains the results of the GRASP approach from [3] and the GRASP approach from [14], both re-implemented by the authors of MA. Note that all techniques were applied with a computation time limit of 600 CPU seconds to each problem instance. The main observations are the following ones. First, both MA and $\text{ACO}^+_{\text{neg}}$ clearly outperform the two GRASP approaches. Second, while MA seems to perform better than $\text{ACO}^+_{\text{neg}}$ in the case of the smaller problem instances, that is, $n = 100$ and ($n = 200, m = 300$), $\text{ACO}^+_{\text{neg}}$ seems to outperform MA with growing problem instance size. This observation holds for both considered threshold values.

6   Conclusions and Outlook

In this work we have described our application of negative learning ant colony optimization to a notoriously difficult combinatorial optimization problem from bioinformatics: the so-called far from most string problem. Our algorithm utilizes negative learning which is based on the results obtained from the MILP solver CPLEX when solving sub-instances of the tackled problem instances at each iteration. In comparison to the state-of-the-art GRASP approaches from [4] we were able to show that our algorithm can considerably improve over the results for instances when solved with a threshold value of $t = 0.8\,\text{m}$. On the other side, our algorithm seems to work less well for a threshold value of $t = 0.85\,\text{m}$, especially in the context of large problem instances. In comparison to the state-of-the-art memetic algorithm from [7], we were able to show that, although our algorithm has slight disadvantages for smaller problem instances, it is able to outperform the memetic algorithm with growing problem instance size. This last finding is independent of the considered threshold value.

In the future, we plan to re-implement both the GRASP approaches and the memetic algorithm from the literature in order to perform a comprehensive comparison on a large and diverse set of problem instances. One of the current shortcomings is, for example, that all considered problem instances are based on an alphabet of size four. However, this is very limiting. Therefore, we plan to generate problem instances over a whole range of different alphabet sizes. Moreover, we will consider a wider range of threshold values in order to study, for each alphabet size, the range of threshold values that make the problem complicated to be solved.

Another line for future research consists in testing different ant colony optimization approaches from the literature that include some type of negative learning; see, for example, [17,18].

References

1. Blum, C., Dorigo, M.: The hyper-cube framework for ant colony optimization. IEEE Trans. Man Syst. Cybern. Part B **34**(2), 1161–1172 (2004)
2. Blum, C., Festa, P.: A hybrid ant colony optimization algorithm for the far from most string problem. In: Blum, C., Ochoa, G. (eds.) EvoCOP 2014. LNCS, vol. 8600, pp. 1–12. Springer, Heidelberg (2014). https://doi.org/10.1007/978-3-662-44320-0_1

3. Ferone, D., Festa, P., Resende, M.G.C.: Hybrid metaheuristics for the far from most string problem. In: Blesa, M.J., Blum, C., Festa, P., Roli, A., Sampels, M. (eds.) HM 2013. LNCS, vol. 7919, pp. 174–188. Springer, Heidelberg (2013). https://doi.org/10.1007/978-3-642-38516-2_14

4. Ferone, D., Festa, P., Resende, M.G.: Hybridizations of grasp with path relinking for the far from most string problem. Int. Trans. Oper. Res. **23**(3), 481–506 (2016)

5. Festa, P.: On some optimization problems in mulecolar biology. Math. Biosci. **207**(2), 219–234 (2007)

6. Festa, P., Pardalos, P.: Efficient solutions for the far from most string problem. Ann. Oper. Res. **196**(1), 663–682 (2012)

7. Gallardo, J.E., Cotta, C.: A GRASP-based memetic algorithm with path relinking for the far from most string problem. Eng. Appl. Artif. Intell. **41**, 183–194 (2015)

8. Kennelly, P.J., Krebs, E.G.: Consensus sequences as substrate specificity determinants for protein kinases and protein phosphatases. J. Biol. Chem. **266**(24), 15555–15558 (1991)

9. Lanctot, J., Li, M., Ma, B., Wang, S., Zhang, L.: Distinguishing string selection problems. Inf. Comput. **185**(1), 41–55 (2003)

10. Liu, X., Liu, S., Hao, Z., Mauch, H.: Exact algorithm and heuristic for the closest string problem. Comput. Oper. Res. **38**(11), 1513–1520 (2011)

11. López-Ibáñez, M., Dubois-Lacoste, J., Cáceres, L.P., Birattari, M., Stützle, T.: The irace package: iterated racing for automatic algorithm configuration. Oper. Res. Perspect. **3**, 43–58 (2016)

12. Meneses, C.N., Oliveira, C.A., Pardalos, P.M.: Optimization techniques for string selection and comparison problems in genomics. IEEE Eng. Med. Biol. Mag. **24**(3), 81–87 (2005)

13. Mousavi, S.R.: A hybridization of constructive beam search with local search for far from most strings problem. Int. J. Comput. Inf. Eng. **4**(8), 1200–1208 (2010)

14. Mousavi, S., Babaie, M., Montazerian, M.: An improved heuristic for the far from most strings problem. J. Heuristics **18**, 239–262 (2012)

15. Nurcahyadi, T., Blum, C.: Adding negative learning to ant colony optimization: a comprehensive study. Mathematics **9**(4), 361 (2021)

16. Nurcahyadi, T., Blum, C., Manyà, F.: Negative learning ant colony optimization for maxsat. Int. J. Comput. Intell. Syst. **15**(1), 1–19 (2022)

17. Rojas-Morales, N., Riff, M.C., Montero, E.: Opposition-inspired synergy in sub-colonies of ants: the case of focused ant solver. Knowl.-Based Syst. **229**, 107341 (2021)

18. Ye, K., Zhang, C., Ning, J., Liu, X.: Ant-colony algorithm with a strengthened negative-feedback mechanism for constraint-satisfaction problems. Inf. Sci. **406–407**, 29–41 (2017)

19. Zörnig, P.: Reduced-size integer linear programming models for string selection problems: application to the farthest string problem. J. Comput. Biol. **22**(8), 729–742 (2015)

Monte Carlo Tree Search with Adaptive Simulation: A Case Study on Weighted Vertex Coloring

Cyril Grelier[ID], Olivier Goudet[ID], and Jin-Kao Hao[(✉)][ID]

LERIA, Université d'Angers, 2 Boulevard Lavoisier, 49045 Angers, France
{cyril.grelier,olivier.goudet,jin-kao.hao}@univ-angers.fr

Abstract. This work presents a hyper-heuristic approach to online learning, which combines Monte Carlo Tree Search with multiple local search operators selected on the fly during the search. The impacts of different operator policies, including proportional bias, one-armed bandit, and neural network, are investigated. Experiments on well-known benchmarks of the Weighted Vertex Coloring Problem are conducted to highlight the advantages and limitations of each dynamic selection strategy.

Keywords: Monte Carlo Tree Search · Local Search · Hyper-heuristic · Weighted Vertex Coloring · Learning-driven optimization

1 Introduction

Given a graph $G = (V, E, w)$ with vertex set V, edge set E, and a weight function $w : V \rightarrow \mathbb{R}^+$, assigning a positive weight $w(v)$ to each node in V, the goal of the Weighted Vertex Coloring Problem (WVCP) is to find a partition $S = \{V_1, \ldots, V_k\}$ of the vertex set V, into k non-empty independent subsets V_i (also called color classes or groups, k is not fixed) such that the function $f(S) = \sum_{i=1}^{k} \max_{v \in V_i} w(v)$ is minimized.

A set V_i is an independent set if and only if $\forall u, v \in V_i, \{u, v\} \notin E$.

The WVCP generalizes the popular NP-hard graph coloring problem, which corresponds to solving this problem when all weights $w(v)$ are equal for $v \in V$, and is therefore itself NP-hard. The WVCP belongs to the larger family of grouping problems whose main task is to partition a set of elements into mutually disjoint subsets such that additional constraints and/or optimization objectives are satisfied. The WVCP has many practical applications such as matrix decomposition [19] and batch job scheduling in a multiprocessor environment [18].

In addition to the exact methods for the WVCP, such as [5,16], dedicated heuristics have been proposed in the literature recently, including local search algorithms [17,19,22,24], and memetic algorithms [10]. These heuristics can provide approximate solutions of good quality for medium and large instances, which

L. Pérez Cáceres and T. Stützle (Eds.): EvoCOP 2023, LNCS 13987, pp. 98–113, 2023.
https://doi.org/10.1007/978-3-031-30035-6_7

cannot be solved in a reasonable time by exact methods. However, given the difficulty of the WVCP, no simple heuristic is able to obtain the best-known results for all instances of the literature. This can be explained by the fact that these instances have different characteristics (average degree, degree distribution, weight distribution). Even by fine-tuning the hyperparameters or using an adaptive selection of these parameters (e.g. the algorithm AFISA [22]), it is difficult for a single heuristic to provide enough flexibility to solve all instances successfully.

One possible way to overcome this difficulty is to find on the fly the best heuristic to use during the search for a given instance, hyper-heuristics have been proposed in the literature for this purpose. Hyper-heuristics are search methods that explore the space formed by low-level heuristics, instead of the space of direct solutions [3]. We refer the reader to [3,7] for an overview of existing hyper-heuristics proposed in the literature to solve various combinatorial optimization problems. Specifically, hyper-heuristics have already been used to solve grouping problems, with applications to scheduling and graph coloring [8,20], but never to the WVCP to our knowledge.

Most hyper-heuristics used for grouping problems are single-point hyper-heuristics in the sense that they apply a chosen low-level heuristic to the current solution at each step of the search before deciding to accept or reject the newly created solution [8]. However, these iterative hyper-heuristics are prone to get stuck in local optima, especially when tackling graph coloring problems that are generally characterized by a rough fitness landscape [2]. This occurs when the hyper-heuristic does not incorporate efficient low-level components that can be chosen to diversify the search into another area of the search space. Moreover, most of the hyper-heuristics proposed in the literature use a high-level strategy taking into account the gain in fitness obtained by the different low-level components (e.g. [9]) but do not take into account the current state of the iteratively improved solution.

To overcome these two limitations, we propose a hyper-heuristic approach with online learning, which combines the Monte Carlo tree search (MCTS) framework proposed in [11] and a high-level learning heuristic selector that takes into account the raw state of the current solution to be improved during the simulation phase of MCTS. This MCTS framework can continuously learn new promising starting points for a local search heuristic, which could help an iterative hyper-heuristic to avoid local minima traps.

Note that two previous works [1,21] have also used an MCTS in a hyper-heuristic for combinatorial optimization problems. In [21], the search space of low-level heuristics is explored by an MCTS, which searches for the best sequence of low-level heuristics to apply to a given solution, but without guarantee of optimality. Unlike this approach, the MCTS used in this work is directly constructed to explore the tree of legal solutions. Therefore, it has the property of being able to provide a theoretical guarantee that an optimal solution can be found if enough time is given to the algorithm. In [1], an MCTS was used to create initial sequences, which are improved in a separate second step by a hyper-heuristic using a wide variety of local moves, selected by a high-level strategy. In this work, we aim to push further the coupling between the selection by the MCTS and the choice of the low-level heuristic.

We summarize the contributions of the present paper as follows. First, we propose an integration between the MCTS, which is used to discover new starting points and the high-level operator selection strategy, which takes as input this starting point to make its choice. The score obtained by a given low-level heuristic is used to update both the MCTS learning strategy and the operator selector. Secondly, we investigate the impact of different online learning strategies, including a neural network taking into account not only the past scores of the low-level components as in [6], but also the raw state of the initiating solution to be improved by a local search heuristic. This neural network is made invariant by permuting the color groups by taking inspiration from the deep-set architecture [25], which is a desirable property for handling the WVCP.

2   Related Works on the WVCP

Most of the best heuristics for the WVCP are local search based algorithms and constructive algorithms. This section presents a review of these methods.

2.1   Local Search Algorithms

We described the four most effective local search based algorithms for the WVCP: TW [11], AFISA [22], RedLS [24] and ILS-TS [17]. These four heuristics will constitute the set of low-level heuristics that we will manipulate in the hyper-heuristic proposed in this paper.

Legal Tabu Search. TabuWeight (TW) [11] is inspired by the classical TabuCol algorithm for the GCP [12]. TW explores the space of legal colorings. It uses the *one-move* operator (which displaces a vertex from its color group to another color group) without creating conflicts. The best move not forbidden by the tabu list is applied at each iteration. The move is then added to the tabu list and is forbidden for the next tt iterations, tt being a parameter called tabu tenure.

Adaptive Feasible and Infeasible Tabu Search. AFISA [22] is a tabu search algorithm using the *one-move* operator and explores both the legal and illegal search spaces.[1] AFISA uses an adaptive coefficient to oscillate between legal and illegal solutions during the search.

Local Search with Selection Rules. RedLS [24] explores both the illegal and legal search spaces. It uses the configuration checking strategy [4] to avoid cycling in neighborhoods solutions with the *one-move* operator. When the solution is legal, RedLS moves all the heaviest vertices from a color group to other colors to lower the WVCP score. Then it resolves the conflicts with different selection rules.

---

[1] A solution is said illegal if two adjacent vertices share the same color.

Iterated Local Search with Tabu Search. ILSTS [17] explores both the legal and partial search spaces. From a complete solution, ILSTS iteratively performs two steps: (i) it removes the heaviest vertices from the several color groups and places them in the set of uncolored vertices; (ii) it minimizes the score $f(S)$ by applying different variants of the *one-move* operator and a *grenade* operator[2] until the set of uncolored vertices becomes empty.

2.2 Constructive Heuristics

Two constructive heuristics have been proposed for the WVCP. These heuristics start with an empty or a partial solution and then color the nodes one by one until a complete legal solution is obtained. They have been used in combination with local search heuristics.

Reactive Greedy Randomized Adaptive Search Procedure. RGRASP [19] is an algorithm that iterates over two steps. First, it initializes a solution with a greedy algorithm. Secondly, it improves the solution with a local search procedure. At each iteration, it randomly removes the color of a part of the vertices and starts again at the first step. The choice of the vertices to recolor is managed by an adaptive parameter that evolves according to the quality of the solution.

Monte Carlo Tree Search Algorithm. The MCTS algorithm proposed in [11] is a constructive heuristic for the WVCP, where a search tree is built to explore the partial and legal search space. The search tree is built incrementally and asymmetrically. For each iteration of the MCTS, a selection strategy balancing exploration and exploitation is used to construct a partial solution that is further completed and improved by a simulation strategy. To build this partial solution, the vertices of the graph are considered in a predefined order (they are sorted by decreasing order of their weight, then by degree) and each uncolored vertex $u \in U$ is assigned a particular color i. Such a move is denoted as $< u, U, V_i >$ and applying this move to a partial solution S being constructed gives a new solution $S \oplus < u, U, V_i >$. After the simulation phase, a legal solution is obtained and the search tree is updated with the WVCP score of this solution. In [11], the authors investigated the impact of various local search procedures used during this simulation phase of this MCTS algorithm, instead of the classical random simulation which performs badly for this problem. It appears from this study that the choice of the local search procedure has an important impact which depends on the type of instance considered.

To make the algorithm more robust to each specific instance, we propose to use this same MCTS framework, but to select dynamically the local search procedure during the search in a set $\mathcal{O} = \{o_1, \ldots, o_d\}$ of d different local search heuristics. In the proposed framework, the choice of the local search strategy

---

[2] The *grenade* operator consists in moving a vertex u to a new group (or color) V_i, but first, each adjacent vertex of u in V_i is relocated to another color group or to a set of unassigned vertices to keep a legal solution.

may depend on the raw state of the initiating solution built by the MCTS at each iteration of the algorithm.

3   MCTS with Adaptive Simulation Strategy

This section presents the general framework of the adaptive selection of local search operators combined with the MCTS algorithm presented in [11].

3.1   Main Scheme

The proposed hyper-heuristic approach is detailed in Algorithm 1. It takes as input a weighted graph $G = (V, E, w)$, a set of local search operators $\mathcal{O} = \{o_1, \dots, o_d\}$ and a high-level selection strategy $\pi_\theta : \Omega \to \mathcal{O}$ taking as input a complete and legal solution S and giving in output a local search operator $o \in \mathcal{O}$ to apply to S.

The selection strategy π_θ is parametrized by a vector of parameters θ initialized at random at the beginning of the search. Then the algorithm repeats a loop (iteration) until a cutoff time limit is met. Each iteration of the MCTS involves the execution of five steps:[3]

1. **Selection**: starting from the root node of the tree, the most promising children nodes are iteratively selected until a leaf node is reached (a node without all children opened). The selection of a child node at each level corresponds to the choice of color for the next vertex of the graph[4]. The most promising node is selected based on an exploitation/exploration trade-off UCT (Upper Confidence bounds for Trees).
2. **Expansion**: the MCTS tree grows by adding a new child node to the leaf node reached during the selection phase.
3. **Adaptive Simulation**: the current partial solution is completed with greedy legal moves to obtain a complete solution S. Then the selected operator $o = \pi_\theta(S)$ is applied to the current solution S to improve it for a fixed time. This leads to a new solution S' and a new learning example (S, o, r), and the reward $r = -f(S')$ is stored in a database D. For every nb iterations, the selection strategy is learned online on this database D.
4. **Update**: after the simulation, the average score and the number of visits of each node on the explored branch are updated.
5. **Pruning**: if a new best score is found, some branches of the MCTS tree may be pruned if it is not possible to improve the best current score with it.

The fact that symmetries are cut in the search tree by restricting the set of legal moves considered during each step 1 and 2, and that pruning rules are

---

[3] One notices that compared to the MCTS method from [11], only the simulation phase (step 3) changes.

[4] The vertices are considered in the decreasing order of their weight then of their degree.

applied in step 5, allows the whole tree to be explored in a reasonable time for small instances. This characteristic of the algorithm allows to provide proof of optimality for such instances. We refer the reader to [11] for more details on

---

Algorithm 1. MCTS with adaptive simulation strategy

1: **Input:** Weighted graph $G = (V, E, w)$, $\mathcal{O}$ a set of local search operators and π_θ a selection strategy.
2: **Output:** The best legal coloring S^* found
3: $S^* = \emptyset$ and $f(S^*) = MaxInt$
4: $D = \emptyset$.
5:
6: **while** stop condition is not met **do**
7:     $C \leftarrow R$ ▷ Current node corresponding to the root node of the tree
8:     $S \leftarrow \{V_1, U\}$ with $V_1 = \{v_1\}$ and $U = V \backslash V_1$ ▷ first vertex in first color group
9:
10:    /* Step 1 - Selection */
11:    **while** C is not a leaf **do**
12:        $C \leftarrow$ select_best_child(C) with legal move $< u, U, V_i >$
13:        $S \leftarrow S \oplus < u, U, V_i >$
14:    **end while**
15:
16:    /* Step 2 - Expansion */
17:    **if** C has a potential child, not yet open **then**
18:        $C \leftarrow$ open_first_child_not_open(C) with legal move $< u, U, V_i >$
19:        $S \leftarrow S \oplus < u, U, V_i >$
20:    **end if**
21:
22:    /* **Step 3 - Adaptive simulation strategy** */
23:    $S \leftarrow$ greedy(S) ▷ **Complete current solution with a greedy algorithm**
24:    $o = \pi_\theta(S)$ ▷ **Select a local search operator**
25:    $S' \leftarrow o(S)$ ▷ **Improve the solution with the selected operator.**
26:    $D \leftarrow D \cup (S, o, -f(S'))$ ▷ **Store a learning example in the database.**
27:    $\theta \leftarrow learning(D, \theta)$ ▷ **Online learning of the adaptive selection strategy**
28:
29:    /* Step 4 - Update */
30:    **while** $C \neq R$ **do**
31:        update(C,f(S'))
32:        $C \leftarrow$ parent(C)
33:    **end while**
34:    **if** $f(S') < f(S^*)$ **then**
35:        $S^* \leftarrow S'$
36:    **end if**
37:
38:    /* Step 5 - Pruning */
39:    apply pruning rules
40:
41: **end while**
42: return S^*

---

the different steps 1, 2, 4, and 5 of this algorithm. In what follows, only step 3, with the adaptive simulation strategy written in bold in Algorithm 1, will be described in detail.

3.2 Adaptive Simulation Strategy Framework

As shown in [11], no simulation strategy dominates the others for the WVCP, and some operators are more successful for some types of instances than others.

Dynamic Operator Selection. To choose the best possible local search operator during the search, a high-level strategy π_θ automatically selects an operator $o = \pi_\theta(S)$ in $\mathcal{O}$ to be applied to the current solution S. Note that the choice of this operator may depend on the solution S to be improved. The set $\mathcal{O}$ of low-level heuristic components considered for the simulation strategy are the four local search procedures presented in Sect. 2.1, namely TW, AFISA, RedLS, and ILS-TS. The different high-level strategies π_θ used in this work will be presented in Sect. 4.

Collecting Learning Examples. Even if some of these local search operators o make transitions between different search spaces, it always returns the best legal solution (with the smallest WVCP score) encountered during the search denoted as $S' = o(S)$. According to this new solution found, a new learning example (S, o, r), associated with the reward $r = -f(S')$, is stored in a database D. The reward r is negative because the WVCP is a minimization problem. One notices that in the literature on hyper-heuristics, the reward of a given operator is often proportional to $f(S) - f(S')$ (for a minimization problem), namely the difference between the score before and after the search (see for example [9] and [6]). However, in our algorithm, we experimentally found that is better that this reward does not depend on the score of the solution S from which the local search starts as it may introduce some noise in the learning process. Indeed, the MCTS can produce solutions of different quality at each iteration and if the score of the solution S from which the local search starts is taken into account in the evaluation of the reward, operators starting from an already good solution (low $f(S)$ value) will be at a disadvantage compared to those starting from a bad solution (high $f(S)$ value), as it is comparatively easier to improve a solution of poor quality than a solution that is already of good quality.

Online Learning of the High Level Operator Selector. Every nb iterations of the MCTS, the policy π_θ is trained on the database D and the set of its parameters θ is updated. The database D is modeled as a queue of size N corresponding to the last N examples obtained during the past iterations. This queue of limited size allows to better adapt to the potential variations of the operators' results, in case some operators are better at the beginning of the search than at the end.

4 Operator Selectors

We investigate the impact of five different operator selection policies π_θ of different level of complexity: a neural network, three fitness-based criteria and one baseline random selector, where each operator has $\frac{1}{|\mathcal{O}|}$ chance to be selected.

4.1 Neural Network Selector

In this case, the function $\pi_\theta : \Omega \rightarrow \mathcal{O}$, is modeled by a neural network, parametrized by a vector θ (initialized at random at the beginning).

This neural network π_θ takes as input a coloring S as a set of k binary vectors $\mathbf{v_j}$ of size n, $S = \{\mathbf{v_1}, \ldots, \mathbf{v_k}\}$, where each $\mathbf{v_j}$ indicates the vertices belonging to the color group j. From such an entry the neural network outputs a vector with $|\mathcal{O}|$ values in $\mathbb{R}$ corresponding to the expected reward of each operator $o \in \mathcal{O}$. Then, with a 90% chance, the operator associated with the maximum expected reward is selected. We still keep a 10% chance to select an operator at random, to ensure a minimum of diversity in the operator selection (exploration).

Using the *deep set* architecture [10,15,25], the output of this neural network is made invariant by the permutation of the color groups in S, so that the neural network selector will take the same decision for two input colors S and S_σ which are equivalent up to color group permutation σ. Specifically, for any permutation σ of the input color groups we have

$$\pi_\theta(\mathbf{v}_{\sigma(1)}, \ldots, \mathbf{v}_{\sigma(k)}) = \pi_\theta(\mathbf{v_1}, \ldots, \mathbf{v_k}). \tag{1}$$

For a coloring $S = \{\mathbf{v_1}, \ldots, \mathbf{v_k}\}$, the color group invariant network π_θ is defined as

$$\pi_\theta(S) = \frac{1}{k} \sum_{i=1}^{k} (\phi_{\theta_P} \circ \phi_{\theta_{P-1}} \circ \cdots \circ \phi_{\theta_0}(S))_i, \tag{2}$$

where each ϕ_{θ_j} is a permutation invariant function from $\mathbb{R}^{k \times l_{j-1}}$ to $\mathbb{R}^{k \times l_j}$ with l_j being the layer sizes. Note that we have $l_{j-1} = |V|$ for the first layer and $l_{j-1} = |V|$ and $l_j = |\mathcal{O}|$ for the last layer. See [15] for more details on the permutation invariant function ϕ_{θ_j}.

In this work, we use a neural network with two hidden layers of size $h1 = |V|$ and $h2 = \frac{|V|+|\mathcal{O}|}{2}$, and a non-linear activation function defined as $\mu(x) = \max(0.2 \times x, x)$ (LeakyReLU).

During this training phase, each learning example (S, o, r) of the database D is converted into a supervised learning example (X, y), with X an input matrix of size $k \times |V|$ corresponding to the set of k vectors $S = \{\mathbf{v_1}, \ldots, \mathbf{v_k}\}$, and y is a real vector of size $|\mathcal{O}|$ (number of candidate operators), such that y is initialized to the value $\pi_\theta(X)$ for each operator and then its value $y[o]$ for the chosen operator o is replaced by the expected reward r: $y[o] = r$.

Every $nb = 10$ iterations, this neural network is trained with the average mean square error loss on the $|\mathcal{O}|$ outputs (supervised learning) on the dataset D during 15 epochs with Adam optimizer [14] and learning rate 0.001.

4.2 Classic Fitness-Based Selectors

We compare the neural network selector above with three *basic* selectors, which do not take as input the raw solution to make their choice. These three selectors are two criteria based on a proportional bias (roulette wheel and pursuit) and one criterion based on the one-arm bandit method. These three criteria have extensively been used in the literature of hyperheuristics (e.g. [9,13,23]).

For these three operators, the reward r of each training example (S, o, r) is normalized between 0 and 1 over the sliding window consisting of the last N examples stored in the database D, 0 being the worst value obtained during the N last iterations (corresponding to the highest value of the fitness $f(S')$) and 1 being the best value. This normalized average reward computed on this sliding window and associated to an operator o is written r_o. n_o denotes the number of times the operator o have been selected during the last N iterations.

Adaptive Roulette Wheel and Pursuit. For these two probability based strategies inspired from [9], π_θ is a random procedure driven by a vector of $|\mathcal{O}|$ parameters $\theta = [p_1, \ldots, p_{|\mathcal{O}|}]$, such that p_i corresponds to the probability that the operator i is chosen this turn. We have $\sum_{i=1}^{|\mathcal{O}|} p_i = 1$. In the beginning, all the probabilities are set to equal value. Therefore, $p_i = \frac{1}{|\mathcal{O}|}$, for $i = 1, \ldots, |\mathcal{O}|$.

The more an operator o achieves good rewards r_o, the more its associated probability p_i increases and the more it is chosen in the following iterations.

Adaptive Roulette Wheel. In this strategy, the probabilities p_o for $o = 1, \ldots, |\mathcal{O}|$ are updated at each iteration according to the formula

$$p_o = p_{min} + (1 - |\mathcal{O}| * p_{min}) * \frac{r_o}{\sum_{o'}^{|\mathcal{O}|} r_{o'}},$$

where p_{min} is the minimum selection probability for each operator (which is set to a strictly positive value to ensure a minimum level of exploration). It is a hyperparameter of the methods set to the value of $\frac{1}{5 \times |\mathcal{O}|}$ in this work.

Pursuit. In this strategy, the probabilities are updated during the learning process as

$$\begin{cases} p_{o*} = p_{o*} + \beta(p_{max} - p_{o*}) \\ p_o = p_o + \beta(p_{min} - p_o), \end{cases} \tag{3}$$

where $o*$ is the index of the best operator on the sliding window, $p_{min} = \frac{1}{5 \times |\mathcal{O}|}$, $p_{max} = 1 - (|\mathcal{O}| - 1) * p_{min}$ and $\beta = 0.7$ is a coefficient introduced to control this winner-take-all strategy.

Note that unlike the adaptive roulette strategy, which gives more balanced chances to all low-level operators, the Pursuit strategy is more elitist as it gives more chances to the best strategy applied in the previous iterations.

One-Arm Bandit. This policy π_θ is based on the one-arm bandit method. At each iteration, according to the UCB (Upper Confidence Bound) formula, the operator with the highest score $s_o = r_o + c * \sqrt{\frac{2*\log|D|}{n_o+1}}$ is selected, where c is a hyperparameter set to the value of 1.

5 Experimentation

This section presents experiments to assess the different adaptive simulation strategies.

5.1 Experimental Settings and Benchmark Instances

We consider the 188 WVCP benchmark instances in the literature. These instances come from various problems, 35 pxx instances and 30 rxx instances from matrix decomposition [19] and 123 from DIMACS/COLOR competitions [22].

Among these instances, 124 have a known optimal score. Therefore, when evaluating a given method on such an instance, we stop it when the optimal score is reached. We perform 20 independent runs of each method to solve each instance on a computer equipped with Intel Xeon ES 2630, 2,66 GHz CPU. The time limit for each local search algorithm was one hour. For the combined methods, MCTS with local search, with and without adaptive selection, a time limit of one hour was set for local search, in order to compare all methods with the same overall time spent in the local search solvers. At each iteration of the MCTS algorithm, the time allowed in a local search (during the simulation phase) is set to $|V| * 0.02$ s.

The algorithm is coded in C++[5], compiled, and optimized with a g++ 12.1 compiler.

In the following, when a method is said to be better than another on a given instance, it means that the difference between the average scores computed over 20 runs is in favor of the first method and that this difference is significant (a non-parametric Wilcoxon signed-rank test with a p-value ≤ 0.001).

5.2 Adaptive Operator Selection During the Search

Figure 1 shows the mean of the cumulative selection of each operator with error bars during the 20 runs on four difficult instances for each criterion: Random, Roulette wheel, Pursuit, One-armed Bandit (UCB), and Neural network. The four local search algorithms that can be selected at each iteration of the MCTS (see Sect. 2.1) are TW (dash-dotted line in light blue), ILST-TS (solid line in red), RedLS (dashed line in green) and AFISA (dotted line in purple).

---

[5] The source code is available at https://github.com/Cyril-Grelier/gc_wvcp_adaptive_mcts.

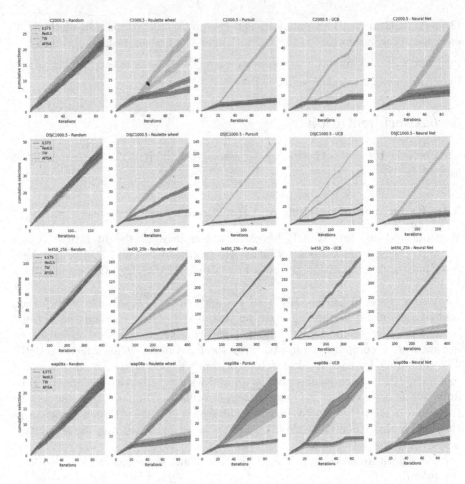

Fig. 1. Each plot represents the mean of the cumulative selection of operators for one criterion based on the 20 runs on one instance, with error bar. One line per instance (4 different instances), one column per adaptive method.

We first observe that during the first 20 iterations of the algorithm, the choices made by the various high-level strategies are almost random. It corresponds to the time spent collecting enough examples to learn.

Figure 1 shows that it is not always the same strategy that is favored for the different instances. For example, the method TW is preferred for the instance C2000.5, RedLS for DSJC1000.5 and ILSTS for le450_25b. The small error bars show that the selection is generally consistent over the different runs, except for wap08a for which the different strategies oscillate a lot between the ReldLS and ILSTS methods (which are the two best performing methods for this instance [17,24]) and thus for this instance the cumulative selection can vary a lot from one run to another. These curves highlight that the different operator selectors (except the Random selector) are in general able to identify the best performing method for each given instance.

One can observe different trends in the selection of operators depending on the policy. First, the random selection proceeds as expected and we observe that the average cumulative selection numbers are almost equal for all operators at different time steps. Roulette Wheel still keeps a lot of diversity in the selection of the different operators but learns a bias toward the best methods during the search. The UCB strategy behaves much like the Roulette wheel strategy. Note that a lower exploration vs. exploitation coefficient would reduce the number of times the worst operators are selected. The Pursuit and Neural network strategies have similar behaviors: in general the best operators for a given instance are selected in priority, the other ones are picked randomly. These two criteria are more elitists than the others.

5.3   Performance Comparisons on the Different Benchmark Instances

Table 1 displays performance comparisons between each pair of methods on the 188 benchmark instances considered in this work. The different types of methods are (i) the four standalone local search solvers AFISA, TW, RedLS, and ILSTS; (ii) the MCTS versions with a fixed local search solver used for simulation: MCTS+AFISA, MCTS+TW, MCTS+RedLS, and MCTS+ILSTS; (iii) the MCTS versions with adaptive simulation strategies: Random, Roulette Wheel, Pursuit, UCB, and Neural network.

The MCTS combined with a local search solver improves the results of the local search alone except for ILSTS. Among the different MCTS+LS versions, MCTS+RedLS is the most efficient.

Looking at the adaptive selection strategy, we observe that the random strategy always performed the worst compared to the other adaptive strategies. This highlights the relevance of dynamically learning the best strategy for each given instance during the search.

When comparing the different adaptive strategies, it appears that the methods Roulette wheel and UCB obtained almost the same results, and are significantly better than the random strategy. The best-performing strategies among all the compared strategies are Pursuit and Neural network. These methods are the most elitist. This indicates that it is beneficial to identify rapidly which method is efficient for each given instance and thus favor more intensification in the choice of the low-level heuristics to use at each iteration of the algorithm.

In general, although adaptive selection with the neural network performs well, comparative advantage over the more basic Pursuit method does not appear in these experiments. There seems to be no obvious advantage to choosing an operator depending on the raw state of the solution from which the search begins. A simple fitness-based selection strategy can be just as effective as with more information in this context.

Table 2 displays more detailed results for several methods tested in this work. Column 1 indicates the name of the instance. Due to space limitation, we show here the results for a set of 20 instances (over 188) of different types: size of the graph $|V|$ ranging from 125 (instance DSJC125.5gb) to 2000 (instance C2000.5),

Table 1. Comparison of all local search solvers and MCTS variants with and without adaptive selection. Each value corresponds to the number of instances where the row method is significantly better than the column method over the 188 benchmark instances considered (Wilcoxon signed-rank test with a p-value ≤ 0.001). A number is written in bold if the number of instances is higher for the method in the row.

	AFISA	MCTS+AFISA	TW	MCTS+TW	RedLS	MCTS+RedLS	ILSTS	MCTS+ILSTS	Random	Roulette Wheel	Pursuit	UCB	Neural network
AFISA	–	35	**48**	16	**56**	0	9	13	0	0	0	0	0
MCTS+AFISA	**40**	–	**72**	10	**86**	0	1	1	0	0	0	0	0
TabuWeight	39	38	–	26	47	2	16	23	2	2	2	2	2
MCTS+TabuWeight	**68**	71	**78**	–	**99**	5	8	17	0	0	0	0	0
RedLS	44	38	47	29	–	13	23	25	15	15	15	15	14
MCTS+RedLS	**102**	79	**105**	68	**102**	–	42	43	21	7	6	4	4
ILSTS	**87**	80	**86**	56	**83**	13	–	25	8	5	1	2	4
MCTS+ILSTS	**82**	78	**83**	48	**81**	11	0	–	5	2	0	1	2
Random	**103**	81	**103**	72	**98**	12	34	39	–	0	0	0	1
Roulette Wheel	**103**	81	**104**	73	**100**	13	35	40	4	–	1	0	1
Pursuit	**103**	81	**105**	75	**101**	14	35	40	11	0	–	1	2
UCB	**103**	81	**104**	75	**100**	13	36	40	4	0	1	–	1
NN	**103**	81	**104**	75	**101**	13	35	41	9	2	0	0	–

degree ranging from 0.1 (instance DSJC250.1) to 0.9 (instance DSJC250.9), as well as different weight and degree distributions. Column 2 shows the best-known score (BKS) for each given instance reported in the literature. A score in bold in this column indicates that the score has been proven optimal. Note that some of these best-known scores have been obtained under specific and relaxed conditions (computation time of more than one day and parallel computation on GPU device), while the results reported here are reached on a single CPU and with a time limit of one hour of local search. Column 3–26 display the results for one part of the different methods compared in this work. For the sake of space, the results of the versions AFISA, TW, MCTS+AFISA, MCTS+TW, and MCTS+ILSTS are not reported here. For each method are displayed the best and the average scores obtained over 20 runs, and the time spent in seconds to obtain the best scores.

The detailed results for all the 188 instances reached by the compared methods are available at the GitHub site given in footnote 5. The last lines of this Table 2 summarize the results obtained by all methods for all the 188 instances: (1) "# BKS" is the number of times a method finds the best know score of the literature on each given instance, (2) "# best average" is the number of times the method gets the best average score compared to the other methods. We first observe that the two best local search solvers RedLS and ILSTS perform well but not for the same type of instance. More specifically, the RedLS algorithm

Table 2. Part of the table reporting detailed results of the different methods for each instance of the benchmarks. For the sake of space, the results of the methods AFISA, TW, MCTS+AFISA, MCTS+TabuWeight, and MCTS+ILSTS are not reported here, and only the results for 20 instances over 188 are shown.

instance	BKS	RedLS			MCTS+RedLS			ILSTS			Random			Roulette Wheel			Pursuit			UCB			NN		
		best	avg	time	best	avg	time	best	avg	time	best	avg	time	best	avg	time	best	avg	time	best	avg	time	best	avg	time
C2000.5	2144	2175	2196	1624	2494	2508	2665	2459	2501	3365	2456	2463	1640	2439	2453	3567	2439	2453	1148	2438	2450	574	2441	2452	4529
DSJC125.5gb	240	245	260	0	241	241	906	240	241	1160	240	240	2477	240	240	2042	240	240	2073	240	240	1666	240	240	2120
DSJC250.1	127	130	131	4	127	127	2187	130	131	1361	127	128	1494	127	128	2154	127	128	1830	127	128	1598	127	127	2418
DSJC250.5	392	398	401	248	396	398	1980	398	408	3591	398	400	2046	397	399	4050	397	399	894	396	399	822	397	398	1238
DSJC250.9	934	934	935	1039	935	936	144	936	944	3214	937	938	2988	935	937	3234	936	937	822	937	937	1932	936	937	2728
DSJC500.1	184	187	202	1448	188	188	777	199	202	2174	189	189	1133	187	188	3124	188	188	1727	187	189	539	188	188	2672
DSJR500.1	169	171	184	0	169	169	222	169	169	0	169	169	19	169	169	18	169	169	13	169	169	21	169	169	15
GEOM120	72	72	75	0	72	72	12	72	72	0	72	72	5	72	72	4	72	72	4	72	72	4	72	72	3
le450.15a	212	214	236	604	213	214	2578	222	225	967	214	214	1188	213	214	126	213	214	1278	213	214	2043	213	214	1095
le450.15b	216	218	225	105	217	217	2700	225	227	2341	218	218	967	217	217	1395	217	217	2106	217	217	1953	217	217	1743
le450.25a	306	306	307	223	306	306	220	306	307	1743	306	306	771	306	306	663	306	306	590	306	306	710	306	306	1012
le450.25b	307	307	313	130	307	307	235	307	307	1444	307	307	688	307	307	610	307	307	609	307	307	988	307	307	771
queen15.15	223	227	229	1110	225	226	460	233	237	2493	225	227	480	226	226	2241	225	226	2196	226	227	1812	226	226	2203
wap07a	555	729	745	0	574	644	175	627	636	1942	574	629	1575	640	644	3290	637	643	1435	584	640	490	578	639	1129
wap08a	529	537	614	2257	549	551	2376	600	614	2563	551	554	432	551	553	1332	551	556	756	551	553	2124	551	553	3956
p42	2466	2466	2522	0	2466	2466	109	2466	2466	16	2466	2466	111	2466	2466	76	2466	2466	62	2466	2466	55	2466	2466	63
r28	9407	9410	9563	126	9407	9427	1917	9407	9407	16	9407	9407	111	9407	9407	155	9407	9407	101	9407	9407	133	9407	9407	223
r29	8693	8696	8817	2	8693	8693	972	8693	8693	206	8693	8693	585	8693	8693	714	8693	8693	403	8693	8693	459	8693	8693	590
r30	9816	9836	9988	4	9816	9823	2827	9816	9816	20	9816	9816	79	9816	9816	78	9816	9816	82	9816	9816	90	9816	9816	112
# BKS	112/188				153/188			151/188			154/188			155/188			156/188			156/188			157/188		
# best avg	46/188				144/188			141/188			139/188			142/188			155/188			146/188			152/188		

is better for large instances such as DSJC1000.5 and C2000.5, while ILSTS is better for smaller instances (DSJC250.1, DSJC250.5, and rxx instances). This shows that these methods can be complementary and that it may be beneficial to choose the right one on the fly for a given instance.

Secondly, we observe that the combination of the MCTS framework with e.g. the RedLS method (MCTS + RedLS version) improves the average results comparison to the RedLS solver alone for many instances, but not for the largest ones like DSJC1000.5 and C2000.5. This can be explained by the fact that the search space is very large for these two instances, and in this case it seems more advantageous to favor intensification in a limited area of the search space to get good results, rather than favoring more diversity with multiple restarts but with less time spent in different areas of the search space.

Thirdly, it appears that the two MCTS versions with adaptive simulation, such as Pursuit and Neural network, reach the largest number of BKS over the whole set of 188 instances, with good results for a wide variety of instances (such as DSJC250_X, DSJC500_X, le450_xx, and rxx instances).

6 Conclusions

A Monte Carlo Tree Search framework with adaptive simulation strategy was presented and tested on the Weighted Graph Coloring Problem. Different high-level operator selectors have been introduced in this work.

The results show that the MCTS versions with adaptive operator selection reach the highest number of best-known scores for the set of 188 benchmark instances of the literature compared to the state-of-the-art methods in one hour of computation time on a CPU (except compared to the DLMCOL algorithm [10] which uses a GPU and was run with an extended computation time).

Analysis of the operator selections during the search for each particular instance shows that, in general, the choice of operators does not change during the search. In fact, once the best solver for each instance has been identified, it is usually still chosen for the rest of the search. This lack of variation during the search may be explained by the fact that for a given instance, there is usually a dominant solver and we do not observe complementarity in the use of the different operators during the search for this problem. This may explain why the neural network selector taking into account the raw state of the current solution does not bring better results than a more basic fitness-based selector such as the Pursuit strategy.

A future work would be to introduce this new neural network operator selection strategy into a memetic algorithm for graph coloring problems where the different operators selected can be more complementary during the search (e.g. local search and crossover operators).

Acknowledgment. We would like to thank Dr. Wen Sun [22], Dr. Yiyuan Wang, [24] and Pr. Bruno Nogueira [17] for sharing their codes. This work was granted access to the HPC resources of IDRIS (Grant No. 2020-A0090611887, 2022-A0130611887) from GENCI and the Centre Régional de Calcul Intensif des Pays de la Loire (CCIPL). We are grateful to the reviewers for their comments.

Contribution Statement. C. Grelier developed the code and performed the tests. O. Goudet and J.K. Hao planned and supervised the work. All the authors contributed to the analysis and the writing of the article.

References

1. Asta, S., Karapetyan, D., Kheiri, A., Özcan, E., Parkes, A.J.: Combining Monte-Carlo and hyper-heuristic methods for the multi-mode resource-constrained multi-project scheduling problem. Inf. Sci. **373**, 476–498 (2016)
2. Bouziri, H., Mellouli, K., Talbi, E.G.: The k-coloring fitness landscape. J. Comb. Optim. **21**(3), 306–329 (2011)
3. Burke, E.K., et al.: Hyper-heuristics: a survey of the state of the art. J. Oper. Res. Soc. **64**(12), 1695–1724 (2013)
4. Cai, S., Su, K., Sattar, A.: Local search with edge weighting and configuration checking heuristics for minimum vertex cover. Artif. Intell. **175**(9), 1672–1696 (2011)
5. Cornaz, D., Furini, F., Malaguti, E.: Solving vertex coloring problems as maximum weight stable set problems. Discret. Appl. Math. **217**, 151–162 (2017)
6. Dantas, A., Rego, A.F.d., Pozo, A.: Using deep Q-network for selection hyper-heuristics. In: Proceedings of the Genetic and Evolutionary Computation Conference Companion, pp. 1488–1492 (2021)
7. Drake, J.H., Kheiri, A., Özcan, E., Burke, E.K.: Recent advances in selection hyper-heuristics. Eur. J. Oper. Res. **285**(2), 405–428 (2020)
8. Elhag, A., Özcan, E.: A grouping hyper-heuristic framework: application on graph colouring. Expert Syst. Appl. **42**(13), 5491–5507 (2015)
9. Goëffon, A., Lardeux, F., Saubion, F.: Simulating non-stationary operators in search algorithms. Appl. Soft Comput. **38**, 257–268 (2016)

10. Goudet, O., Grelier, C., Hao, J.K.: A deep learning guided memetic framework for graph coloring problems. Knowl.-Based Syst. **258**, 109986 (2022)

11. Grelier, C., Goudet, O., Hao, J.-K.: On Monte Carlo tree search for weighted vertex coloring. In: Pérez Cáceres, L., Verel, S. (eds.) EvoCOP 2022. LNCS, vol. 13222, pp. 1–16. Springer, Cham (2022). https://doi.org/10.1007/978-3-031-04148-8_1

12. Hertz, A., Werra, D.: Using tabu search techniques for graph Coloring. Computing **39**, 345–351 (1987)

13. Hoos, H.H.: Automated algorithm configuration and parameter tuning. In: Hamadi, Y., Monfroy, E., Saubion, F. (eds.) Autonomous Search, pp. 37–71. Springer, Heidelberg (2011). https://doi.org/10.1007/978-3-642-21434-9_3

14. Kingma, D.P., Ba, J.: Adam: a method for stochastic optimization. arXiv preprint arXiv:1412.6980 (2014)

15. Lucas, T., Tallec, C., Ollivier, Y., Verbeek, J.: Mixed batches and symmetric discriminators for GAN training. In: International Conference on Machine Learning, pp. 2844–2853 (2018)

16. Malaguti, E., Monaci, M., Toth, P.: Models and heuristic algorithms for a weighted vertex coloring problem. J. Heuristics **15**(5), 503–526 (2009)

17. Nogueira, B., Tavares, E., Maciel, P.: Iterated local search with tabu search for the weighted vertex coloring problem. Comput. Oper. Res. **125**, 105087 (2021)

18. Pemmaraju, S.V., Raman, R.: Approximation algorithms for the max-coloring problem. In: Caires, L., Italiano, G.F., Monteiro, L., Palamidessi, C., Yung, M. (eds.) ICALP 2005. LNCS, vol. 3580, pp. 1064–1075. Springer, Heidelberg (2005). https://doi.org/10.1007/11523468_86

19. Prais, M., Ribeiro, C.C.: Reactive GRASP: an application to a matrix decomposition problem in TDMA traffic assignment. INFORMS J. Comput. **12**(3), 164–176 (2000)

20. Sabar, N.R., Ayob, M., Qu, R., Kendall, G.: A graph coloring constructive hyper-heuristic for examination timetabling problems. Appl. Intell. **37**(1), 1–11 (2012)

21. Sabar, N.R., Kendall, G.: Population based Monte Carlo tree search hyper-heuristic for combinatorial optimization problems. Inf. Sci. **314**, 225–239 (2015)

22. Sun, W., Hao, J.K., Lai, X., Wu, Q.: Adaptive feasible and infeasible tabu search for weighted vertex coloring. Inf. Sci. **466**, 203–219 (2018)

23. Thierens, D.: An adaptive pursuit strategy for allocating operator probabilities. In: Proceedings of the 7th Annual Conference on Genetic and Evolutionary Computation, pp. 1539–1546 (2005)

24. Wang, Y., Cai, S., Pan, S., Li, X., Yin, M.: Reduction and local search for weighted graph coloring Problem. In: Proceedings of the AAAI Conference on Artificial Intelligence, vol. 34, no. 0303, pp. 2433–2441 (2020)

25. Zaheer, M., Kottur, S., Ravanbakhsh, S., Poczos, B., Salakhutdinov, R.R., Smola, A.J.: Deep sets. In: Advances in Neural Information Processing Systems, vol. 30 (2017)

Evolutionary Strategies for the Design of Binary Linear Codes

Claude Carlet[1,2], Luca Mariot[3(✉)], Luca Manzoni[4], and Stjepan Picek[5]

[1] Department of Mathematics, Université Paris 8, 2 Rue de la Liberté,
93526 Saint-Denis Cedex, France
[2] University of Bergen, Bergen, Norway
[3] Semantics, Cybersecurity & Services Group, University of Twente, Drienerlolaan 5,
7522, NB Enschede, The Netherlands
l.mariot@utwente.nl
[4] Department of Mathematics and Geosciences, University of Trieste,
Via Valerio 12/1, Trieste, Italy
lmanzoni@units.it
[5] Digital Security Group, Radboud University, PO Box 9010, Nijmegen,
The Netherlands
stjepan.picek@ru.nl

Abstract. The design of binary error-correcting codes is a challenging optimization problem with several applications in telecommunications and storage, which has been addressed with metaheuristic techniques such as evolutionary algorithms. Still, all these efforts are focused on optimizing the minimum distance of unrestricted binary codes, i.e., with no constraints on their linearity, which is a desirable property for efficient implementations. In this paper, we present an Evolutionary Strategy (ES) algorithm that explores only the subset of linear codes of a fixed length and dimension. We represent the candidate solutions as binary matrices and devise variation operators that preserve their ranks. Our experiments show that up to length $n = 14$, our ES always converges to an optimal solution with a full success rate, and the evolved codes are all inequivalent to the Best-Known Linear Code (BKLC) given by MAGMA. On the other hand, for larger lengths, both the success rate of the ES as well as the diversity of the codes start to drop, with the extreme case of $(16, 8, 5)$ codes which all turn out to be equivalent to MAGMA's BKLC.

Keywords: Error-correcting codes · Boolean functions · Algebraic normal form · Evolutionary strategies · Variation operators

1 Introduction

A central problem in information theory is the transmission of messages over noisy channels. To this end, error-correcting codes are usually employed to add redundancy to a message before sending it over a channel. A common setting is

L. Pérez Cáceres and T. Stützle (Eds.): EvoCOP 2023, LNCS 13987, pp. 114–129, 2023.
https://doi.org/10.1007/978-3-031-30035-6_8

to consider messages over the binary alphabet $\mathbb{F}_2 = \{0, 1\}$, under the hypothesis of a binary symmetric channel [1]. To be useful in practice, a binary code must have the following properties: (a) a high minimum Hamming distance, (b) a high number of codewords, and (c) an efficient decoding algorithm. While (a) and (b) induce a direct trade-off, property (c) is usually addressed by requiring that the code is linear, i.e., that it forms a k-dimensional subspace of $\mathbb{F}_2^n$.

The design of a good binary code is a combinatorial optimization problem where the objective is to maximize the minimum distance of a set of codewords, and it is equivalent to finding a maximum clique in a graph [2]. Several optimization algorithms have been applied to solve this problem, including evolutionary algorithms [3–6] and other metaheuristics [5,7–10]. Most of these works target the construction of unrestricted binary codes, without any linearity requirement. The only exception is [4], where genetic algorithms are used to evolve the generator matrices of linear codes, but without preserving their ranks; thus, the dimension of the evolved codes can vary.

In this paper, we propose for the first time an Evolutionary Strategy (ES) algorithm for the design of binary linear codes with the best possible minimum distance d for a given combination of length n and dimension k. We adopt a combinatorial representation that allows the ES to explore only the set of k-dimensional subspaces of $\mathbb{F}_2^n$. The ES encodes a candidate solution as a $k \times n$ binary generator matrix of full rank k. On account of a recent result proved in [11], the fitness of a matrix is evaluated as the number of monomials of degree less than d in the Algebraic Normal Form (ANF) of the indicator Boolean function of the linear code. Next, for the variation operators, we consider both a classical ES using only mutation and a variant combining mutation and crossover. Both types of operators are designed so that the rank of the offspring matrices is preserved.

We evaluate experimentally our approach over five different instances of linear codes for lengths ranging from $n = 12$ to $n = 16$ and dimension set to $k = \lfloor \frac{n}{2} \rfloor$, seeking to reach the bounds on the minimum distance reported in [12]. We test four different versions of our ES algorithm, depending on the replacement strategy (either (μ, λ) or $(\mu+\lambda)$) and whether the crossover is applied or not. All variants achieve a full success rate up to length $n = 14$. Somewhat surprisingly, for the larger instances $n = 15$ and $n = 16$, the simple (μ, λ) without crossover scores the best performance. Indeed, we observe that the average fitness and distance of the population in the $(\mu + \lambda)$ variants quickly converge to a highly fit, low-diversity area of the search space.

Finally, we investigate the diversity of the codes evolved by our ES algorithm up to isomorphism. In particular, we test how many of our codes are inequivalent to the Best-Known Linear Code (BKLC) constructed through the MAGMA computer algebra system, and we group them into equivalence classes. Interestingly, for lengths up to $n = 14$, all codes turn out to be inequivalent to MAGMA's BKLC, and they belong to a high number of equivalence classes. The situation is, however, reversed for the larger instances: while for $n = 15$, there is still a good proportion of inequivalent codes, for $n = 16$, all codes are equivalent to MAGMA's BKLC. Therefore, our ES essentially converges to the same solution.

2   Background .

In this section, we cover all background information related to binary linear codes and Boolean functions that we will use in the paper. The treatment is far from exhaustive, and we refer the reader to [1,13] and [14] for a more complete survey of the main results on error-correcting codes and Boolean functions, respectively.

2.1   Binary Linear Codes

Let $\mathbb{F}_2 = \{0, 1\}$ denote the finite field with two elements. For any $n \in \mathbb{N}$, the n-dimensional vector space over $\mathbb{F}_2$ is denoted by $\mathbb{F}_2^n$, where the sum of two vectors $x, y \in \mathbb{F}_2^n$ corresponds to their bitwise XOR, while the multiplication of x by a scalar $a \in \mathbb{F}_2$ is defined as the logical AND of a with each coordinate of x. The Hamming distance $d_H(x, y)$ of two vectors $x, y \in \mathbb{F}_2^n$ is the number of coordinates where x and y disagree. Given $x \in \mathbb{F}_2^n$, the Hamming weight of x is the number of its nonzero coordinates, or equivalently the Hamming distance $d_H(x, \underline{0})$ between x and the null vector $\underline{0}$.

A binary code of length n is any subset C of $\mathbb{F}_2^n$. The elements of C are also called codewords, and the size of C is usually denoted by M. The minimum distance d of C is the minimum Hamming distance between any two codewords $x, y \in C$. One of the main problems in coding theory is to determine what is the maximum number of codewords $A(n, d)$ that a code C of length n and minimum distance d can have. Several theoretical bounds exist on $A(n, d)$. For example, the Gilbert-Varshamov bound, and the Singleton bound, respectively, give a lower and an upper bound on $A(n, d)$ as follows:

$$\frac{2^n}{\sum_{i=0}^{d-1} \binom{n}{i}} \leq A(n, d) \leq 2^{n-d+1} . \tag{1}$$

More refined bounds exist, such as the Hamming bound, for which we refer the reader to [13]. A slightly different but equivalent problem is to fix the length n and size M of a code and then maximize the resulting minimum distance d according to an analogous upper bound.

A binary code $C \subseteq \mathbb{F}_2^n$ is called linear if it forms a vector subspace of $\mathbb{F}_2^n$. In this case, the size of the code can be compactly described by the dimension $k \leq n$ of the subspace. Indeed, the encoding process amounts to multiplying a k-bit vector $m \in \mathbb{F}_2^k$ by a $k \times n$ generator matrix G_C which spans the code C (and, therefore, G_c has rank k). The resulting n-bit vector $c = mG_C \in C$ will be the codeword corresponding to the message m. Thus, the size of C is $M = 2^k$. A linear code of length n, dimension k, and minimum distance d is also denoted as an (n, k, d) code. Some of the bounds mentioned above can be simplified if one is dealing with a linear code. For example, the Singleton bound for a (n, k, d) linear code becomes

$$k \leq n - d + 1 , \tag{2}$$

which also gives an upper bound on the minimum distance d as $d \leq n - k + 1$.

2.2 Boolean Functions

A Boolean function of $n \in \mathbb{N}$ variables is a mapping $f : \mathbb{F}_2^n \to \mathbb{F}_2$, i.e., a function associating to each n-bit input vector a single output bit. The support of f is defined as $supp(f) = \{x \in \mathbb{F}_2^n : f(x) \neq 0\}$, i.e., the set of all input vectors that map to 1 under f. The most common way to represent a Boolean function is by means of its truth table: assuming that a total order is fixed on $\mathbb{F}_2^n$ (e.g., the lexicographic order), then the truth table of f is the 2^n-bit vector specifying for each input vector $x \in \mathbb{F}_2^n$ the corresponding output value $f(x) \in \mathbb{F}_2$.

A second representation method for Boolean functions commonly used in cryptography and coding theory is the Algebraic Normal Form (ANF). Given $f : \mathbb{F}_2^n \to \mathbb{F}_2$, the ANF of f is the multivariate polynomial in the quotient ring $\mathbb{F}_2[x_1, \ldots, x_n]/(x_1 \oplus x_1^2, \ldots, x_n \oplus x_n^2)$ defined as follows:

$$P_f(x) = \bigoplus_{I \in 2^{[n]}} a_I \left(\prod_{i \in I} x_i \right) , \tag{3}$$

with $2^{[n]}$ being the power set of $[n] = \{1, \cdots, n\}$, and I being any element of $2^{[n]}$. The coefficients $a_I \in \mathbb{F}_2^n$ of the ANF can be computed from the truth table of f via the Möbius transform:

$$a_I = \bigoplus_{x \in \mathbb{F}_2^n : supp(x) \subseteq I} f(x) , \tag{4}$$

where $supp(x) = \{i \in [n] : x_i \neq 0\}$ denotes the support of x, or equivalently the set of nonzero coordinates of x.

The degree of a coefficient a_I corresponds to the size of I (that is, to the number of variables in the corresponding monomials). Then, the algebraic degree of f is defined as the largest monomial occurring in the ANF of f, or equivalently as the cardinality of the largest $I \in 2^{[n]}$ such that $a_I \neq 0$.

The algebraic normal form of Boolean functions can be used to characterize the minimum distance of binary linear codes, as shown by C. Carlet [11]:

Proposition 1. *Let* $C \subseteq \mathbb{F}_2^n$ *be a* (n, k, d) *binary linear code, and define the indicator of* C *as the Boolean function* $1_C : \mathbb{F}_2^n \to \mathbb{F}_2$ *whose support coincides with the code, i.e.,* $supp(1_C) = C$. *Then,*

$$d = \min\{|I| \in 2^{[n]} : a_I = 0\} , \tag{5}$$

where a_I *denotes the coefficients of the ANF of* 1_C.

In other words, one can check if a binary linear code of length n and dimension k has minimum distance d by verifying that all monomials of degree strictly less than d appear in the ANF of the indicator function 1_C, while the smallest non-occurring monomial has degree d. This observation will be used in the next sections to define a fitness function for our optimization problem of interest.

3   Related Works

El Gamal et al. [7] were the first to investigate the design of unrestricted codes by simulated annealing (SA). Their results showed that SA was capable of finding many new constant-weight and spherical codes, in some cases improving on the known bounds for $A(n, d)$. The first application of Genetic Algorithms (GAs) to evolve error-correcting codes with maximal distance was proposed by Dontas and De Jong [3]. The authors encoded a candidate solution as a bitstring of length $n \cdot M$, representing the concatenation of M codewords of length n, and maximized two fitness functions based on the pairwise Hamming distance between codewords. Using the same encoding and fitness, Chen et al. [5] developed a hybrid algorithm combining GA and SA to design error-correcting codes.

McGuire and Sabin [4] employed a GA to search for linear codes. To enforce the linearity of the evolved codes, the genotype of the candidate solutions were $k \cdot n$ bitstrings, which represented the concatenation of the rows of $k \times n$ generator matrices. However, the authors used crossover and mutation operators that do not preserve the ranks of the resulting matrices. Therefore, the dimension of the codes evolved by their GA is not fixed, contrary to what is claimed in the paper.

Alba and Chicano [8] designed a so-called Repulsion Algorithm (RA) for the error-correcting codes problem that takes inspiration from electrostatic phenomena. In particular, the codewords are represented as particles obeying Coulomb law, and their next position on the Hamming cube is determined by computing the resultant force vectors. Cotta [9] experimented with several combinations of Scatter Search (SS) and Memetic Algorithms (MAs) to evolve codes of length up to $n = 20$ and $M = 40$ codewords. The results indicated that both SS and MAs could outperform other metaheuristics on this problem.

Blum et al. [10] investigated an Iterated Local Search (ILS) technique and combined it with a constructive heuristic to design error-correcting codes, showing that it achieved state-of-the-art performances. McCarney et al. [6] considered GAs and Genetic Programming (GP) to evolve binary codes, where the chromosome's genes represent the entire codeword rather than a single symbol, as in most other approaches. The results on codes of length $n = 12, 13$, and 17 and minimum distance 6 suggested that GP can outperform GA on this problem.

A few works address the construction of combinatorial designs that are analogous to error-correcting codes through evolutionary algorithms. For example, Mariot et al. [15] employed GA and GP to evolve binary orthogonal arrays, which are equivalent to binary codes. Knezevic et al. [16] considered using Estimation of Distribution Algorithms (EDAs) to design disjunct matrices, which can be seen as superimposed codes. More recently, Mariot et al. [17] proposed an evolutionary algorithm for the incremental construction of a permutation code, where the codewords are permutations instead of binary vectors.

Although in this paper we use Boolean functions to compute the fitness, we note that their construction via metaheuristics is also a solid research thread, especially concerning the optimization of their cryptographic properties. We refer the reader to [18] for a survey of the main results in this area.

4   Evolutionary Strategy Algorithm

In this section, we describe the main components of our Evolutionary Strategy (ES) algorithm used to evolve binary linear codes. As a reference for later, we formally state the optimization problem that we are interested in as follows:

Problem 1. Let $n, k \in \mathbb{N}$ with $k \leq n$. Find a (n, k, d) binary linear code $C \subseteq \mathbb{F}_2^n$ reaching the highest possible minimum distance d.

Remark that the upper and lower bounds on the highest minimum distance mentioned in Sect. 2.1 are not tight in general. However, for binary codes of length up to $n = 256$, one can use the tables provided by Grassl [12] to determine the best-known values. In particular, for the combinations of n and k that we will consider in our experiments in Sect. 5, the lower and upper bounds on d coincide. So, Problem 1 is well-defined for all instances considered in our tests.

4.1   Solutions Encoding and Search Space

As we discussed in Sect. 3, most of the works addressing the design of error-correcting codes via metaheuristic algorithms usually target generic codes without any constraint on their linearity. The only exception seems to be the paper by McGuire and Sabin [4] where a GA evolves a generator matrix, but there is no control on the dimension k of the corresponding code. As a matter of fact, if one applies unrestricted variation operators on a matrix of rank k (such as one-point crossover or bit-flip mutation), then the vectors in the resulting matrix might not be linearly independent, and thus the associated code could have a lower dimension. This might, in turn, pose issues because the optimal value of the fitness function (which is related to the best-known minimum distance for a given combination of n and k) can change during the evolution process.

In our approach, we consider the genotype of a candidate solution S as a $k \times n$ binary matrix G of full rank k. Accordingly, the phenotype corresponding to G is the code C which is the image of the linear map defined by G. Formally, we can define the phenotype code as:

$$C = \{c \in \mathbb{F}_2^n : c = x \cdot G, \ x \in \mathbb{F}_2^k\} \ . \tag{6}$$

Therefore, the search space $\mathcal{S}_{n,k}$ for a given combination of code length n and dimension k is the set of all $k \times n$ binary matrices of rank k, or equivalently the set of all k-dimensional subspaces of $\mathbb{F}_2^n$, also called the Grassmannian $Gr(\mathbb{F}_2^n, k)$ of $\mathbb{F}_2^n$. It is known that the size of this space equals the Gaussian binomial coefficient $\binom{n}{k}_2$, defined as [19]:

$$\binom{n}{k}_2 = \frac{(2^n - 1)(2^{n-1} - 1) \cdots (2^{n-k+1} - 1)}{(2^k - 1)(2^{k-1} - 1) \cdots (2^{k-(k-2)} - 1)} \ . \tag{7}$$

It is clear from the expression above that the size of the Grassmannian grows very quickly, and thus exhaustive search in this space becomes unfeasible already for small values of n and k. For example, setting $n = 12$ and $k = 6$, the corresponding Grassmannian is composed of about $2.3 \cdot 10^{11}$ (230 billion) subspaces. This observation provides a basic argument motivating the use of heuristic optimization algorithms to solve Problem 1.

In what follows, we will endow the Grassmannian $Gr(\mathbb{F}_2^n, k)$ with a distance, turning it into a metric space. This will be useful to study the diversity of the population evolved by the ES. In particular, let $A, B \subseteq \mathbb{F}_2^n$ be two k-dimensional subspaces of $\mathbb{F}_2^n$. Then, the distance between A and B equals:

$$d(A, B) = dim(A) + dim(B) - 2dim(A \cap B) = 2(k - dim(A \cap B)) . \qquad (8)$$

This distance was introduced by Kötter and Kschischang in [20] to study error-correction in the setting of random network linear coding, where the transmitted codewords are not vectors of symbols, but rather vector subspaces themselves. Hence, the problem becomes to find a set of subspaces in the projective space of $\mathbb{F}_2^n$ (that is, the set of all subspaces of $\mathbb{F}_2^n$), which are far apart from each other under the distance defined in Eq. (8). In particular, this distance is the length of a geodesic path joining the two subspaces A and B when seen as points on the poset (partially ordered set) of the projective space of $\mathbb{F}_2^n$, where the elements are ordered with respect to subset inclusion. In our case, the search space $\mathcal{S}_{n,k} = Gr(\mathbb{F}_2^n, k)$ corresponds to the antichain of this poset that includes all k-dimensional subspaces of $\mathbb{F}_2^n$.

4.2   Fitness Function

Many of the works surveyed in Sect. 3 optimize binary codes by maximizing a fitness function that directly measures the pairwise Hamming distance between the codewords. In this work, on the other hand, we define a new fitness function that is based on the characterization of the minimum distance in terms of the ANF of the indicator function proved in [11]. In particular, we use the fitness function to count the number of coefficients of degree strictly less than d that occur in the ANF of the indicator, with the objective of maximizing it. Formally, given a code C of length n and dimension k, and denoting by a_I the monomial in the ANF of the indicator function 1_C for $I \in 2^{[n]}$, the fitness of C is defined as:

$$fit(C) = \{ I \in 2^{[n]} : |I| < d, \ a_I \neq 0 \} . \qquad (9)$$

By Proposition 1, C is a (n, k, d) linear code if and only if all coefficients of degree less than d are in the ANF of its indicator, and the optimal value for fit is:

$$fit_{n,d}^* = \sum_{i=0}^{d-1} \binom{n}{i} . \qquad (10)$$

To summarize, the fitness value of a solution $G \in \mathcal{S}_{n,k}$ is computed as follows:

1. Generate the linear code C as the subspace spanned by the matrix G.
2. Write the truth table of the indicator function $1_C : \mathbb{F}_2^n \to \mathbb{F}_2$ by setting $f(x) = 1$ if $x \in C$, and zero otherwise.

3. Compute the ANF of 1_C using the Fast Möbius Transform algorithm.
4. Compute $fit(C)$ using Eq. (9).

Although the Fast Möbius Transform yields a significant improvement in the time complexity required to compute the ANF over the naive procedure, it is still computationally cumbersome to compute it for Boolean functions with a relatively high number of variables. For this reason, in our experiments, we limit ourselves to linear codes of length up to $n = 16$.

4.3 Rank-Preserving Mutation and Crossover

We now describe the variation operators that we employed to generate new candidate solutions from a population of k-dimensional subspaces of $\mathbb{F}_2^n$, represented by their generator matrices.

For mutation, we adopt the same operator proposed in [21]: there, the authors were interested in evolving an invertible binary matrix that was used to define an affine transformation for a bent Boolean function. The operator can be straightforwardly adapted to our problem, even though we are not dealing with invertible matrices. Indeed, the basic principle of [21] is to preserve the invertibility of a square $n \times n$ binary matrix by keeping its rank equal to n. In our case, we use the same idea to maintain the rank of a rectangular matrix. Specifically, given a $k \times n$ binary matrix G of full rank k and a row $i \in [k]$, the mutation operator samples a random number $r \in (0, 1)$ and checks whether it is smaller than the mutation probability p_{mut}. If this is the case, G is mutated as follows:

1. Remove the i-th row of G, obtaining a $(k-1) \times n$ matrix G' of rank $k-1$.
2. Generate the subspace spanned by G', denoted as $span(G')$, and compute the complement $\mathcal{C} = \mathbb{F}_2^n \setminus span(G')$.
3. Pick a random vector $v \in \mathcal{C}$ and insert it in G' as a row in position i, obtaining the mutated $k \times n$ matrix H.

By construction, the random vector v sampled in step 3 is linearly independent with all vectors in the span of G'. Therefore, the mutated matrix H has the same rank k as the original matrix G. The process is then repeated for all rows $i \in [k]$. One can notice that the two matrices G and H are at distance 2 under the metric of Eq. (8), which is the minimum possible for distinct points in the Grassmannian $Gr(\mathbb{F}_2^n, k)$ [20]. Therefore, this operator effectively perturbs a candidate solution by transforming it into one of its closest neighbors.

Given two $k \times n$ parent matrices G_1, G_2 of rank k, our crossover operator generates an offspring matrix H in the following way:

1. Concatenate the rows of G_1 and G_2, thus obtaining a $2k \times n$ matrix J.
2. Perform a random shuffle of the rows in J.
3. Generate H by selecting a subset of k linearly independent vectors from J.

Step 3 is performed incrementally: the offspring matrix H is filled by adding the rows of J from top to bottom, checking if the current row is linearly independent with all previously added ones. If it is not, then the row is discarded, and the next one is attempted. Notice that it is always possible to find a set of k linearly independent vectors in J to construct H since both G_1 and G_2 have rank k. Thus, the worst case arises when all rows of one of the parents are in the span of the other (i.e., G_1 and G_2 generate the same subspace). In this situation, the offspring will also end up spanning the same subspace, although the generator matrix might look different from the parents. From a linear algebraic point of view, this eventuality corresponds to a change of basis on the same subspace. By the above argument, it follows that also the crossover operator preserves the rank k of the parents in the offspring.

5   Experiments

5.1   Experimental Setting

Evolutionary strategies are specified by two main parameters, the population size λ and the reproduction pool size μ [22]. At each generation, the μ best parents in the population are selected for reproduction (truncation selection). Then, in the (μ, λ)-ES variant each selected individual generates μ/λ offspring individuals, and their fitness is evaluated. In this case, the new offspring entirely replaces the old population, and the process is then iterated. The $(\mu + \lambda)$-ES variant differs from the fact that the μ parents from the old population are brought into the new population. To keep the population size fixed to λ in the $(\mu + \lambda)$-ES variant in our experiments, we generate $(\lambda/\mu) - 1$ offspring individuals for each selected parent. In classical ES, the parents usually create the offspring only by applying a mutation operator. We also considered a variant of ES augmented with crossover, which works as follows: each parent generates an offspring matrix by first performing crossover with a random mate selected from the reproduction pool of the μ best individuals. Then, mutation is applied as usual. Therefore, in our experiments we considered four variants of ES, depending on the replacement mechanism ((μ, λ) or $(\mu + \lambda)$), and whether crossover (χ) is applied or not.

Concerning the combinations of length n, dimension k, and minimum distance d of the codes, we experimented over five problem instances: $(12, 6, 4)$, $(13, 6, 4)$, $(14, 7, 4)$, $(15, 7, 5)$, and $(16, 8, 5)$. In particular, we always set $k = \lfloor n/2 \rfloor$ since this gives the largest search space possible for a given n. Starting from $n = 12$ yields the smallest instance that is not amenable to exhaustive search. The corresponding minimum distance d (that represents the optimization objective) has been taken from the tables reported in [12]. In all these cases, the lower and upper bounds on d coincide, so these are the best minimum distances one can get for these combinations of n and k. For the ES parameters, we set the population size λ equal to the length n, and $\mu = \lfloor n/3 \rfloor$. The mutation probability was set to $p_{mut} = 1/n$. These are the same parameters settings adopted for the ES in [21] to evolve invertible binary matrices, and after a preliminary tuning phase we noticed that they also worked well on Problem 1. The fitness budget was set to 20 000 generations of the ES algorithm, since we remarked that the best fitness seldom improved after this threshold. Finally, we repeated each experiment over 100 independent runs to get statistically sound results.

Table 1. Summary of the parameter settings, search space size, and best fitness value for each problem instance.

(n, k, d)	$\#\mathcal{S}_{n,k}$	$fit^*_{n,d}$	λ	μ	p_{mut}
$(12, 6, 4)$	$2.31 \cdot 10^{11}$	299	12	4	0.083
$(13, 6, 4)$	$1.49 \cdot 10^{13}$	378	13	4	0.077
$(14, 7, 4)$	$1.92 \cdot 10^{15}$	470	14	4	0.071
$(15, 7, 5)$	$2.47 \cdot 10^{17}$	1941	15	5	0.067
$(16, 8, 5)$	$6.34 \cdot 10^{19}$	2517	16	5	0.063

Table 1 summarizes our experimental design with all relevant parameters, along with the size of the corresponding search space $\mathcal{S}_{n,k}$ and the best fitness value for each considered problem instance.

5.2 Results

Table 2 reports the success rates of the four ES variants over 100 independent runs for the five considered problem instances, that is, how many times they converged to an optimal linear code. We denote a crossover-augmented ES variant by appending $+\chi$ to it.

Table 2. Success rates (over 100 runs) of the four considered ES variants.

(n, k, d)	(μ, λ)-ES	$(\mu, \lambda)+\chi$-ES	$(\mu + \lambda)$-ES	$(\mu + \lambda)+\chi$-ES
$(12, 6, 4)$	100	100	100	100
$(13, 6, 4)$	100	100	100	100
$(14, 7, 4)$	100	100	100	100
$(15, 7, 5)$	100	100	77	81
$(16, 8, 5)$	92	76	18	17

The first interesting remark is that all ES variants always converge to an optimal solution up to length $n = 14$, seemingly indicating that Problem 1 is rather easy on these problem instances, independently of the replacement mechanism and the use of crossover. For $(15, 7, 5)$, the (μ, λ) variants still converge in all runs, while the $(\mu + \lambda)$-ES achieve a lower success rate, although still quite high. The biggest difference can be seen on the largest problem instance $(16, 8, 5)$. In this case, the only variant reaching a very high success rate of 92% is the (μ, λ)-ES. Somewhat surprisingly, adding crossover to this variant actually worsens the performance. On the other hand, the $(\mu + \lambda)$-ES variants reach a very low success rate on this instance, independently of crossover. Therefore, in general the main factor influencing the performance is the replacement mechanism, rather than crossover. Apparently, letting the parents directly compete with their children as in the $(\mu + \lambda)$ variant is detrimental for this particular optimization problem.

To investigate more in detail the effects of the replacement mechanism and the crossover operator, we plotted the distributions of the number of fitness evaluations in Fig. 1.

In general, one can see that the number of fitness evaluations necessary to converge to an optimal solution is not directly correlated with the length of the code,

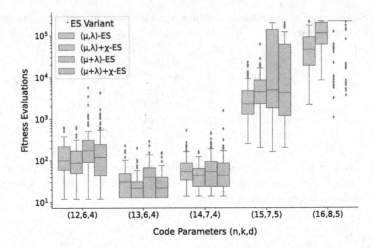

Fig. 1. Fitness evaluation distributions for the four considered ES variants.

and consequently with the size of the search space. As a matter of fact, the median number of evaluations of $(12, 6, 4)$ is always higher than that required for $(13, 6, 4)$ and $(14, 7, 4)$. Indeed, the most evident correlation is with the minimum distance, since for the two largest instances with $d = 5$ the number of fitness evaluations is significantly higher. This is reasonable, since as reported in Table 1, the optimal fitness values for $d = 5$ are consistently greater than for $d = 4$.

As expected, for the larger instances $(15, 7, 5)$ and $(16, 8, 5)$, a clear difference emerges between the two replacement mechanisms, as already indicated by the success rates. The (μ, λ) variants converge to an optimal solution more quickly than the $(\mu + \lambda)$ ones. On the other hand, up to $(14, 7, 4)$ it is not possible to distinguish the performances of the four evolutionary strategies by just looking at the respective boxplots. For this reason, we used the Mann-Whitney-Wilcoxon statistical test to compare two ES variants, with the alternative hypothesis that the corresponding distributions are not equal, and setting the significance value to $\alpha = 0.05$. The obtained p-values show that the (μ, λ)-ES variants give an advantage over the $(\mu + \lambda)$-ES without crossover for $(12, 6, 4)$, while for $(13, 6, 4)$, only $(\mu, \lambda) + \chi$-ES is significantly better than $(\mu + \lambda)$-ES ($p = 0.007$). For $(14, 7, 4)$, there is no significant difference between any two combinations of ES. Another interesting insight from the statistical test concerns the effect of the crossover operator. While for the instances up to $(14, 7, 4)$ there is no significant difference whether the ES is augmented with crossover or not (with the exception of $(13, 6, 4)$ where $(\mu + \lambda) + \chi$ is better than its counterpart without crossover, $p = 0.031$), the situation is different with $(15, 7, 5)$ and $(16, 8, 5)$ for the (μ, λ) variants. In these cases, using crossover actually worsens the convergence speed of the ES algorithm. This is somewhat surprising, as one would expect that crossover allows to exploit the local search space more efficiently. Overall, our results show that the simplest (μ, λ)-ES variants without crossover is the best performing one over this optimization problem.

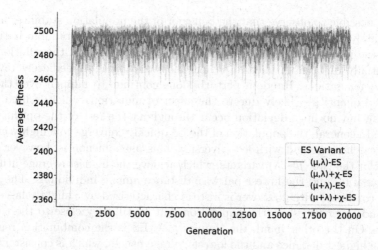

Fig. 2. Average population fitness for $(16, 8, 5)$.

5.3 Solutions Diversity

To analyze more deeply the influence of the replacement mechanism and the crossover operator on the performances of the ES algorithm, we ran again the experiments for 30 independent repetitions on the $(16, 8, 5)$ instance, where the effects are more evident. We set the stopping criterion to 20 000 generations, independently of the fact that an optimal solution might be found before. In each run, we recorded every 40 generations the average fitness of the population and the average pairwise distance between individuals, using Eq. (8). The corresponding lineplots are displayed in Figs. 2 and 3, respectively.

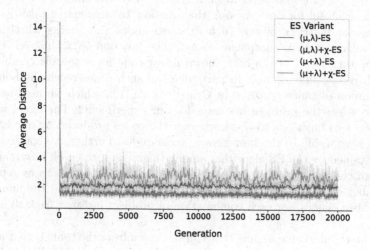

Fig. 3. Average pairwise distance distributions for $(16, 8, 5)$.

It is possible to observe that the behavior of the population stabilizes almost immediately for all four ES variants. In particular, after the random initialization of the population where the fitness is relatively low and the solutions are substantially different from each other, the situation is immediately reversed in a few generations. Random perturbations continue to happen over the two measured quantities (likely due to the effect of mutation, which is used in all variants), but no huge deviation occur throughout the rest of the optimization process. In general, the population of the ES quickly converge to a highly fit area of the search space and with low diversity. This phenomenon is, however, more evident for the two $(\mu + \lambda)$ variants, which achieve the highest average fitness in the population and the lowest pairwise distance among individuals. The (μ, λ) variant combined with crossover is instead characterized by a slightly larger distance and lower fitness in the population, but is still very close to the $(\mu + \lambda)$ variants. On the other hand, the simple (μ, λ)-ES is the combination reaching both the highest distance and the lowest average fitness, which is consistent with our earlier observation that this variant is the best performing one. In particular, having a higher diversity might hamper the average fitness in the population, but at the same time can help the population to escape local optima. The low diversity observed in the $(\mu + \lambda)$ variants indicates that the convex hull defined by the population under the distance of Eq. (8) shrinks very quickly, and does not grow anymore throughout the evolutionary process. Thus, if this convex hull represents a highly fit area of the search space, which however does not contain a global optimum, chances are that the population will remain stuck in that area. Likely, this effect is further strengthened by the use of crossover.

As a final analysis, we investigated the diversity of the optimal codes produced by the four ES variants in terms of code isomorphism. Two codes $C, D \subseteq \mathbb{F}_2^n$ are called isomorphic if there exists a sequence of permutations on the coordinates of the codewords and on the symbols set that transforms C into D [13]. We used the computer algebra system MAGMA since it has two built-in functions useful for our purpose: the function IsIsomorphic takes as input the generator matrices of two (n, k, d) linear codes and checks whether they are equivalent up to isomorphism or not. The function BKLC, instead, returns the generator matrix of the best known linear code for a specific combination of length n and dimension k. In particular, all such codes reach the bound on the minimum distance reported in Grassl's table [12], which we used as a reference to select the problem instances for our experiments. Therefore, we first used these two functions to check whether the codes produced by our ES variants are isomorphic to the best known linear codes. Further, we compared the codes obtained by the ES algorithm among themselves, to check how many isomorphism classes they belong to. Table 3 summarizes this analysis by reporting the number of codes that are not isomorphic to the BKLC and the number of isomorphism classes, for each combination of problem instance (n, k, d) and ES variant.

The first remarkable finding that can be drawn from the table is that all four ES variants always discover codes that are inequivalent to the BKLC for the

Table 3. Number of non-isomorphic codes to the BKLC (#non-iso) and equivalence classes (#eq) found by the four considered ES variants.

(n, k, d)	(μ, λ)-ES		$(\mu, \lambda)+\chi$-ES		$(\mu + \lambda)$-ES		$(\mu + \lambda)+\chi$-ES	
	#non-iso	#eq	#non-iso	#eq	#non-iso	#eq	#non-iso	#eq
$(12, 6, 4)$	100	23	100	22	100	22	100	22
$(13, 6, 4)$	100	85	100	81	100	78	100	79
$(14, 7, 4)$	100	89	100	94	100	95	100	93
$(15, 7, 5)$	72	5	63	6	51	5	44	5
$(16, 8, 5)$	0	1	0	1	0	1	0	1

smaller instances with minimum distance $d = 4$. Moreover, such codes belong to a high number of equivalence classes, so they are also quite diverse among themselves. From this point of view, there is also no particular difference between different ES variants. For $(15, 7, 5)$ one can remark a lower diversity since more codes turn out to be equivalent to the BKLC. Moreover, there is a noticeable difference between the (μ, λ) and the $(\mu + \lambda)$ variants, with the former scoring a higher number of codes inequivalent to the BKLC than the latter. Further, in general, the number of isomorphism classes drops substantially, with only 5 or 6 classes grouping all evolved codes. This phenomenon is even more extreme for the $(16, 8, 5)$ instance: in this case, all discovered codes are equivalent to the BKLC provided by MAGMA, and thus they all belong to the same equivalence class. This fact is independent of the underlying ES variant.

6 Conclusions and Future Work

To conclude, we summarize our experimental findings and discuss their relevance concerning the design of binary linear codes using evolutionary algorithms:

- The proposed ES algorithm easily converges to an optimal solution for the smaller problem instances of $(12, 6, 4)$, $(13, 6, 4)$, and $(14, 7, 4)$, with no significant differences among the four tested variants. On the other hand, there is a huge increase in the difficulty of the problem for the larger instances of $(15, 7, 5)$ and $(16, 8, 5)$, although the simple (μ, λ)-ES variant is able to maintain a very high success rate.
- Contrary to our initial expectation, the crossover operator that we augmented our ES with either does not make any difference on the performances of the algorithm, or it even deteriorates them over the harder instances. We speculate that this is due to the small variability offered by the crossover, since it is based on the direct selection of the vectors from the parents, rather than on the vectors spanned by their generator matrices.
- The optimal codes obtained by the ES are quite interesting from a theoretical point of view, as most of them for small instances are not equivalent to the best-known linear code produced by MAGMA, and moreover they belong to a high number of isomorphism classes. The fact that all codes instead turn out to be equivalent to the BKLC for $(16, 8, 5)$ is curious, and we hypothesize that

this is related to the specific structure of the search space for this instance, where the global optima might be very sparse.

Overall, our results suggest that ES represent an interesting tool to discover potentially new linear codes, and prompt us to multiple ideas for future research. One obvious direction is to apply the ES algorithm over larger instances. However, the computation of the fitness function could become a significant bottleneck in this case. Indeed, our Java implementation of the ES algorithm takes around 20 min to perform 20 000 generations on a Linux machine with an AMD Ryzen 7 processor, running at 3.6 GHz. Therefore, it makes sense to explore also with other fitness functions, maybe without relying on the characterization through the ANF of the indicator function. A second interesting direction for future research concerns the study of the variation operators proposed in this paper, especially with respect to their topological properties. In particular, we believe that both operators can be proved to be geometric in the sense introduced by Moraglio and Poli [23]. This might in turn give us some insights related to the structure of the Grassmannian metric space under the distance defined in Eq. (8), and thus help us in designing better crossover operators for this problem. One idea, for instance, could be to follow an approach similar to those adopted for fixed-weight binary strings in [24].

References

1. McEliece, R.J.: The Theory of Information and Coding. Number 86. Cambridge University Press (2004)
2. Karp, R.M.: Reducibility among combinatorial problems. In Miller, R.E., Thatcher, J.W., eds.: Proceedings of a symposium on the Complexity of Computer Computations, held March 20–22, 1972, pp. 85–103. The IBM Research Symposia Series, Plenum Press, New York (1972)
3. Dontas, K., Jong, K.A.D.: Discovery of maximal distance codes using genetic algorithms. In: Proceedings of IEEE TAI 1990, pp. 805–811. IEEE Computer Society (1990)
4. McGuire, K.M., Sabin, R.E.: Using a genetic algorithm to find good linear error-correcting codes. In George, K.M., Lamont, G.B., eds. Proceedings of the 1998 ACM symposium on Applied Computing, SAC 1998, Atlanta, GA, USA, February 27 - March 1, 1998, pp. 332–337. ACM (1998)
5. Chen, H., Flann, N.S., Watson, D.W.: Parallel genetic simulated annealing: a massively parallel SIMD algorithm. IEEE Trans. Parallel Distrib. Syst. 9(2), 126–136 (1998)
6. McCarney, D.E., Houghten, S.K., Ross, B.J.: Evolutionary approaches to the generation of optimal error correcting codes. In Soule, T., Moore, J.H., eds. Proceedings of GECCO 2012, pp. 1135–1142. ACM (2012)
7. Gamal, A.A.E., Hemachandra, L.A., Shperling, I., Wei, V.K.: Using simulated annealing to design good codes. IEEE Trans. Inf. Theory 33(1), 116–123 (1987)
8. Alba, E., Chicano, J.F.: Solving the error correcting code problem with parallel hybrid heuristics. In: Haddad, H., Omicini, A., Wainwright, R.L., Liebrock, L.M., (eds.) Proceedings of SAC 2004), pp. 985–989. ACM (2004)

9. Cotta, C.: Scatter search and memetic approaches to the error correcting code problem. In: Gottlieb, J., Raidl, G.R. (eds.) EvoCOP 2004. LNCS, vol. 3004, pp. 51–61. Springer, Heidelberg (2004). https://doi.org/10.1007/978-3-540-24652-7_6
10. Blum, C., Blesa, M.J., Roli, A.: Combining ILS with an effective constructive heuristic for the application to error correcting code design. In: Metaheuristics International Conference (MIC-2005), Vienna, Austria, pp. 114–119 (2005)
11. Carlet, C.: Expressing the minimum distance, weight distribution and covering radius of codes by means of the algebraic and numerical normal forms of their indicators. In: Advances in Mathematics of Communications (2022)
12. Grassl, M.: Bounds on the minimum distance of linear codes and quantum codes (2007). http://www.codetables.de. Accessed 13 Nov 2022
13. Huffman, W.C., Pless, V.: Fundamentals of Error-correcting Codes. Cambridge University Press (2010)
14. Carlet, C.: Boolean functions for cryptography and coding theory (2021)
15. Mariot, L., Picek, S., Jakobovic, D., Leporati, A.: Evolutionary search of binary orthogonal arrays. In: Auger, A., Fonseca, C.M., Lourenço, N., Machado, P., Paquete, L., Whitley, D. (eds.) PPSN 2018. LNCS, vol. 11101, pp. 121–133. Springer, Cham (2018). https://doi.org/10.1007/978-3-319-99253-2_10
16. Knezevic, K., Picek, S., Mariot, L., Jakobovic, D., Leporati, A.: The design of (Almost) disjunct matrices by evolutionary algorithms. In: Fagan, D., Martín-Vide, C., O'Neill, M., Vega-Rodríguez, M.A. (eds.) TPNC 2018. LNCS, vol. 11324, pp. 152–163. Springer, Cham (2018). https://doi.org/10.1007/978-3-030-04070-3_12
17. Mariot, L., Picek, S., Jakobovic, D., Djurasevic, M., Leporati, A.: On the difficulty of evolving permutation codes. In: Jiménez Laredo, J.L., Hidalgo, J.I., Babaagba, K.O. (eds.) EvoApplications 2022. LNCS, vol. 13224, pp. 141–156. Springer, Cham (2022). https://doi.org/10.1007/978-3-031-02462-7_10
18. Mariot, L., Jakobovic, D., Bäck, T., Hernandez-Castro, J.: Artificial intelligence for the design of symmetric cryptographic primitives. In: Security and Artificial Intelligence, pp. 3–24 (2022)
19. Mullen, G.L., Panario, D. (eds.): Handbook of Finite Fields, Discrete Mathematics and its Applications. CRC Press (2013)
20. Koetter, R., Kschischang, F.R.: Coding for errors and erasures in random network coding. In: IEEE International Symposium on Information Theory, ISIT 2007, Nice, France, June 24–29, 2007, pp. 791–795. IEEE (2007)
21. Mariot, L., Saletta, M., Leporati, A., Manzoni, L.: Heuristic search of (semi-)bent functions based on cellular automata. Nat. Comput. **21**(3), 377–391 (2022)
22. Luke, S.: Essentials of Metaheuristics. 2nd edn. Lulu (2015)
23. Moraglio, A., Poli, R.: Topological crossover for the permutation representation. In: Rothlauf, F., ed. Genetic and Evolutionary Computation Conference, GECCO 2005, Workshop Proceedings, Washington DC, USA, June 25–26, 2005, pp. 332–338. ACM (2005) ACM (2005)
24. Manzoni, L., Mariot, L., Tuba, E.: Balanced crossover operators in genetic algorithms. Swarm Evol. Comput. **54**, 100646 (2020)

A Policy-Based Learning Beam Search for Combinatorial Optimization

Rupert Ettrich, Marc Huber[(✉)], and Günther R. Raidl

Algorithms and Complexity Group, Institute of Logic and Computation, Wien, TU,
Austria
{mhuber,raidl}@ac.tuwien.ac.at

Abstract. Beam search (BS) is a popular incomplete breadth-first
search widely used to find near-optimal solutions to hard combinatorial optimization problems in limited time. Its central component is an
evaluation function that estimates the quality of nodes encountered on
each level of the search tree. While this function is usually manually
crafted for a problem at hand, we propose a Policy-Based Learning Beam
Search (P-LBS) that learns a policy to select the most promising nodes
at each level offline on representative random problem instances in a
reinforcement learning manner. In contrast to an earlier learning beam
search, the policy function is realized by a neural network (NN) that is
applied to all the expanded nodes at a current level together and does
not rely on the prediction of actual node values. Different loss functions
suggested for beam-aware training in an earlier work, but there only
theoretically analyzed, are considered and evaluated in practice on the
well-studied Longest Common Subsequence (LCS) problem. To keep P-
LBS scalable to larger problem instances, a bootstrapping approach is
further proposed for training. Results on established sets of LCS benchmark instances show that P-LBS with loss functions "upper bound" and
"cost-sensitive margin beam" is able to learn suitable policies for BS such
that results highly competitive to the state-of-the-art can be obtained.

Keywords: Beam Search · Machine Learning · Reinforcement
Learning · Longest Common Subsequence Problem

1 Introduction

Beam search (BS) is a prominent incomplete, i.e., heuristic, graph search algorithm widely used to tackle hard planning and discrete optimization problems
in limited time. Starting from a root node r, BS traverses a state graph in a
breadth-first search manner but restricts the search by selecting at each level
only up to β most promising nodes to pursue further and discards the others.
The subset of selected nodes at the current level is referred to as the *beam*, and

This project is partially funded by the Doctoral Program "Vienna Graduate School
on Computational Optimization", Austrian Science Foundation (FWF), grant W1260-
N35.

L. Pérez Cáceres and T. Stützle (Eds.): EvoCOP 2023, LNCS 13987, pp. 130–145, 2023.
https://doi.org/10.1007/978-3-031-30035-6_9

parameter β as *beam width*. To select the β most promising nodes at each level, every node v from this level is typically evaluated by an evaluation function $f(v) = g(v) + h(v)$, where $g(v)$ represents the cost of the path from the root node r to the current node v and $h(v)$, called heuristic function, is an estimate for the cost of a best path from the current node v to a goal node. The β nodes with the best values according to this evaluation then form the beam.

Clearly, the quality of the solution BS obtains depends in general fundamentally on the evaluation function f. This function is usually developed manually for a specific problem, and its effectiveness relies on a good understanding and careful exploitation of the problem structure and possibly properties of expected instances. In practice, it is sometimes difficult to come up with an effective evaluation function that strikes the right balance between good BS guidance and reasonable computational effort.

The main contribution of this work is the investigation of a *Policy-Based Learning Beam Search* (P-LBS) that learns a *policy* for BS to select the β most promising nodes at each level of a BS, replacing the traditional approach of evaluating each node independently with the hand-crafted evaluation function f and afterwards selecting the nodes remaining on the beam based on their f-values. It builds upon our earlier Learning Beam Search (LBS) [11] framework, in which a machine learning model is used as heuristic function h as part of f and trained offline in a reinforcement learning manner on a large number of representative randomly generated problem instances to approximate specifically the expected cost-to-go from a node to a goal node. By learning a policy that is applied to all the nodes at a current level together in order to do the node selection and not insisting on approximating the actual cost-to-go, we allow now for greater flexibility and alternative modeling and training approaches.

In earlier work, Negrinho et al. [16] already described the learning of beam search policies for structured prediction problems by imitation learning and analyzed different variants from a purely theoretical perspective. In their approach, an abstract *scoring function* replaces the classical evaluation function f, which is not expected to approximate real solution costs anymore but shall just express how promising a node is in relation to the others at the current level. These scores thus induce a policy over the nodes, nodes are ranked accordingly, and the best-ranked nodes are accepted for the beam—just as in classical BS. Imitation learning is done on representative problem instances for which exact solutions, i.e., optimal paths, are assumed to be known. While Negrinho et al. [16] suggested and studied different loss functions for training with respect to theoretical convergence, no practical experiments were done.

Building on LBS, P-LBS again relies on *reinforcement learning* and does not need problem instances with known optimal solutions for training. We iteratively apply a BS with an initially randomly initialized neural network (NN) model as scoring function on many randomly generated representative problem instances. In each BS iteration, a subset of the BS tree levels is selected for generating training data. A training sample consists of all nodes encountered on a selected BS level. Two different approaches, *beam-unaware* and *beam-aware*, are investigated to label training data. In *beam-unaware* training, the node in a training sample that lies on the path from the root node to the finally best solution node obtained by BS is labeled with one and all other nodes with zero. In *beam-aware* training, we perform a *nested beam search* (NBS) on each subinstance induced by each node of a training sample in order to approximate values for the true cost-to-go. Based on these values, we rank the nodes of a sample accordingly, and consider this ranking as the training target. Following Negrinho et al. [16], we consider different surrogate loss functions for the actual training.

To achieve reasonable scalability to larger problem instances, we stop the NBS executions when they reach a maximum level $d \in \mathbb{N}$ in their search trees, and evaluate the returned nodes by the so far trained NN to obtain suitable training targets for the new training data. This approach resembles a form of bootstrapping as known in reinforcement learning [22]. Produced training samples are stored in a FIFO replay buffer and used to continuously train the NN, intertwined with the P-LBS's further training data production.

While the general principle of P-LBS is quite generic, we test its effectiveness on the well-known NP-hard Longest Common Subsequence (LCS) problem [6]. Experiments show that policies trained by P-LBS are able to guide BS on established LCS benchmark instances well such that results being competitive to the state-of-the-art can be obtained.

Section 2 reviews related work. In Sect. 3 we present the general P-LBS framework, different loss functions, and the bootstrapping approach for speeding up the training data generation. The general NN architecture used as scoring function in P-LBS is described in Sect. 4. Section 5 introduces the LCS problem, its specific state graph, and the features used for the NN. Results of computational experiments are discussed in Sect. 6. Finally, we conclude in Sect. 7, where we also outline promising future work.

2   Related Work

In recent years there has been a growing interest at leveraging machine learning (ML) techniques to better solve discrete optimization problems. Under the umbrella term *learning to search* much work has been done in different directions for improving classical tree search [3]. We focus here particularly on beam search, which is a conceptually simple and classical incomplete search strategy for obtaining a heuristic solution in controlled time. It was originally introduced in the context of speech recognition [14], but since then has been widely applied to many combinatorial problems including scheduling, packing, and various string problems from bioinformatics such as the LCS problem, for which it frequently yields state-of-the-art results [6].

In the context of prediction tasks and sequence to sequence learning, BS is frequently used to derive better or feasible solutions than just by applying a simple greedy solution construction, see, e.g., [8,21]. These approaches rely on ML models that are trained independently of the BS beforehand on the basis of given labeled data, imitation learning, or occasionally reinforcement learning. The BS is then applied as a decoder in the actual application (test time). Typically, such approaches suffer from ignoring the existence of the beam during training.

In contrast, beam-aware learning algorithms use BS at both, training and application/test time. A first approach by Collins and Roark [4] is perceptron-based and updates the parameters when the best node does not score first in the beam. On the other hand, Daumé et al. [5] described an approach that updates the parameters only when the best node falls out of the beam.

While further work on beam-aware algorithms exists in the context of prediction tasks and sequence to sequence learning, see, e.g., [23], most approaches do not expose the learned model to its own consecutive mistakes at train time: when a transition leads to a beam where the assumed best node is excluded, the algorithms either stop or reset to a beam with the best node included. To our knowledge, only Negrinho et al. [16] described an approach to learn beam search policies that addresses this issue. They formulate the task as learning a policy to traverse the combinatorial search space of beams. A scoring function is learned to match the ranking induced by given oracle costs from an assumed expert strategy. The authors proposed and analyzed several loss functions and data collection strategies that consider the beam also at train time and proved novel no-regret guarantees for learning BS policies.

In the context of classical combinatorial optimization, we are only aware of our LBS [11] sketched already in the introduction as a method where a guidance function is learned and used within a BS. This approach also exposes its learned model to its own mistakes by using the model in the BS for further training data generation and performing training in an interleaved way. However, it cannot be considered an actual beam-aware approach, as the model is specifically trained to approximate the cost to go, and the respective labels are obtained by independent NBS calls. In [12], we refined the original LBS specifically for the LCS problem by making the model independent of the number of strings and relying on a relative value function in which a cut-off is applied to the values of nodes at the same level.

LBS as well as the new P-LBS are both based on principles inspired by AlphaZero [20], although AlphaZero relies on Monte Carlo Tree Search (MCTS) and not BS. AlphaZero has proven to be very successful in the board games Go, chess, and shogi, with its predecessor AlphaGo being the first computer program that was able to beat a human Go champion. In the MCTS a neural network is used to evaluate game states and to provide a policy over possible moves. Training data is continuously produced by self-play in a reinforcement learning manner and stored in a replay buffer for training. AlphaZero has also been adapted to solve various combinatorial optimization problems like 3D packing problems [13], minimum vertex cover and maximum cut [1], or graph coloring [10].

3   Policy-Based Learning Beam Search

Solving a combinatorial optimization problem can be formulated as search in a state graph $G = (V, \mathcal{A})$ with nodes V and arcs $\mathcal{A}$. Each node $v \in V$ represents a problem-specific state, e.g., a partial assignment of values to the decision variables. Nodes $u, v \in V$ are connected by an arc $(u, v) \in \mathcal{A}$ if there is a valid problem-specific action that can be performed to transform state u into state v, for example, the assignment of a specific value to a so far unassigned decision variable of state u. Let label $\tau(u, v)$ denote this action transforming state u into state v. We assume each arc $(u, v) \in \mathcal{A}$ has associated cost $c_a(u, v)$ that are induced by the action w.r.t. the objective function of the problem. State graph G has a dedicated root node $r \in V$ representing the initial state, in which typically all decision variables are unassigned. Moreover, there are one or more goal nodes $T \subset V$, which have no outgoing arcs and represent valid final states, e.g., in which all decision variables have feasible values. A complete solution is represented by a path from r to a goal node $t \in T$, referred to as r–t path, and we assume that the arc costs are defined in such a way that the objective value of the solution corresponds to the sum of the path's arc costs.

As already pointed out in the introduction, classical BS explores such a state space in an incomplete breadth-first search manner to find one or more heuristic solutions. Nodes are considered level by level, and at each level only up to β nodes are selected as beam to continue with. Now, let V_{ext} be the set of all nodes that have been derived as successors of the current beam. Moreover, let $f_s \colon (V, 2^V) \to \mathbb{R}$ be a *scoring function* so that $f_s(v, V_{\text{ext}})$ assigns each node $v \in V$ a real-valued score *in relation to all the other nodes in V_{ext}*. Thus, the score of a node is not determined independently for each node but under consideration of V_{ext}. The score obtained by evaluating $f_s(v, V_{\text{ext}})$ for each $v \in V_{\text{ext}}$ induces a policy over the nodes in V_{ext}, where higher values shall indicate a higher probability of a node leading to a best goal node. In P-LBS this scoring function f_s replaces the classical node-individual evaluation function f of BS and is realized in the form of a neural network that will be described in Sect. 4.

The core idea of P-LBS is to train function f_s via "self-play" similarly as in AlphaZero [20] by iterated application on many random instances generated according to the properties of the instances expected in the future application. A pseudocode of the P-LBS framework is shown in Algorithm 1. It starts with a randomly initialized NN as scoring function f_s, and an initially empty replay buffer R which will contain the training data. The buffer is realized as first-in-first-out (FIFO) queue of maximum size ρ. The idea hereby is to remove older, outdated training samples when the scoring function has already been improved. After initialization, a certain number z of iterations is performed. In each iteration, a new independent random problem instance I with root node r is created and a BS with the current scoring function f_s is applied. This BS returns a best goal node t, and also the set L of node sets V_{ext} encountered at each level during the search. Next, from each set $V_{\text{ext}} \in L$ a training sample is derived with probability α/L, where parameter α controls the expected number of samples produced per instance.

Algorithm 1. Policy-Based Learning Beam Search (P-LBS)

1: **Input:** nr. of iterations z, beam width β, NBS beam width β', replay buffer size ρ, min. buffer size for training γ, nr. of training samples per instance α
2: **Output:** trained scoring function f_s
3: $f_s \leftarrow$ scoring function (randomly initialized NN)
4: $R \leftarrow \emptyset$ // replay buffer: FIFO of max. size ρ
5: **for** z iterations **do**
6:     $I, r \leftarrow$ create representative random problem instance with root node r
7:     $t, L \leftarrow$ BeamSearch(I, β, f_s) // best found goal node t,
8:                         // set L of node sets V_{ext} encountered at each level
9:     **for** $V_{\text{ext}} \in L$ **do**
10:         **if** rand() $< \alpha/|L|$ **then** // generate training sample
11:             **for** $v \in V_{\text{ext}}$ **do**
12:                 **if** beam-unaware **then**
13:                     $c_v \leftarrow \begin{cases} 1, & \text{if node } v \text{ lies on } r\text{–}t \text{ path} \\ 0, & \text{otherwise} \end{cases}$
14:                 **else if** beam-aware **then**
15:                     $t'_v \leftarrow$ BeamSearch$(I(v), \beta', f_s)$ // NBS call $\rightarrow$ best goal node
16:                     $c_v \leftarrow g(t'_v)$
17:                 **end if**
18:             **end for**
19:             add training sample $(V_{\text{ext}}, \{c_v\}_{v \in V_{\text{ext}}})$ to R
20:         **end if**
21:     **end for**
22:     **if** $|R| \geq \gamma$ **then**
23:         train f_s with batches of randomly sampled data from R
24:     **end if**
25: **end for**
26: **return** f_s

For *beam-unaware* training, target values for the nodes in V_{ext} are derived by mapping a node $v \in V_{\text{ext}}$ to $c_v = 1$ if node v lies on the best solution path r–t (ties are broken randomly in case there are multiple such paths with equal cost) and to $c_v = 0$ otherwise. For *beam-aware* training, each node $v \in V_{\text{ext}}$ is mapped to a target value (i.e., approximate oracle cost) c_v obtained by performing an independent nested beam search (NBS) with beam width β' and scoring function f_s on the problem subinstance $I(v)$ induced by node v. However, as each NBS call is in general computationally expensive, we apply a bootstrapping approach (details below) to keep P-LBS scalable. All training samples derived are added to the replay buffer R.

At the end of each P-LBS iteration, if the replay buffer has reached a minimum fill level of γ, the scoring function f_s is incrementally trained with batches of data sampled uniformly at random from R using one of the following loss functions.

3.1 Loss Functions

Let $c = (c_v)_{v \in V_{\text{ext}}}$ be the vector of all target values of the nodes in V_{ext}. Moreover, given a training sample (V_{ext}, c), let $s_v = f_s(v, V_{\text{ext}})$ be the score obtained by evaluating our learnable scoring function f_s for each $v \in V_{\text{ext}}$ and $s = (s_v)_{v \in V_{\text{ext}}}$. Moreover, let $\hat{\sigma}$ be a permutation of V_{ext} that sorts the scores in s in descending order such that $s_{\hat{\sigma}(1)} \geq s_{\hat{\sigma}(2)} \geq \cdots \geq s_{\hat{\sigma}(|V_{\text{ext}}|)}$, and let σ^* be a permutation of V_{ext} that sorts the target values in c in descending order such that $c_{\sigma^*(1)} \geq c_{\sigma^*(2)} \geq \cdots \geq c_{\sigma^*(|V_{\text{ext}}|)}$. We consider the following loss functions originally proposed by Negrinho et al. [16], as well as one introduced by ourselves called cost-sensitive marginal beam (cmb)

perceptron first (pf): $\ell(s, c) = \max(0, s_{\hat{\sigma}(1)} - s_{\sigma^*(1)})$

This loss is positive if the node with the highest target value does not correspond to the highest score node.

perceptron last (pl): $\ell(s, c) = \max(0, s_{\hat{\sigma}(\beta)} - s_{\sigma^*(1)})$

The loss is positive if the node with the highest target value falls out of the beam.

margin last (ml): $\ell(s, c) = \max(0, 1 + s_{\hat{\sigma}(\beta)} - s_{\sigma^*(1)})$

A penalty is given if the highest target value node is not among the β best nodes in s, but also a smaller penalty may be given if the highest target value node is placed low in the beam.

cost-sensitive margin last (cml):

$$\ell(s, c) = (c_{\sigma^*(1)} - c_{\hat{\sigma}(\beta)}) \max(0, 1 + s_{\hat{\sigma}(\beta)} - s_{\sigma^*(1)})$$

The previous ml loss is here weighted by the difference between the highest target value and the target value of the node at place β in the beam according to $\hat{\sigma}$.

cost-sensitive margin beam (cmb):

$$\ell(s, c) = \sum_{i=1}^{\beta-1} \max(0, c_{\sigma^*(i)} - c_{\hat{\sigma}(\beta)}) \max(0, 1 + s_{\hat{\sigma}(\beta)} - s_{\sigma^*(i)})$$

We suggest this additional variant of cml, in which the sum of the weighted ml losses for the first $(\beta - 1)$ elements in the beam is calculated. A penalty is given if any of the $(\beta - 1)$ first nodes in the beam according to c falls out of the beam according to s. This penalty is weighted as in the cml loss for each of the $(\beta - 1)$ first nodes in the beam.

log loss neighbors (lln): $\ell(s, c) = -s_{\sigma^*(1)} + \log\left(\sum_{i=1}^{|V_{\text{ext}}|} \exp(s_i)\right)$

Here we normalizes over all elements in V_{ext}. A higher penalty is given if there are nodes with higher scores than the score of the highest target value node.

log loss beam (llb): $\ell(s,c) = -s_{\sigma^*(1)} + \log\left(\sum_{i\in I}\exp(s_i)\right)$

Here, I denotes the index set that contains the index of the highest target value node and the indices of the β elements with the highest scores in s. This loss function is similar to the lln loss, but normalization is done only over the nodes in the beam according to the scores.

upper bound (ub): $\ell(s,c) = \max(0, \delta_{\beta+1}, \ldots, \delta_{|V_{\text{ext}}|})$

Here, $\delta_j = (c_{\sigma^*(1)} - c_{\sigma^*(j)})(s_{\sigma^*(j)} - s_{\sigma^*(1)})$ for $j = \beta+1, \ldots, |V_{\text{ext}}|$. Negrinho et al. [16] showed that this loss function is a convex upper bound for the expected beam transition cost.

Preliminary tests indicated that it is beneficial to use for the loss calculation not necessarily the beam width for which BS is intended to be finally applied, but an independent value proportional to $|V_{\text{ext}}|$. Therefore, the beam width considered in the loss functions is $\lceil |V_{\text{ext}}| \cdot \xi \rceil$, where $\xi \in (0,1]$ is a control parameter.

3.2 Bootstrapping

In beam-aware training, a training sample for a node set V_{ext} is obtained by executing NBS on each subinstance $I(v)$ induced by a node $v \in V_{\text{ext}}$. Depending on the beam width and the specific instance to be solved, these NBS executions can become computationally expensive. To keep beam-aware training scalable and reduce the computational effort, NBS executions are stopped when they reach a maximum level $d \in \mathbb{N}$. For simplicity, we assume in the following maximization and that goal nodes deeper in the search tree are always better, as it is the case in our benchmark, the LCS problem. An extension to the general case needs to consider the g-values of nodes but is otherwise straight-forward. Each depth-limited NBS call returns then either a set of nodes if level d is reached and the execution stopped or a best goal node otherwise. To determine the target costs for the nodes in V_{ext}, let $M \subseteq V_{\text{ext}}$ be the set of nodes for which the respective NBS calls finish with a goal node before or at level d, and let $N \subseteq V_{\text{ext}}$ be the set of nodes for which the NBS calls are stopped prematurely at level d. Moreover, let $\text{NBS}(I(v))$ be the set of nodes that is returned from level d for $v \in N$. Three cases are now distinguished:

1. $M = V_{\text{ext}}$, $N = \emptyset$. In this case no early stopping occurred, and the target value of node $v \in M$ is set to $c_v = g(\text{NBS}(I(v)))$.
2. $M = \emptyset$, $N = V_{\text{ext}}$. Let $V'_{\text{ext}} = \{\text{argmax}_{u\in\text{NBS}(I(v))} f_s(u, \text{NBS}(I(v))) \mid v \in N\}$ be the set of nodes with highest f_s-values from each node set returned by $\text{NBS}(I(v))$. Moreover, let $v' \in V'_{\text{ext}}$ be the node that corresponds to $v \in V_{\text{ext}}$, i.e., the node $v' = \text{argmax}_{u\in\text{NBS}(I(v))} f_s(u, \text{NBS}(I(v)))$. The target values are then set to $c_v = f_s(v', V'_{\text{ext}})$.
3. $M \neq \emptyset$, $N \neq \emptyset$. Let M' be the set of goal nodes that represent the solutions returned by the NBS calls executed on subinstances $I(v)$ for $v \in M$. Set V'_{ext} is derived analogously to the previous case as the node set representing the

most promising partial solutions for the nodes in N. The target value for a node $v \in V_{\text{ext}}$ is then determined as $c_v = f_s(v', V'_{\text{ext}} \cup M'_v)$, where

$$v' = \begin{cases} \text{NBS}(I(v)) & \text{if } v \in M \\ \text{argmax}_{u \in \text{NBS}(I(v))} f_s(u, \text{NBS}(I(v))) & \text{if } v \in N \end{cases}.$$

Additional post-processing may help in problem-specific scenarios. For example, in case of uniform arc costs as in our LCS benchmark problem, the nodes in N should always be ranked higher than the nodes in M, because the respective NBS calls on the subinstances induced by nodes in M finish at an earlier level than the NBS calls on subinstances induced by nodes in N. The oracle cost corresponding to nodes in N can then simply all be increased by the same value in order to be ranked above all nodes in M while still maintaining their relative positioning among the nodes in N.

4   Neural Network Architecture

The NN used as scoring function f_s in P-LBS must fulfill an important property: It must be able to deal with inputs of variable size as $|V_{\text{ext}}|$ in general varies and we aim at scoring each node in dependence of all nodes in V_{ext}.

Let the input to the NN be a vector of vectors $(x_v)_{v \in V_{\text{ext}}}$, where x_v is a problem-specific feature vector representing the state associated with node v. Moreover, also $g(v)$, the cost from the r–v path, are appended as an additional feature in x_v. Figure 1 illustrates the NN architecture realizing f_s. The NN is a feedforward network with layers $j = 0, \ldots, 3$ described in the following. Hereby, $A^{(j)}$ denotes a weight matrix and $b^{(j)}$ a bias vector for each layer j. Weights and biases are shared within each layer among the components for the individual nodes' feature vectors.

Layer 0: The inputs x_v are first embedded by a linear transformation

$$h_v^{(0)} = A^{(0)} x_v + b^{(0)} \qquad \forall v \in V_{\text{ext}}.$$

Layer 1: The embeddings $h_v^{(0)}$ of the individual nodes are then pooled to obtain a constant-size global embedding for V_{ext}. We do this simply by averaging, i.e.,

$$h^{(1)} = \frac{1}{|V_{\text{ext}}|} \sum_{v \in V_{\text{ext}}} h_v^{(0)}.$$

Layer 2: Now, the node-individual embeddings from layer 0 are combined with the the global embedding from layer 1 by concatenation and used subsequently as inputs for a per-node linear transformation followed by a ReLU activation:

$$h_v^{(2)} = \text{ReLU}(A^{(2)}(h_v^{(0)} \,||\, h^{(1)}) + b^{(2)}) \qquad \forall v \in V_{\text{ext}}.$$

Layer 3: A final linear transformation is used to compute the scores s_v in the form of logits

$$s_v = h_v^{(3)} = A^{(3)} h_v^{(2)} + b^{(3)} \qquad \forall v \in V_{\text{ext}}.$$

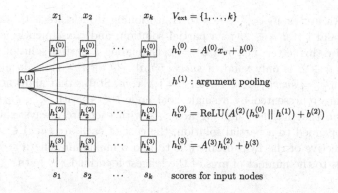

Fig. 1. Four-layer feedforward NN architecture for P-LBS.

5 Case Study: Longest Common Subsequence Problem

A string is a finite sequence of symbols taken from an alphabet Σ. A subsequence of a string is a string that is obtained by deleting zero or more symbols. A *common* subsequence of a set of strings $S = \{S_1, \ldots, S_m\}$ is a string that is a subsequence of every string in S. The longest common subsequence (LCS) problem aims at finding a common subsequence of maximum length for S. For example, the LCS of strings AGACT, GTAAC, and GTACT is GAC. The LCS problem is well-studied and has many applications, in particular in bioinformatics [18], database query optimization [17], and image processing [2]. For a fixed number m of strings the LCS problem is polynomially solvable by dynamic programming in time $O(n^m)$ [9], where n denotes the length of the longest input string, while for general m it is NP-hard [15]. The current state-of-the-art heuristic approaches for large m and n are based on BS with a theoretically derived function EX that approximates the expected length of the result of random strings from a partial solution [6] and also on our LBS [11].

Notations. We denote the length of a string S by $|S|$, and the maximum length of all input strings in S by n. The j-th letter of a string shall be $S[j]$. By $S[j, j']$ we denote the substring of S starting with $S[j]$ and ending with $S[j']$ if $j \leq j'$ or the empty string ε otherwise. As in previous works [6,11], the following data structure is prepared in preprocessing to enable an efficient "forward stepping" in the strings. For each $i = 1, \ldots, m$, $j = 1, \ldots, |S_i|$, and $a \in \Sigma$, $succ[i, j, a]$ stores the minimal position j' such that $j' \geq j \wedge S_i[j'] = a$ or 0 if letter a does not occur in S_i from position j onward.

State Graph for the LCS Problem. In the state graph $G = (V, \mathcal{A})$ for the LCS problem, a node $v \in V$ represents a state by a position vector $p^v = (p_i^v)_{i=1,\ldots,m}$ with $p_i^v \in 1, \ldots, |S_i|+1$, indicating the remaining relevant substrings $S_i[p_i^v, |S_i|]$ for $i = 1, \ldots, m$. These substrings form the LCS subproblem instance $I(v) = S_i[p_i^v, |S_i|]$, for $i = 1, \ldots, m$ induced by node v. The root node $r \in V$ has position vector $p_i^v = (1, \ldots, 1)$. Hence, $I(v)$ corresponds to the original LCS instance. An arc $(u, v) \in \mathcal{A}$ refers to transitioning from state u to state v by adding a valid letter $a \in \Sigma$ to a partial solution, and consequently, arc (u, v) is labeled by this letter, i.e., $\tau(u, v) = a$. Extending a partial solution at state u by letter $a \in \Sigma$ is only valid, if $succ[i, p_i^u, a] > 0$ for $i = 1, \ldots, m$ and yields state v with $p_i^v = succ[i, p_i^u, a] + 1$ for $i = 1, \ldots, m$. States that allow no feasible extension are represented by a single goal node $t \in V$ with $p_i^t = |S_i| + 1$ for $i = 1, \ldots, m$ that has no outgoing arcs. As with each arc always exactly one letter is appended to a partial solution, the cost of each arc $(u, v) \in \mathcal{A}$ is one. As the objective of the LCS problem is to find a maximum length string, $g(v)$ corresponds to the number of arcs of the longest identified r–v path.

Node Features and Training. As features to represent a state we only use the *remaining string lengths* $|S_i| - p_i^v + 1$, $i = 1, \ldots, m$, on which also the heuristic from [6] is based. To prevent possible difficulties in learning symmetries, the remaining string lengths are always sorted according to non-decreasing values before providing them as feature vector x_v. In the NN, the hidden vectors $h_v^{(j)}$ for $j = 0, 1$, have size ten, whereas the hidden vectors $h_v^{(2)}$ have size 20, and the weight matrices and bias vectors were dimensioned accordingly.

As in previous work [11], the ADAM optimizer with step size 0.001 and exponential momentum decay rates 0.9 and 0.999 is applied for training. In each P-LBS iteration, two mini-batches of eight random samples are selected from the replay buffer R and used for learning. The loss of a single sample is obtained by one of the loss functions from Sect. 3.1, and the loss of a batch is determined by the mean loss of the individual batches.

6   Experimental Evaluation

We implemented P-LBS in Julia 1.7 using Flux for the NN. All experiments were executed on an Intel Xeon E5-2640 processor with 2.40 GHz and a memory limit of 20 GB. Two in the literature commonly used benchmark sets for the LCS problem are considered to empirically analyze and evaluate P-LBS. The first set referred to as rat was introduced in [19], and consists of 20 single instances composed of sequences from rat genomes. Each of these instances differs in the combination of the alphabet size $|\Sigma| \in \{4, 20\}$, number of input strings $m \in \{10, 15, 20, 25, 40, 60, 80, 100, 150, 200\}$. The length of the strings is $n = 600$. The second benchmark set denoted as ES is from [7] and consists of 50 random instances for each combination of $|\Sigma| \in \{2, 10, 25\}$, $m \in \{10, 50, 100\}$, where $n = 1000$ for instances with $|\Sigma| \in \{2, 10\}$, and $n = 2500$ for instances with $|\Sigma| = 25$. Preliminary tests led to the following P-LBS configuration that turned out to be suitable for all our benchmark sets unless stated otherwise: nr. of P-LBS iterations $z = 2000$, P-LBS and NBS beam widths $\beta = \beta' = 50$, NBS depth limit $d = 5$, beam width parameter for loss calculation $\xi = 0.1$, max. buffer size $\rho = 500$, min. buffer size for learning $\gamma = 250$, nr. of training samples generated per instance $\alpha = 5$, ten restarts with randomly initialized NN weights and final adoption of the NN yielding the best result on 30 independent random instances.

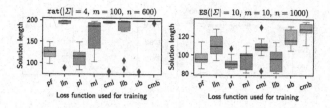

Fig. 2. Impact of the loss function in P-LBS on the solution lengths of BS on `rat` and ES instances.

Loss Functions. One of our main goals is to analyze the impact of the different loss functions from Sect. 3.1 in practice, as they were so far only theoretically considered in [16]. For this purpose, ten P-LBS runs were performed for ($|\Sigma| = 4$, $m = 100$, $n = 600$) and ($|\Sigma| = 10$, $m = 10$, $n = 1000$), and the learned scoring functions were used in BS to solve the respective `rat` and ES instances. Figure 2 shows the obtained solution lengths for each loss function pf, lln, pl, ml, cml, llb, ub, and cmb as box plots. Loss functions pf and lln were used in the conjunction with beam-unaware training, whereas all other loss functions were used with beam-aware training. We can clearly see that loss functions pf, pl, ml, and llb perform significantly worse than cml, cmb, lln, and ub. Therefore, we use only loss functions cml, cmb, lln, and ub in the further experiments.

NBS Depth Limit. The choice of depth limit d in the NBS calls has a considerable impact on the runtime of P-LBS. Thus, we want to use a depth limit in the NBS calls that is as small as possible, but at the same time, large enough to produce robust models leading to high-quality predictions. In order to examine this aspect, we performed ten P-LBS runs each for different depth limits d in the NBS calls. Figure 3 shows exemplary box plots for final LCS lengths and training times on a representative `rat` instance, obtained by BS with scoring functions trained via P-LBS using the different depth limits. As one may expect, higher values for d lead to a more stable convergence of the NN, reflected by the smaller standard deviation and generally larger solution lengths seen in the left subfigure. The right subfigure shows the runtimes of P-LBS with $z = 2000$ iterations. We can see that our bootstrapping approach works well already for quite moderate depth limits $d \geq 5$ and can save much time. We therefore apply $d = 5$ in the remaining experiments.

Evaluation on Benchmark Instances. Finally, we evaluate our approach with each of the remaining loss function alternatives on all instances from benchmark sets `rat` and ES and also compare it to state-of-the-art methods from the literature. For this purpose, NNs were trained for each combination of $|\Sigma|$, m, and n occurring in the benchmark instances on random instances using P-LBS with each loss function. All training with P-LBS was done using beam width

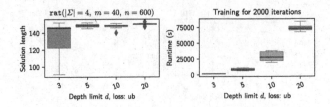

Fig. 3. Impact of depth limit d in NBS calls on the solution length of BS on a `rat` instance.

$\beta = \beta' = 50$, except for `rat` instances with $|\Sigma| = 20$, for which $\beta = \beta' = 20$ and $z = 1000$ were used due to computational budget limitations. Concerning the final testing, we followed [6] and applied BS on all benchmark instances using $\beta = 50$ to aim at low (computation) time and $\beta = 600$ to aim at high-quality solutions, respectively. Table 1 shows the obtained results. Columns $|g_{\text{lln}}|$, $|g_{\text{cml}}|$, $|g_{\text{ub}}|$, and $|g_{\text{cmb}}|$ list the average solution lengths obtained by BS with the NNs trained by P-LBS with loss functions lln, cml, ub, and cmb, respectively. Additionally, respective average solution qualities of LBS from [12] are shown in columns $|g_{\text{LBS}}|$. So far best-known average solution lengths reported in [6] are listed in columns $|g_{\text{lit-best}}|$. Average runtimes of the BS with the trained NNs (with loss function cmb) and corresponding ones from LBS are provided in columns $t_{\text{cmb}}[s]$ and $t_{\text{LBS}}[s]$.

The results show that BS with the trained NNs with loss function ub yields for both, low time and high-quality experiments, in six out of nine instance groups on benchmark set ES higher average solution lengths than lln, cml and cmb, while BS with the trained NNs with loss functions ub and cmb achieves on many instance groups on benchmark set `rat` higher average solution lengths than lln and cml. This coincides with our previous loss function analysis, where loss functions ub and cmb yielded higher solution lengths than lln and cml. Furthermore, the high variance in the results obtained by BS with the trained NNs with loss functions lln and cml on benchmark set `rat` indicates that these loss functions produce less robust models than ub and cmb. We conclude that loss functions ub and cmb are most suitable for training NNs to guide BS on the LCS problem. While BS with the trained NNs with loss function ub yields in six out of nine instance groups for low time and in seven out of nine for high-quality experiments higher average solution lengths compared to LBS on benchmark set ES, loss functions ub and cmb perform slightly worse than LBS on benchmark set `rat`.

Comparing BS with the trained NNs with loss functions ub and cmb to the so far best-known solutions, ub and cmb yield results being competitive on benchmark set ES but perform slightly worse on benchmark set `rat`. In total, BS with the trained NNs with loss functions lln, cml, ub and cmb could achieve in five out of 29 instance groups new best solutions for low time and in two out of 29 for high-quality. Concerning runtimes, we can conclude that they are lower than those of LBS, as more node features were used in LBS.

Table 1. LCS results on benchmark sets rat and ES obtained by BS with NNs trained by P-LBS (with loss functions lln, cml, ub, and cmb), and LBS [12]. Respective so far best-known average solution lengths reported in [6] are listed in columns $|g_{\text{lit-best}}|$.

Set	$|\Sigma|$	m	n	low time ($\beta = 50$)								high quality ($\beta = 600$)							
				$|g_{\text{lln}}|$	$|g_{\text{cml}}|$	$|g_{\text{ub}}|$	$|g_{\text{cmb}}|$	$t_{\text{cmb}}[s]$	$|g_{\text{LBS}}|$	$t_{\text{LBS}}[s]$	$|g_{\text{lit-best}}|$	$|g_{\text{lln}}|$	$|g_{\text{cml}}|$	$|g_{\text{ub}}|$	$|g_{\text{cmb}}|$	$t_{\text{cmb}}[s]$	$|g_{\text{LBS}}|$	$t_{\text{LBS}}[s]$	$|g_{\text{lit-best}}|$
rat	4	10	600	198.0	199.0	199.0	199.0	0.280	199.0	0.550	201.0	**205.0**	201.0	204.0	203.0	7.542	**205.0**	8.591	205.0
rat	4	15	600	*186.0	178.0	182.0	182.0	0.318	184.0	0.660	184.0	183.0	182.0	184.0	184.0	7.485	**185.0**	9.097	185.0
rat	4	20	600	162.0	159.0	*170.0	167.0	0.298	169.0	0.620	169.0	**173.0**	167.0	172.0	171.0	7.388	**173.0**	8.082	173.0
rat	4	25	600	*169.0	167.0	166.0	166.0	0.344	166.0	0.766	167.0	169.0	167.0	170.0	170.0	8.392	**171.0**	9.295	171.0
rat	4	40	600	147.0	143.0	150.0	150.0	0.204	**152.0**	0.844	152.0	145.0	150.0	153.0	153.0	8.649	**156.0**	10.064	156.0
rat	4	60	600	145.0	144.0	147.0	149.0	0.284	149.0	0.868	150.0	150.0	149.0	151.0	149.0	9.027	**152.0**	12.129	152.0
rat	4	80	600	110.0	134.0	132.0	132.0	0.304	**138.0**	1.056	138.0	132.0	137.0	140.0	138.0	9.451	140.0	12.564	142.0
rat	4	100	600	122.0	119.0	129.0	134.0	0.508	**135.0**	0.483	135.0	128.0	134.0	**138.0**	131.0	9.521	137.0	13.650	138.0
rat	4	150	600	101.0	117.0	123.0	126.0	0.746	**127.0**	1.176	127.0	117.0	122.0	114.0	128.0	10.159	**130.0**	11.625	130.0
rat	4	200	600	105.0	104.0	115.0	115.0	1.438	121.0	1.572	123.0	111.0	115.0	**124.0**	118.0	10.529	**123.0**	14.117	123.0
rat	20	10	600	70.0	*71.0	70.0	70.0	1.420	70.0	1.108	70.0	**71.0**	**71.0**	**71.0**	**71.0**	11.774	**71.0**	10.104	71.0
rat	20	15	600	**62.0**	61.0	**62.0**	60.0	1.788	**62.0**	1.117	**62.0**	62.0	62.0	**63.0**	62.0	12.457	**63.0**	12.048	63.0
rat	20	20	600	53.0	52.0	52.0	52.0	1.184	**54.0**	1.059	**54.0**	*55.0	54.0	54.0	*55.0	11.557	54.0	13.704	54.0
rat	20	25	600	50.0	50.0	**51.0**	50.0	0.983	**51.0**	1.152	**51.0**	**52.0**	**52.0**	**52.0**	51.0	7.662	**52.0**	13.073	52.0
rat	20	40	600	43.0	43.0	44.0	45.0	1.206	**49.0**	0.529	**49.0**	47.0	47.0	47.0	47.0	10.368	**49.0**	16.005	49.0
rat	20	60	600	43.0	43.0	43.0	45.0	1.948	**46.0**	1.945	**46.0**	46.0	45.0	45.0	46.0	16.698	**47.0**	19.734	47.0
rat	20	80	600	40.0	38.0	39.0	**38.0**	1.858	42.0	1.953	43.0	40.0	40.0	40.0	40.0	16.986	43.0	24.741	44.0
rat	20	100	600	37.0	37.0	36.0	**38.0**	1.711	**38.0**	2.007	**38.0**	38.0	39.0	38.0	39.0	13.746	39.0	24.441	40.0
rat	20	150	600	34.0	34.0	35.0	**37.0**	1.828	**37.0**	2.457	**37.0**	**37.0**	**37.0**	**37.0**	**37.0**	15.408	**37.0**	28.719	37.0
rat	20	200	600	33.0	33.0	33.0	33.0	1.956	**34.0**	2.048	**34.0**	**34.0**	**34.0**	**34.0**	**34.0**	18.582	**34.0**	32.118	34.0
ES	2	10	1000	608.74	605.06	608.84	604.62	0.41	606.80	0.71	**609.80**	614.42	612.18	614.68	611.64	8.107	613.35	11.248	**615.06**
ES	2	50	1000	518.54	529.24	531.46	529.30	0.67	529.76	1.02	**535.02**	523.38	532.96	535.76	533.24	13.046	534.29	16.684	**538.24**
ES	2	100	1000	503.64	508.40	511.36	508.54	1.09	514.62	1.72	**517.38**	507.96	512.34	513.94	512.32	17.777	516.85	22.103	**519.84**
ES	10	10	1000	198.70	198.56	198.80	199.00	0.94	198.94	1.17	**199.38**	202.60	202.42	202.42	202.60	18.081	202.10	23.249	**203.10**
ES	10	50	1000	134.60	133.80	134.22	133.64	1.15	134.02	1.89	**134.74**	136.28	135.64	136.12	135.58	16.080	135.56	19.720	**136.32**
ES	10	100	2500	119.46	120.82	120.94	120.88	1.39	121.20	1.20	**122.10**	121.26	122.12	122.38	122.14	17.258	122.67	24.349	**123.32**
ES	25	10	2500	*230.90	230.49	230.49	230.76	4.62	229.39	5.15	230.28	235.38	235.28	235.69	*235.69	76.820	235.20	77.550	235.22
ES	25	50	2500	136.62	136.84	137.18	136.84	8.09	133.88	7.82	**137.9**	138.56	138.92	139.04	138.66	74.902	137.44	105.956	**139.5**
ES	25	100	2500	119.76	120.56	120.64	120.46	10.33	119.70	16.74	**121.74**	121.22	121.92	122.12	121.94	116.728	121.71	159.843	**122.88**

7   Conclusions and Future Work

We proposed a general Policy-Based Learning Beam Search (P-LBS) framework for learning BS policies to solve combinatorial optimization problems. Instead of the traditional approach of evaluating each node independently with a hand-crafted evaluation function in BS, we learn a policy for selecting the nodes to continue with in the next BS level. Learning is performed by utilizing concepts from reinforcement learning, in particular the self-play of AlphaZero: P-LBS generates training data on its own by executing BS with the so far trained policy on many representative randomly generated problem instances. While different loss functions for learning a BS policy have been suggested but only studied from a theoretical point of view in the literature, we compare and evaluate them in the practical scenario of solving the prominent LCS problem. Reasonable scalability to larger problem instances could be achieved by utilizing bootstrapping. Our case study on the LCS demonstrates that P-LBS with loss functions ub and cmb is able to learn BS policies such that highly competitive results can be obtained.

One weakness we recognized in P-LBS using beam-unaware training is that the BS in our implementation returns exactly one best goal node and r–t path disregarding the fact that multiple best goal nodes with equal objective values and different r–t paths may exist. As a result, nodes in a training sample are labeled with zeroes, although these nodes possibly lie on a path of another best goal node. In future work, it would be promising to adapting the BS so that all found equally good goal nodes and corresponding r–t paths are considered. General improvement potential of P-LBS lies in using a more advanced graph neural network as policy to get rid of the dependency of specific instance sizes. Finally, we are interested in applying P-LBS to further problems of different nature to investigate the full potential of P-LBS.

References

1. Abe, K., Xu, Z., Sato, I., Sugiyama, M.: Solving NP-hard problems on graphs with extended alphago zero. arXiv:1905.11623 [cs, stat] (2020)
2. Bezerra, F.: A longest common subsequence approach to detect cut and wipe video transitions. In: Proceedings of the 17th Brazilian Symposium on Computer Graphics and Image Processing, pp. 154–160. IEEE Press (2004)
3. Chang, K.W., Krishnamurthy, A., Agarwal, A., Daumé, H., Langford, J.: Learning to search better than your teacher. In: Proceedings of the 32nd International Conference on Machine Learning, vol. 37, pp. 2058–2066 (2015)
4. Collins, M., Roark, B.: Incremental parsing with the perceptron algorithm. In: Proceedings of the 42nd Annual Meeting on Association for Computational Linguistics, pp. 111-es (2004)
5. Daumé, H., Marcu, D.: Learning as search optimization: approximate large margin methods for structured prediction. In: Proceedings of the 22nd International Conference on Machine Learning, pp. 169–176. ACM Press (2005)
6. Djukanovic, M., Raidl, G.R., Blum, C.: A beam search for the longest common subsequence problem guided by a novel approximate expected length calculation. In: Nicosia, G., Pardalos, P., Umeton, R., Giuffrida, G., Sciacca, V. (eds.) LOD 2019. LNCS, vol. 11943, pp. 154–167. Springer, Cham (2019). https://doi.org/10.1007/978-3-030-37599-7_14

7. Easton, T., Singireddy, A.: A large neighborhood search heuristic for the longest common subsequence problem. J. Heuristics **14**(3), 271–283 (2008)
8. Graves, A., Jaitly, N.: Towards end-to-end speech recognition with recurrent neural networks. In: Proceedings of the 31st International Conference on Machine Learning, pp. 1764–1772. PMLR (2014)
9. Gusfield, D.: Algorithms on Strings, Trees, and Sequences - Computer Science and Computational Biology. Cambridge University Press, Cambridge (1997)
10. Huang, L., et al.: Linearfold: linear-time approximate RNA folding by 5'-to-3' dynamic programming and beam search. Bioinformatics **35**(14), i295–i304 (2019)
11. Huber, M., Raidl, G.R.: Learning beam search: utilizing machine learning to guide beam search for solving combinatorial optimization problems. In: Nicosia, G., et al. (eds.) Machine Learning, Optimization, and Data Science. LNCS, vol. 13164, pp. 283–298. Springer, Cham (2022). https://doi.org/10.1007/978-3-030-95470-3_22
12. Huber, M., Raidl, G.R.: A relative value function based learning beam search for the longest common subsequence problem. In: Moreno-Díaz, R., Pichler, F., Quesada-Arencibia, A. (eds.) EUROCAST 2022. LNCS, vol. 13789, pp. 87–95. Springer, Cham (2022). https://doi.org/10.1007/978-3-031-25312-6_10
13. Laterre, A., et al.: Ranked reward: enabling self-play reinforcement learning for combinatorial optimization. In: AAAI 2019 Workshop on Reinforcement Learning on Games. AAAI Press (2018)
14. Lowerre, B.T.: The harpy speech recognition system. Ph.D. thesis, Carnegie Mellon University, Pittsburgh, PA (1976)
15. Maier, D.: The complexity of some problems on subsequences and supersequences. J. ACM **25**(2), 322–336 (1978)
16. Negrinho, R., Gormley, M., Gordon, G.J.: Learning beam search policies via imitation learning. In: Bengio, S., et al. (eds.) Advances in Neural Information Processing Systems, vol. 31, pp. 10652–10661. Curran Associates, Inc. (2018)
17. Ning, K., Ng, H.K., Leong, H.W.: Analysis of the relationships among longest common subsequences, shortest common supersequences and patterns and its application on pattern discovery in biological sequences. Int. J. Data Min. Bioinf. **5**(6), 611–625 (2011)
18. Ossman, M., Hussein, L.F.: Fast longest common subsequences for bioinformatics dynamic programming. Int. J. Comput. Appl. **975**, 8887 (2012)
19. Shyu, S.J., Tsai, C.Y.: Finding the longest common subsequence for multiple biological sequences by ant colony optimization. Comput. Oper. Res. **36**(1), 73–91 (2009)
20. Silver, D., et al.: A general reinforcement learning algorithm that masters chess, shogi, and Go through self-play. Science **362**(6419), 1140–1144 (2018)
21. Sutskever, I., Vinyals, O., Le, Q.V.: Sequence to sequence learning with neural networks. In: Advances in Neural Information Processing Systems, vol. 27. Curran Associates, Inc. (2014)
22. Sutton, R.S., Barto, A.G.: Reinforcement Learning: An Introduction. MIT Press, Cambridge (2018)
23. Xu, Y., Fern, A.: On learning linear ranking functions for beam search. In: Proceedings of the 24th International Conference on Machine Learning, pp. 1047–1054. ACM Press (2007)

Cooperative Coevolutionary Genetic Programming Hyper-Heuristic for Budget Constrained Dynamic Multi-workflow Scheduling in Cloud Computing

Kirita-Rose Escott(✉) [iD], Hui Ma[iD], and Gang Chen[iD]

Victoria University of Wellington, Wellington 6012, New Zealand
{kirita-rose.escott,hui.ma,aaron.chen}@ecs.vuw.ac.nz

Abstract. Dynamic Multi-workflow Scheduling (DMWS) in cloud computing is a well-known combinatorial optimisation problem. It is a great challenge to tackle this problem by scheduling multiple workflows submitted at different times and meet user-defined quality of service objectives. Scheduling with user-defined budget constraints is becoming increasingly important due to cloud dynamics associated with on-demand provisioning, instance types, and pricing. To address the Budget-Constrained Dynamic Multi-workflow Scheduling (BC-DMWS) problem, a novel Cooperative Coevolution Genetic Programming (CCGP) approach is proposed. Two heuristic rules, namely VM Selection/Creation Rule (VMR) and Budget Alert Rule (BAR), are learned automatically by CCGP. VMR is used to allocate ready tasks to either existing or newly rented VM instances, while BAR makes decisions to downgrade VM instances so as to meet the budget constraint. Experiments show significant performance and success rate improvement compared to state-of-the-art algorithms.

Keywords: Cloud computing · Dynamic workflow scheduling · Genetic programming · Hyper heuristic · Cooperative coevolution · Budget constraint

1 Introduction

Workflow scheduling aims to allocate workflow tasks to cloud resources so as to minimise the average response time involved in executing dynamically arriving tasks in workflows. Such a workflow scheduling problem is widely known as NP-Hard [14]. *Dynamic Multi-workflow Scheduling (DMWS)* brings new challenges of scheduling multiple workflows arriving dynamically over time. Dynamic resource provisioning enables new VMs to be provisioned when there are no suitable available VMs. Therefore, VM provisioning is sometimes considered during VM selection to help meet budget constraints [1,12,24]. However, the introduction of VM provisioning and additional features, such as VM speed, VM cost, and budget used, significantly increases the problem complexity. In the *Budget Constrained Dynamic Multi-workflow Scheduling* (BC-DMWS) problem, the

© The Author(s), under exclusive license to Springer Nature Switzerland AG 2023
L. Pérez Cáceres and T. Stützle (Eds.): EvoCOP 2023, LNCS 13987, pp. 146–161, 2023.
https://doi.org/10.1007/978-3-031-30035-6_10

objective is to allocate a set of workflows to VMs for execution, in order to minimise the makespan within a user-defined budget.

Current works that address budget constrains are mostly for *Static Workflow Scheduling* problems and are often inapplicable for *DMWS* [12,23,24]. This is due to the inability of static approaches to make scheduling decisions at runtime; instead decisions must be made a priori. Furthermore, many existing works rely on simple, man-made heuristics that do not consider many problem-centric features, such as workflow pattern, VM cost and VM speed, which are important for effective scheduling [3,5]. In [9–11], the limitations of the above works are addressed by proposing a *Genetic Programming Hyper-Heuristic* (GPHH) approach for dynamic workflow scheduling. However, previous GPHH studies aim to evolve a single heuristic rule to minimise makespan and ignore budget constraints. Since most real-world scientific applications are data or computation intensive, scientists are more interested in minimising the makespan and prefer to set budget limits for the whole workflow execution [18]. In view of this, it is vital to design scheduling approaches that are capable of satisfying budget constraints.

In this work, we propose a novel Cooperative Coevolution Genetic Programming (CCGP)-based approach that jointly evolves a *Budget Alert Rule* (BAR) as well as a *VM selection/creation rule* (VMR). Several reasons motivate us to propose this CCGP approach. Firstly, the new requirement for VM creation increases the complexity of the scheduling decision. Therefore, we need to evolve two rules by following a divide-and-conquer strategy. Secondly, VMs are rented on an hourly basis. It is important to maximise VM utilisation during its renting period. Therefore, we require a second indicator to identify potential risk of breaking the budget constraint. Finally, it is difficult to meet user-defined budget and minimise makespan simultaneously. Therefore, we introduce a *budget alert indicator* to achieve a desirable trade-off between performance and budget constraint by downgrading expensive VMs.

The overall goal of this paper is to propose a CCGP-based approach to jointly design both BAR and VMR to effectively tackle the BC-DMWS problem. To achieve this goal, we have the following objectives:

1. Propose a new *Budget Constrained Cooperative Coevolutionary Genetic Programming (BC-CCGP)* approach that can evolve BAR and VMR simultaneously for *DMWS* in consideration of the budget constraint.
2. Design a new terminal set with budget-centric features for the BC-CCGP approach. To the best of our knowledge, this is the first study in the literature on CCGP to generate heuristic rules to address budget constraint.
3. In order to evaluate the effectiveness of the CCGP approach we experimentally compare with human-designed heuristic rules and an existing GPHH algorithm [9] on multiple benchmark datasets.

2 Background and Related Work

2.1 Background

Hyper-heuristics are defined as automated methods for selecting or generating heuristics to solve hard computational problems [6]. In other words, hyper-heuristics aim to search the heuristic space to find the best heuristic for a situation, rather than trying to solve the problem directly. Hyper-Heuristic algorithms can be categorised as: *selective* and *generative* [6] methods. Selective hyper-heuristics rely on existing heuristics, whereas generative hyper-heuristics generate new heuristics based on experts' domain knowledge. In our problem, the applicability and effectiveness of existing heuristics is limited, therefore we focus on generative hyper-heuristics.

Recently, Genetic Programming (GP) has become a popular approach for generative hyper-heuristics, known as *GP Hyper-Heuristic* (GPHH) [17], which has been shown to have strong ability to represent and evolve complex heuristics [29–31]. The aim of GPHH is to evolve a heuristic that performs well on user-defined tasks.

Coevolution creates an effective evolutionary process that enables mutual adaption of two or more sub-populations [20]. Cooperative Coevolutionary Genetic Programming (CCGP) is an approach that combines GP with a cooperative coevolutionary framework such that multiple heuristics can be evolved to jointly solve a problem [22]. CCGP maintains $N > 0$ sub-populations for generating N heuristics respectively. GPHH and CCGP have been successfully applied to a variety of combinatorial optimisation problems such as Job Shop Scheduling (JSS) [31], Uncertain Capacitated Arc Routing (UCARP) [21], and Resource Allocation [26].

2.2 Related Work

Due to the dynamic nature of *DMWS*, heuristic rules are employed to allocate tasks to VMs at runtime. In more complex scheduling environments, there may be a number of interrelated problems that have to be resolved, i.e., the allocation of tasks to VMs in environments while satisfying budget constraints. This raises the question of how to best deal with such complex scenarios. The most straightforward approach is to design a set of heuristics, one for each decision [4]. However, it is challenging to design effective rules capable of meeting user-defined budget constraints. To satisfy budget constraints, some existing works attempt to improve an initial schedule. Wu et al. [27] proposed a budget-constrained workflow mapping for Minimum End-to-End Delay in IaaS Multi-Cloud environments (BCMED-MC). Gao et al. [13] proposed a Workflow Mapping algorithm for Financial Cost Optimisation. The limitation of these approaches is that they are designed to solve static problems and may be too time consuming to handle dynamic problems.

Existing approaches for dynamic resource provisioning, such as Greedy Resource Provisioning HEFT (GRP-HEFT) [12] and Just-in-time Algorithm (JIT-C) [24], rely on prior knowledge such as the sequence of workflows, the

workflow arrival time and available VMs, in order to make VM provision decisions. During the execution, the schedule is adjusted to dynamically acquire and
release VMs. However, for *DMWS*, such information is not known in advance.

GPHH and CCGP have recently been successfully applied to Workflow
Scheduling problems [11,29,30]. In these research works, rules evolved by GPHH
and CCGP can outperform human-designed rules. The automatic learning process substantially reduces the complexity of the heuristic design, allowing GPHH
to design effective heuristics. Yu et al. [30] proposed a Multi-Objective Scheduling Genetic Programming (MOSGP) that employs GP to evolve a set of non-
dominated heuristics to schedule workflows with different preferences over conflicting objectives, i.e., minimising makespan and minimising cost. MOSGP
focuses on the static workflow scheduling problem and did not consider the
arrival of multiple workflows over time, making the approach inapplicable for
DMWS. Yang et al. [29] proposed a GPHH approach called DWSGP to evolves
an optimal heuristic with the objective of minimising SLA penalties and VM
rental fees (cost) for dynamic workflow scheduling problems. However, DWSGP
does not aim to minimise makespan.

Xiao et al. [28] proposed a CCGP approach to minimise makespan for static
workflow scheduling. However, constraints are not considered and a single workflow is scheduled one at a time in this work. The addition of constraints further
complicates the already complex *DMWS* problem. For example, continuously
launching VMs can lead to low VM utilisation, high scheduling overhead and
inefficient resource provisioning policies that waste user budget [7]. Furthermore,
constraints can be described as soft or hard constraint. Since it is difficult to meet
hard budget constraints, it is advantageous to design an approach that is able to
adapt to complex scheduling environments and improve the chance for budget
satisfaction. CCGP is suitable for solving the *BC-DMWS* problem as it is able
to design multiple cooperative rules and these rules are able to adapt to the
changing workload patterns.

3   Problem Model

In the *BC-DMWS* problem, a sequence of workflows $W = \{w_1, ..., w_m\}$ arrives
to the cloud at a constant rate per minute to be allocated i.e., each workflow
w_i arrives at rate ar. A sequence of workflows W needs to be processed within
a user defined budget W_b. Each workflow w_i is depicted as a directed acyclic
graph $DAG(\mathbb{N}, \mathbb{E})$, where $\mathbb{N}$ is a set of nodes representing n dependent tasks, $\mathbb{E}$
indicates the data flow dependencies between tasks. Each workflow has a task
with no predecessors named *entry* task and a task with no successors named *exit*
task. There is a set of VM types $\Gamma = \{\tau_1, ..., \tau_m\}$ that can be selected to execute the workflow tasks. Each VM type τ_j has Compute Units $\pi^{cu}(\tau_j)$ and hourly
price $\pi^{price}(\tau_j)$. When a VM v_k is provisioned, it is assigned a VM type τ_j which
determines the speed m_k and hourly cost c_k. Furthermore, VMs are leased on an
hourly basis and charged for the full hour, whether or not the VM is utilised for
the entirety of that hour [2]. Therefore, in order to meet budget constraints, it is
for the best of interest of users to maximise utilisation of the provisioned VMs

and/or remove under-performing VMs. In the *BC-DMWS* problem it is necessary to directly handle the dynamic nature of the cloud computing environment through provisioning and de-provisioning VMs.

In line with the above, the *BC-DMWS* problem is formulated as follows:

$$\textbf{Minimise: } Makespan(W), \tag{1a}$$

$$\textbf{Subject to: } TotalCost(W) \leq Budget(W_b), \tag{1b}$$

The execution time of task t_i on virtual machine v_k is obtained by dividing the task size s_i by the speed m_k of virtual machine v_k below.

$$ET_{ik} = \frac{s_i}{m_k} \tag{2}$$

A task t_i of w_i becomes ready for execution whenever all of its parents $aParent(t_i)$ have completed processing. The remaining lease time RLT_k of a virtual machine v_k is the time until the end of the current lease. All leases are an hour long and can be renewed. The execution cost of task t_i on virtual machine v_k is given by Equation (3). EC_{ik} is obtained by determining if the execution time ET_{ik} is less than the remaining lease time RLT_k. If so, there is no extra cost. Otherwise, the cost c_k for a new lease of v_k is charged. If the ET_{ik} exceeds the hour lease period, the expected multiples of c_k are charged to reflect the cost.

$$EC_{ik} = \begin{cases} 0, & if\,ET_{ik} < RLT_k \\ c_k & otherwise \end{cases} \tag{3}$$

The expected completion time ECT_i of task t_i is the total execution time of t_i and its children $aChild(t_i)$ on virtual machine v_k, and is given by Equation (4).

$$ECT_i = total_{c \in aChild(t_i)} ET_{ik} \tag{4}$$

The current accumulated cost is determined by summing the accumulated cost AC_k of all $\mathcal{N}$ virtual machines. The *remaining budget RB* is obtained by subtracting the current accumulated cost from the given budget W_b below.

$$RB = W_b - \sum_{k=1}^{\mathcal{N}} AC_k \tag{5}$$

4   Methodology

The *Budget Constrained Dynamic Multi-workflow Scheduling (BC-DMWS)* problem consists of four decision-making procedures: (1) *Task selection*, (2) *VM selection/creation*, (3) *Budget alert*, and (4) *VM downgrade*. Task selection chooses the next ready task to be allocated to a VM. VM selection/creation determines which VM (either existing or newly leased VM) is the most appropriate to allocate the selected task to. VMs can be of different types, sizes and costs, according to cloud providers' decisions. VMs are leased on an hourly basis. The price for an hourly lease is fixed for the hour whether or not the VM is utilised the entire time [2]. The budget alert process determines whether it is necessary to downgrade VMs to meet the user-defined budget constraint [25].

4.1 Budget Constrained Cooperative Coevolution Genetic Programming Hyper-Heuristic

This subsection introduces the proposed *Budget Constrained Cooperative Coevolutionary Genetic Programming Hyper-Heuristic (BC-CCGP)* approach for the BC-DMWS problem. We first give an overview of our proposed *BC-CCGP* approach, then introduce the representation, terminal sets and fitness function. Finally, we describe the algorithm in detail.

Overview. We propose *BC-CCGP*, a CCGP based approach, to automatically generate two rules. In *BC-CCGP*, the first sub-population of CCGP is used to evolve the VM selection/creation Rule (VMR), which can be used to select appropriate VMs to execute the next ready task. The second sub-population of CCGP is used to evolve the Budget Alert Rule (BAR), which can be used to determine whether or not a rented VM should be deprovisioned at the end of its current lease.

Our proposed *BC-CCGP* approach is designed to evolve two rules simultaneously. Figure 1 presents an overview of the training process of *BC-CCGP*. *BC-CCGP* initialises two sub-populations of rules randomly. The rules are then cooperatively evolved by applying genetic operators, including parent selection, crossover and mutation. Upon completion of the training process, the generated rules are then used to solve unseen BC-DMWS problem instances during testing.

Representation, Terminal Set, and Function Set. In GP, a feasible solution is represented by a tree that consists of both function and terminal nodes [19]. In this paper, we design a new terminal set T_{set} containing budget-centric features of the BC-DMWS problem. As two rules are being evolved cooperatively, we design two separate terminal sets summarized in Table 1. The function set F_{set} contains a combination of arithmetic operators and mathematical functions. The terminal and function sets are summarized in Table 1.

Fitness Function. The objective of BC-DMWS is to minimise the makespan subject to a budget constraint with respect to a sequence of workflows arriving dynamically over time. A training instance ti_i consists of a sequence of Workflows W with different arrival rates. To evaluate a pair of VMR and BAR, the rules are applied to a set of T training instances ti_i with different arrival rates, to evaluate the generality of the evolved rules. This evaluation is completed through a series of simulations using the Workflowsim simulator [8].

Inspired by [25], we propose a cost-fitness function to evaluate the rules' effectiveness of controlling budgets. Specifically, whenever the cost of a candidate rule pair exceeds the given budget, their fitness is penalised. F_{ti_i} defined in Equation (8) hence encourages *BC-CCGP* to meet the budget constraint.

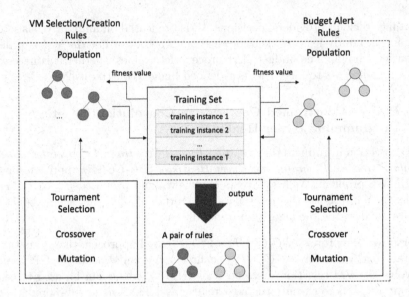

Fig. 1. Overview of the training process of *BC-CCGP*

The cost-fitness $F_{cost_{ti_i}}$ of a training instance defined in (6) is obtained by dividing the *cost* of training instance ti_i by the overall budget B.

$$F_{cost_{ti_i}} = \frac{cost_{ti_i}}{B} \tag{6}$$

For any training instance ti_i, the time-fitness $F_{time_{ti_i}}$ is computed by dividing the *makespan* of ti_i by *maxTime*, which is defined as the largest makespan obtained by the current population. Thus, the fitness depends on the individuals in the population as well as the solution being evaluated. Defined in (7), this encourages the solutions towards minimising the makespan.

$$F_{time_{ti_i}} = \frac{makespan_{ti_i}}{maxTime} \tag{7}$$

The fitness F_{ti_i} of a pair of rules on a training instance is determined by considering both $F_{cost_{ti_i}}$ and $F_{time_{ti_i}}$, according to Equation (8):

$$F_{ti_i} = \begin{cases} F_{cost_{ti_i}} + 1, & \text{if } F_{cost_{ti_i}} > 1 \\ F_{time_{ti_i}} & \text{otherwise} \end{cases} \tag{8}$$

The sum of the fitness of an individual on the training instances $\widetilde{F}$ is divided by the number of training instances T to calculate the overall fitness of a pair of VMR and BAR, as defined in Equation (9).

$$F = \frac{\widetilde{F}}{T} \tag{9}$$

Table 1. The Terminal and Function Sets for the Proposed *BC-CCGP* Approach

Symbol	Description
	Attributes for VM selection
TS	The total size of a task t_i
VS	The speed of a virtual machine v_j
ET	Execution time of a task t_i on vm v_j
EC	Execution cost of a task t_i on vm v_j
ECT	Expected completion time of a task t_i
	Attributes for budget alert
RB	Remaining budget s_b
PB	Percentage of budget this allocation uses
RT	Remaining time on the current lease of v_j
MC	Maximum cost of vm v_j to execute task t_i
Function set	+, -, ×, protected %, min, max

Algorithm. The proposed *BC-CCGP* approach is described in ALGORITHM 1. It starts with the initialisation of two sub-populations. Each sub-population consists of N randomly generated heuristic rules. The *ramped Half-and Half* [16] initialisation method is applied to ensure diversity in each sub-population.

The coevolution process begins on *line 4*, and repeats until a predefined maximum number of generations *maxGen* is reached. Each iteration is counted

---

Algorithm 1: *BC-CCGP* for the BC-DMWS problem

Input: A set of training instance T, Terminal sets and function sets
Output: The best VM selection/creation rule, The best Budget alert rule

1 Initialise population P_r with two sub-populations $P_r = \{P_{vmr}, P_{bar}\}$;
2 $P_r \leftarrow \{p_1^r, p_2^r, ..., p_N^r\}$;
3 $gen \leftarrow 0$;
4 **while** *maxGen is not reached* **do**
5   **for** $r = vmr \rightarrow bar$ **do**
6   **for** $i = 1 \rightarrow N$ **do**
7              $F \leftarrow$ apply p_i^r and $p_{rep}^{r'}$ on the training instance s^{gen} where $r \neq r'$ (see Algorithm 2);
8   calculate $\widetilde{F}$;
9              $p_i^r \leftarrow \frac{\widetilde{F}}{N}$;
10   **end**
11   **end**
12   **for** $r = vmr \rightarrow bar$ **do**
13          $p_r^{selected} \leftarrow$ TournamentSelection(P_r);
14          $p_r \leftarrow$ genetic_operators($p_r^{selected}$);
15   **end**
16      $gen \leftarrow gen + 1$
17 **end**

by *gen*. The rules are evaluated in turns (*line 5* to *line 11*). At the beginning, each heuristic rule p_i^r from P_{vmr} is paired with a representative $p_{rep}^{r'}$ from P_{bar} (*line 6* to *line 11*). In the first generation, a random heuristic rule from P_{bar} is selected as the representative of its sub-population. The pair of rules is then evaluated on *line 7*.

In the evaluation, the pair VMR and BAR are applied to schedule a sequence of workflows on to a set of virtual machines for execution. The detailed scheduling process is given in ALGORITHM 2. The scheduling returns the cost and time of a training instance. The values are then normalised (*line 8*) by Eq. (6) and (7), before calculating the fitness of a training instance Eq. (8). A training instance s^{gen} is comprised of a sequence of workflows, an arrival rate and a budget factor. The set of workflows in each generation is switched to improve generalisation of the evolved rule pairs. Additionally, in each evaluation the application of different arrival rates is considered. The sum of the fitness of the training instances $\widetilde{F}$ is then used to determine the fitness of the given pair of rules by Eq. (9) on *line 9*.

After the rules have been evaluated, the tournament selection and genetic operators are applied on the two sub-populations [26]. Tournament selection guides the evolutionary process [16]. Crossover and mutation generate new solutions from the selected rules. Crossover randomly selects branches of two rules to switch and mutation randomly selects a branch to be replaced with a randomly generated branch [26]. This crossover method is called one-point crossover, and this mutation method is called uniform mutation. After tournament selection and genetic operators are applied, new heuristics rules are added to the new population and the next iteration begins.

Algorithm 2: The procedure of scheduling workflow tasks to VMs

---

Input: VM selection rule *vmr*, Budget alert rule *bar*
Output: cost of training instance *cost*, makespan of scenario *time*

1 **for** *a training instance in S* **do**
2 | $cost = 0$;
3 | $time = 0$;
4 | Order tasks by number of children;
5 | **for** *each ready task r in RT* **do**
6 | | vm = vmSelectionCreation(r, *vmr*);
7 | | **if** *budgetAlert(bar)* **then**
8 | | | vmDowngrade = findVMToDowngrade();
9 | | | flagVMToDowngrade(vmDowngrade);
10 | | **end**
11 | | schedule(r, *vm*);
12 | **end**
13 | $cost$ += calculateCost();
14 | $time$ += calculateMakespan();
15 **end**
16 **return** *time* and *cost*;

ALGORITHM 2 describes the scheduling process using the heuristics generated by *BC-CCGP*. At the beginning, the simulator initialises a data center with no currently available virtual machines. Then, workflows arrive with an arrival rate based on a Poisson distribution [1]. The sequence of workflows need to be processed with a user-defined budget. In this work different budget factors with values ranging from tight (i.e., 1) to relaxed (i.e., 20) are considered, similar to [23]. As the workflows arrive, ready tasks are ordered by the number of child tasks they have (*line 4*). A ready task with a higher number of children is given a higher priority. This is the mechanism for task selection. The *VM selection/creation rule VMR* are used to select an existing VM or create a new VM on *line 6*. The idea of the *VM selection/creation rule* is to include a new VM (one of each type) into the candidate VM list which is only provisioned if selected. Then, the *Budget alert rule BAR* determines if the current allocation of task to VM is likely to result in the violation of the budget constraint (*line 7*). In the case where it is, a man-made heuristic (*line 8*) based on the Less Time-Consuming Task List (LTCTL) step of the Fair Budget-Constrained Workflow Scheduling algorithm (FBCWS) [23] is employed, to select a VM to be turned off at the end of its current lease period. On (*line 9*) the VM downgrade occurs, such that the selected VM will not be renewed at the end of the lease. As VMs are charged for the full hour it makes sense to keep the VM on until the end of the lease to increase the utilisation. Any new ready tasks can still be scheduled to such VMs for execution, unless the execution time exceeds the remainder of the lease. The training process outputs the *cost* and *makespan*, which are then used to determine the fitness.

5   Experiments

To evaluate the performance of our proposed *BC-CCGP* approach for BC-DMWS, we conduct experiments using five benchmark scientific workflows and compare with two state-of-the-art methods. Benchmark workflows obtained from real-world applications are commonly used in literature [15, 29, 30]. These benchmark workflows are summarised in able 4. This section first describes the experiment design, including competing methods, datasets, and test instances, before presenting the full results.

5.1   Experiment Design

To experimentally compare our *BC-CCGP* and other methods, we evaluate the performance of our *BC-CCGP* algorithm on a variety of training instances with different complexities, involving 6 VM types, 15 workflow applications, 3 arrival rates, and 5 budget factors. We perform this evaluation using the WorkflowSim simulator [8] to simulate a real cloud environment and execute a series of heterogeneous workflows using heuristic rules evolved by *BC-CCGP*.

Competing Algorithms. We compared our *BC-CCGP* approach with two state-of-the-art scheduling algorithms summarized below:

- *BCGP* [9] is an existing GPHH for dynamic workflow scheduling, with modification to the fitness function to explicitly consider the budget constraint as well as makespan as the optimisation objective. BCGP generates a single heuristic to select the best VM for the current ready task.
- *GRP-HEFT* [12] is a modified version of HEFT that is designed to tackle the budget constraint while always selecting the fastest available virtual machine to execute a task.

Dataset. We design 15 test instances for the experiments (see Table 2) which are divided into five groups, according to the budget factor. A larger budget factor indicates a larger budget. Each test instance has distinct arrival rates ranging from 1.5 to 10. We have extended the simulator to include dynamic workflow arrival and dynamic VM provisioning. Each test instance contains a set of 10 scientific workflows of different sizes and patterns.

Table 2. Test Instances

Test instance	Budget factor	Arrival rate
1	1	1.5
2	1	5
3	1	10
4	5	1.5
5	5	5
6	5	10
7	10	1.5
8	10	5
9	10	10
10	15	1.5
11	15	5
12	15	10
13	20	1.5
14	20	5
15	20	10

Table 3. VM Types Based on Amazon EC2

Type of VM	vCPU	ECU	Price ($/h)
m5.large	2	10	0.096
m5.xlarge	4	16	0.192
m5.2xlarge	8	37	0.384
m5.4xlarge	16	70	0.768
m5.8xlarge	32	128	1.536
m5.12xlarge	48	168	2.304

To create all the instances, we use five well-known real world scientific workflow patterns: CyberShake, Epigenomics, Inspiral, Montage and Sipht [15,29,30]. We consider small, medium and large workflows that consist of 24-30, 46-60 and 100 tasks, respectively. Each time a simulation is performed, 10 workflows are randomly sampled from the 15 workflows shown in Table 4.

Table 4. Number of Tasks in Workflow Applications

Application Size	CyberShake	Epigenomics	Inspiral	Montage	SIPHT
Small	30	24	30	25	30
Medium	50	46	50	50	60
Large	100	100	100	100	100

Cloud service providers typically offer various types of VMs with varying configurations. This paper considers a data centre equipped with 6 different VM types. Table 3 presents the VM configurations of EC2 that have been studied previously in [29]. As evidenced in the table, large vCPUs correspond to fast processing speed and high hourly rental price.

Parameter Settings. In this work we follow the standard parameter settings that are used commonly in literature [26]. In our experiments we set the population size to 512, and the number of generations to 100. The crossover, mutation and reproduction rates are set to 85%, 10% and 5%. The tournament size for tournament selection is 5 and the maximum depth of a tree is set to 10. Tournament selection is used to encourage the survival of effective heuristics in the population. We run the experiments 30 times to verify our results.

Performance Metrics. We use the following performance metrics in the experiments: Makespan (M), Cost (C) and Success Rate (SR). M and C measure the average cost and makespan achieved by an algorithm over 30 independent runs. The Success Rate (SR) of each algorithm is calculated in Equation (10) as the ratio between the number of simulation runs that were able to meet the budget $n_{W_{b_k}}$ and the total number of simulations n_{total}, multiplied by 100.

$$SR = \frac{n_{W_{b_k}}}{n_{total}} \times 100; \tag{10}$$

5.2   Experiment Results

This subsection first reports the cost, makespan and success rate achieved by $BC\text{-}CCGP$, $BCGP$ [9] and $GRP\text{-}HEFT$ [12]. Subsequently, we investigate the scheduling processes of these methods to identify potential causes of the observed performance differences among all the competing methods.

Overall Results. The comparison of cost, makespan and success rate of the three methods are shown in Table 5. The bold entries in the table indicate the best performance and the entries that are in red indicate the worst performance. The standard deviation is presented for makespan and cost. Although in parts the standard deviation is high, statistically it is clear that rules generated by $BC\text{-}CCGP$ have the advantage over $BCGP$ and $GRP\text{-}HEFT$ for minimising cost and maximising the success rate in all test instances. $BCGP$ is able to achieve slightly shorter makespan than $BC\text{-}CCGP$, at the expense of a substantially higher cost. Meanwhile, the success rate of $BCGP$ suffers. $GRP\text{-}HEFT$ consistently performs the worst across makespan and cost and is unable to meet the budget in any of the test instances.

Detailed Results. As detailed in Table 2, the budget factor increases over the test instances. Table 5 shows that the cost also increases over the test instances across all methods. Thus, in the cases of $BC\text{-}CCGP$ and $BCGP$, an increase in success rate over the test instances as the budget factor increases can be observed.

Table 5. Overall Results of Test Instances

Test Instance	Makespan (x10⁴sec)			Cost ($)			Success Rate (%)		
	BC-CCGP	BCGP[9]	GRPHEFT[12]	BC-CCGP	BCGP[9]	GRPHEFT[12]	BC-CCGP	BCGP[9]	GRP HEFT[12]
1	1.96± 1.21	1.95±1.21	13.33±11.5	15.08±5.09	71.73±45.60	67.69±51.01	20.00	0.00	0.00
2	1.93±1.22	1.92±1.22	13.5±11.66	15.48±5.11	74.43±49.48	65.81±47.36	13.33	0.00	0.00
3	1.92±1.22	1.92±1.22	13.4±11.6	16.39±4.65	82.34±53.14	65.26±46.70	10.00	0.00	0.00
4	1.95± 1.21	1.95±1.21	4.90±2.93	24.98±22.09	73.21±47.14	105.03±74.85	60.00	6.67	0.00
5	1.92±1.22	1.92±1.22	4.81±2.80	25.22±24.42	81.05±54.64	105.58±74.59	60.00	3.33	0.00
6	1.92±1.22	1.92±1.22	4.95±2.91	27.70±27.26	89.97±59.43	104.67±75.64	60.00	3.33	0.00
7	1.96± 1.21	1.95±1.21	3.40±1.41	30.66±36.56	80.16±67.52	171.10±120.41	86.67	50.00	0.00
8	1.93±1.22	1.92±1.22	3.37±1.45	34.29±43.81	83.75±69.14	174.26±120.51	83.33	53.33	0.00
9	1.93±1.22	1.92±1.22	3.34±1.40	34.75±41.21	91.24±71.22	169.96±116.04	70.00	43.33	0.00
10	1.96±1.21	1.95±1.21	2.66±0.79	47.51±49.73	74.96±45.40	332.23±348.04	93.33	83.33	0.00
11	1.93±1.22	1.92±1.22	2.63±0.78	49.47±48.09	77.98±49.75	333.81±363.72	96.67	86.67	0.00
12	1.93±1.22	1.92±1.22	2.62±0.77	52.14±49.97	89.11±58.47	299.53±339.81	96.67	80.00	0.00
13	1.95±1.21	1.95±1.21	2.78±0.94	44.37±48.74	176.21±271.48	665.30±679.48	93.33	73.33	0.00
14	1.93±1.22	1.92±1.22	2.78±0.96	45.66±48.05	176.53±267.08	675.20±727.29	96.67	76.67	0.00
15	1.92±1.22	1.92±1.22	2.76±0.98	51.38±54.82	191.44±273.57	555.98±614.46	93.33	70.00	0.00

GRP-HEFT, however, is unsuccessful in meeting the budget constraint across all test instances. This is due to the fact that *GRP-HEFT* is a fundamentally static method that only partially supports dynamic resource provisioning. First, GRP-HEFT greedily provisions a set of VMs according to the budget, starting from the most expensive VM type. Then, the modified HEFT allocates tasks onto VMs, removing unused VMs during the process. This is time consuming and often loses effectiveness for real-time scheduling.

In *BC-DMWS*, scheduling decisions must be made in real time. This is likely the cause of the inability for *GRP-HEFT* to meet the budget constraint in any test instance. Furthermore, in test instances 1-3, *GRP-HEFT* achieves a much larger makespan. Due to the tight budget, *GRP-HEFT* is unable to provision many VMs. This can lead to a bottleneck of waiting tasks, ultimately increasing the makespan.

It is not surprising that *BCGP* is able to achieve competitive makespan, comparable to previous findings [9]. The fitness function is the same as *BC-CCGP*, however, *BCGP* does not perform as well on cost or success rate. In fact *BCGP* achieves the highest cost for test instances 1-3 where there is a tight budget. This is like due to the fact that the allocation of tasks to VMs, as well as VM creation rely on a singular rule. Moreover, the single rule does not consider budget factors such as remaining budget and maximum cost. Furthermore as a single rule does the allocation, no adjustments are made to improve the probability of meeting the budget constraint after allocation.

During the scheduling process, an initial rule generated by *BC-CCGP* (VMR) is employed to create or select a VM to execute a ready task. A secondary rule (BAR) is then able to determine whether an adjustment needs to be made to improve the probability of meeting the budget constraint. The VM that is selected to be downgraded will be turned off at the end of its current lease. This allows the VM to be considered for scheduling during its entire lease period, improving the utilisation of every leased VM. As a result, *BC-CCGP* is able to achieve a competitive makespan across different arrival rates and budgets consistently, outperforming the other methods in both cost and success rate.

5.3 Analysis

Our *BC-CCGP* approach uses a set of features to evolve rules, including various properties of workflows and VMs as summarised in Table 1. Based on the best rules evolved by *BC-CCGP* and *BCGP*, the frequency of different features/terminals utilised by these rules are summarized in Table 6.

Table 6. Frequency of Each Terminal for Virtual Machine Selection/Creation (VMR) and Budget Alert Rule (BAR)

Rules Terminal	VM Selection/Creation Rule (VMR)						Budget Alert Rule (BAR)			
	EC	ECT	ET	RLT	TS	VS	MC	PB	RB	RT
BC-CCGP	42.1	7.1	8.1	15.3	11.8	15.6	21.7	22.0	27.1	29.1
BCGP	14.9	13.2	14.4	21.5	15.3	20.6				

Comparing VMRs evolved by *BC-CCGP* and *BCGP*, we see that the terminals have different frequency in different approaches. In *BC-CCGP*, the Execution Cost (*EC*) is apparently the most frequently occurring terminal. In comparison, in *BCGP* the frequency of terminals is more evenly distributed with Remaining Lease Time (*RLT*) and VM Speed (*VS*) being the most frequently occurring.

Similarly, we can see that the frequency of terminals is relatively evenly distributed for BAR. The most frequently occurring terminals are the Remaining Time on Current Lease (*RT*) and Remaining Budget (*RB*). We can hence conclude that Execution Cost (*EC*), Remaining Time on Current Lease (*RT*) and Remaining Budget (*RB*) are among the most important features to consider in the *BC-DMWS* problem. *EC* was the fourth commonly occurring terminal in *BCGP*. As *EC* was among the most important features for *BC-DMWS*, the fact that it was less prevalent in the rules generated by *BCGP* may attribute to the poor performance of *BCGP*.

6 Conclusions

In this paper, we present a new CCGP approach, *BC-CCGP*, for *Budget Constrained Dynamic Workflow Scheduling* (BC-DWS). The novelty of our approach is that our approach is capable of coevolving multiple scheduling heuristics, one for VM selection/creation and one for budget alert, to effectively solve the budget-constraint dynamic multi-workflow scheduling problem. The experimental evaluations demonstrate that our approach is able to make adjustments during scheduling to improve the probability of meeting budget constraints. The results show that the rules generated by our approach are able to consistently minimise cost and achieve the best success rate without sacrificing the performance for makespan. In the future, our proposed approach can be extended to coevolve rules for other scheduling decisions such as task selection and VM downgrade to further improve the performance.

References

1. Arabnejad, V., Bubendorfer, K., Ng, B.: Dynamic multi-workflow scheduling: a deadline and cost-aware approach for commercial clouds. Future Gener. Comput. Syst. **100**, 98–108 (2019)
2. AWS: Amazon EC2 on demand pricing (2022). https://aws.amazon.com/ec2/pricing/on-demand/
3. Blythe, J., Jain, S., et al.: Task scheduling strategies for workflow-based applications in grids. In: CCGrid 2005. IEEE International Symposium on Cluster Computing and the Grid, 2005, vol. 2, pp. 759–767. IEEE (2005)
4. Branke, J., Nguyen, S., Pickardt, C.W., Zhang, M.: Automated design of production scheduling heuristics: a review. IEEE Trans. Evol. Comput. **20**(1), 110–124 (2015)
5. Braun, T.D., Siegel, H.J., et al.: A comparison of eleven static heuristics for mapping a class of independent tasks onto heterogeneous distributed computing systems. J. Parallel Distrib. Comput. **61**(6), 810–837 (2001)
6. Burke, E.K., et al.: A classification of hyper-heuristic approaches. In: Gendreau, M., Potvin, J.Y. (eds.) Handbook of Metaheuristics. International Series in Operations Research & Management Science, vol. 146, pp. 449–468. Springer, Boston, MA (2010). https://doi.org/10.1007/978-1-4419-1665-5_15
7. Chakravarthi, K.K., Neelakantan, P., Shyamala, L., Vaidehi, V.: Reliable budget aware workflow scheduling strategy on multi-cloud environment. Cluster Comput. **25**(2), 1189–1205 (2022)
8. Chen, W., Deelman, E.: Workflowsim: a toolkit for simulating scientific workflows in distributed environments. In: 2012 IEEE 8th International Conference on E-Science, pp. 1–8. IEEE (2012)
9. Escott, K.-R., Ma, H., Chen, G.: Genetic programming based hyper heuristic approach for dynamic workflow scheduling in the cloud. In: Hartmann, S., Küng, J., Kotsis, G., Tjoa, A.M., Khalil, I. (eds.) DEXA 2020. LNCS, vol. 12392, pp. 76–90. Springer, Cham (2020). https://doi.org/10.1007/978-3-030-59051-2_6
10. Escott, K.R., Ma, H., Chen, G.: A genetic programming hyper-heuristic approach to design high-level heuristics for dynamic workflow scheduling in cloud. In: 2020 IEEE Symposium Series on Computational Intelligence (SSCI), pp. 3141–3148. IEEE (2020)
11. Escott, K.-R., Ma, H., Chen, G.: Transfer learning assisted GPHH for dynamic multi-workflow scheduling in cloud computing. In: Long, G., Yu, X., Wang, S. (eds.) AI 2022. LNCS (LNAI), vol. 13151, pp. 440–451. Springer, Cham (2022). https://doi.org/10.1007/978-3-030-97546-3_36
12. Faragardi, H.R., Sedghpour, M.R.S., Fazliahmadi, S., Fahringer, T., Rasouli, N.: GRP-HEFT: a budget-constrained resource provisioning scheme for workflow scheduling in IaaS clouds. IEEE Trans. Parallel Distrib. Syst. **31**(6), 1239–1254 (2019)
13. Gao, T., Wu, C.Q., Hou, A., Wang, Y., Li, R., Xu, M.: Minimizing financial cost of scientific workflows under deadline constraints in multi-cloud environments. In: Proceedings of the 34th ACM/SIGAPP Symposium on Applied Computing, pp. 114–121 (2019)
14. Jakobović, D., Jelenković, L., Budin, L.: Genetic programming heuristics for multiple machine scheduling. In: Ebner, M., O'Neill, M., Ekárt, A., Vanneschi, L., Esparcia-Alcázar, A.I. (eds.) EuroGP 2007. LNCS, vol. 4445, pp. 321–330. Springer, Heidelberg (2007). https://doi.org/10.1007/978-3-540-71605-1_30

15. Juve, G., Chervenak, A., Deelman, E., Bharathi, S., Mehta, G., Vahi, K.: Characterizing and profiling scientific workflows. Future Gener. Comput. Syst. **29**(3), 682–692 (2013)

16. Koza, J.R.: Genetic programming as a means for programming computers by natural selection. Statist. Comput. **4**(2), 87–112 (1994)

17. Koza, J.R., Koza, J.R.: Genetic Programming: on the Programming of Computers by Means of Natural Selection, vol. 1. MIT Press (1992)

18. Li, H., Wang, D., Xu, G., Yuan, Y., Xia, Y.: Improved swarm search algorithm for scheduling budget-constrained workflows in the cloud. Soft Comput. **26**(8), 3809–3824 (2022). https://doi.org/10.1007/s00500-022-06782-w

19. Lin, J., Zhu, L., Gao, K.: A genetic programming hyper-heuristic approach for the multi-skill resource constrained project scheduling problem. Expert Syst. Appl. **140**, 112915 (2020)

20. Liu, L., Zhang, M., Buyya, R., Fan, Q.: Deadline-constrained coevolutionary genetic algorithm for scientific workflow scheduling in cloud computing. Concurr. Comput. Pract. Exp. **29**(5), e3942 (2017)

21. MacLachlan, J., Mei, Y.: Look-ahead genetic programming for uncertain capacitated arc routing problem. In: 2021 IEEE Congress on Evolutionary Computation (CEC), pp. 1872–1879. IEEE (2021)

22. Nguyen, S., Zhang, M., Johnston, M., Tan, K.C.: Automatic design of scheduling policies for dynamic multi-objective job shop scheduling via cooperative coevolution genetic programming. IEEE Trans. Evol. Comput. **18**(2), 193–208 (2013)

23. Rizvi, N., Ramesh, D.: Fair budget constrained workflow scheduling approach for heterogeneous clouds. Cluster Comput. **23**(4), 3185–3201 (2020). https://doi.org/10.1007/s10586-020-03079-1

24. Sahni, J., Vidyarthi, D.P.: A cost-effective deadline-constrained dynamic scheduling algorithm for scientific workflows in a cloud environment. IEEE Trans. Cloud Comput. **6**(1), 2–18 (2015)

25. Shi, T., Ma, H., Chen, G., Hartmann, S.: Location-aware and budget-constrained service deployment for composite applications in multi-cloud environment. IEEE Trans. Parallel Distrib. Syst. **31**(8), 1954–1969 (2020)

26. Tan, B., Ma, H., Mei, Y., Zhang, M.: A cooperative coevolution genetic programming hyper-heuristic approach for on-line resource allocation in container-based clouds. IEEE Trans. Cloud Comput. (2020)

27. Wu, C.Q., Cao, H.: Optimizing the performance of big data workflows in multi-cloud environments under budget constraint. In: 2016 IEEE International Conference on Services Computing (SCC), pp. 138–145. IEEE (2016)

28. Xiao, Q.Z., Zhong, J., Feng, L., Luo, L., Lv, J.: A cooperative coevolution hyper-heuristic framework for workflow scheduling problem. IEEE Trans. Serv. Comput. (2019)

29. Yang, Y., Chen, G., Ma, H., Zhang, M., Huang, V.: Budget and SLA aware dynamic workflow scheduling in cloud computing with heterogeneous resources. In: 2021 IEEE Congress on Evolutionary Computation (CEC), pp. 2141–2148. IEEE (2021)

30. Yu, Y., Ma, H., Chen, G.: Achieving multi-objective scheduling of heterogeneous workflows in cloud through a genetic programming based approach. In: 2021 IEEE Congress on Evolutionary Computation (CEC), pp. 1880–1887. IEEE (2021)

31. Zhang, F., Mei, Y., Nguyen, S., Zhang, M.: Collaborative multifidelity-based surrogate models for genetic programming in dynamic flexible job shop scheduling. IEEE Trans. Cybern. (2021)

OneMax Is Not the Easiest Function for Fitness Improvements

Marc Kaufmann, Maxime Larcher, Johannes Lengler, and Xun Zou[✉]

Department of Computer Science, ETH Zürich, Zürich, Switzerland
{marc.kaufmann,maxime.larcher,johannes.lengler,xun.zou}@inf.ethz.ch

Abstract. We study the $(1 : s + 1)$ success rule for controlling the population size of the $(1, \lambda)$-EA. It was shown by Hevia Fajardo and Sudholt that this parameter control mechanism can run into problems for large s if the fitness landscape is too easy. They conjectured that this problem is worst for the ONEMAX benchmark, since in some well-established sense ONEMAX is known to be the easiest fitness landscape.

In this paper we disprove this conjecture. We show that there exist s and ε such that the self-adjusting $(1, \lambda)$-EA with the $(1 : s + 1)$-rule optimizes ONEMAX efficiently when started with εn zero-bits, but does not find the optimum in polynomial time on DYNAMIC BINVAL. Hence, we show that there are landscapes where the problem of the $(1 : s + 1)$-rule for controlling the population size of the $(1, \lambda)$-EA is more severe than for ONEMAX. The key insight is that, while ONEMAX is the easiest function for decreasing the distance to the optimum, it is *not* the easiest fitness landscape with respect to finding fitness-improving steps.

Keywords: parameter control · onemax · self-adaptation · $(1, \lambda)$-EA · one-fifth rule · dynamic environments · evolutionary algorithm

1 Introduction

The ONEMAX function assigns to a bit string x the number of one-bits in x. Despite, or rather because of its simplicity, this function remains one of the most important unimodal benchmarks for theoretical analysis of randomized optimization heuristics, and specifically of Evolutionary Algorithms (EAs). A reason for its special role is the result by Doerr, Johannsen and Winzen [13] that it is the easiest function with a unique optimum for the $(1 + 1)$-EA in terms of expected optimization time. This result was extended to many other EAs [37] and to stochastic dominance instead of expectations [38]. Easiest and hardest functions have become research topics of their own [6,16,18].

M. Kaufmann—The author was supported by the Swiss National Science Foundation [grant number 200021_192079].
X. Zou—The author was supported by the Swiss National Science Foundation [grant number CR-SII5_173721].

L. Pérez Cáceres and T. Stützle (Eds.): EvoCOP 2023, LNCS 13987, pp. 162–178, 2023.
https://doi.org/10.1007/978-3-031-30035-6_11

Whether a benchmark is easy or hard is crucial for parameter control mechanisms (PCMs) [9,17]. Such mechanisms address the classical problem of setting the parameters of algorithms. They can be regarded as meta-heuristics which automatically tune the parameters of the underlying algorithm. The hope is that (i) optimization is more robust with respect to the meta-parameters of the PCM than to the parameters of the underlying algorithm, and (ii) PCMs can deal with situations where different optimization phases require different parameter settings for optimal performance [3,4,10,12,14].

To this end, PCMs often rely on an (often implicit) measure of how easy the optimization process currently is. One of the most famous examples is the $(1 : s + 1)$-*rule* for step size adaptation in continuous optimization [7,25,34,36]. It is based on the heuristic that improving steps are easier to find if the step size is small, but that larger step sizes are better at exploiting improvements, if improvements are found at all. Thus we have conflicting goals requiring small and large step sizes, respectively, and we need a compromise between those goals. The $(1 : s + 1)$-rule resolves this conflict by defining a *target success rate*[1] of $q_s = 1/(s + 1)$, increasing the step size if the success rate (the fraction of steps which find an improvement) is above q_s, and decreasing the step size otherwise. Thus it chooses larger step sizes in environments where improvements are easy to find, and chooses smaller step sizes in more difficult environments.

More recently, the $(1 : s+1)$-rule has been extended to parameters in discrete domains, in particular to the mutation rate [11,15] and offspring population size [19,20] of EAs. For the self-adapting $(1, \lambda)$-EA with the $(1 : s + 1)$-rule, or SA-$(1, \lambda)$-EA for short, Hevia Fajardo and Sudholt showed in [19,20] an interesting collection of results on ONEMAX. They showed that optimization is highly efficient if the success ratio s is less than one. In this case, the algorithm achieves optimal population sizes λ throughout the course of optimization. The optimal λ ranges from constant at early stages to almost linear (in the problem dimension n) values of λ for the last steps. On the other hand, the mechanism provably fails completely if $s \geq 18$.[2] Then the algorithm does not even manage to obtain an 85% approximation of the optimum in polynomial time.

Hevia Fajardo and Sudholt also gave some important insights into the reasons for failure. The problem is that for large values of s, the algorithm implicitly targets a population size λ^* with a rather small success rate.[3] However, the $(1, \lambda)$-EA is a non-elitist algorithm, i.e., the fitness of its population can decrease over time. This is particularly likely if λ is small. So for large values of s, the PCM chooses population sizes that revolve around a rather small target value λ^*. It is still guaranteed that the algorithm makes progress in successful steps, which comprise a $\approx 1/(s + 1)$ fraction of all steps. But due to the small population size, it loses performance in some of the remaining $\approx s/(s+1)$ fraction of steps, and this loss cannot be compensated by the gain of successful steps.

---

[1] Traditionally the most popular value is $s = 4$, leading to the famous *one-fifth rule* [2].
[2] Empirically they found the threshold between the two regimes to be around $s \approx 3.4$.
[3] See [23, Section 2.1] for a detailed discussion of the target population size λ^*.

A counter-intuitive aspect of this bad trade-off is that it only happens when success is too easy. If success is hard, then the target population size λ^* is also large. In this case, the losses in unsuccessful steps are limited: most of the time, the offspring population contains a duplicate of the parent, in which case the loss is zero. So counter-intuitively, *easy fitness landscapes lead to a high runtime*. For ONEMAX, this means that the problems do not occur close to the optimum, but only at linear distance from the optimum. This result is implicitly contained in [19,20] and explicitly in [23]: for every $s > 0$ there is $\varepsilon > 0$ such that if the SA-$(1, \lambda)$-EA with success ratio s starts within distance at most εn from the optimum, then it is efficient with high probability, i.e., with probability $1 - o(1)$.

The results by Hevia Fajardo and Sudholt in [19,20] were for ONEMAX, but Kaufmann et al. [23,24] showed that this holds for all monotone, even all *dynamic monotone functions*[4]. Only the threshold for s changes, but it is a universal threshold: there exist $s_1 > s_0 > 0$ such that for every $s < s_0$, the SA-$(1, \lambda)$-EA is efficient on *every* (static or dynamic) monotone function, while for $s > s_1$ the SA-$(1, \lambda)$-EA fails for *every* (static or dynamic) monotone function to find the optimum in polynomial time. Moreover, for all $s > 0$ there is $\varepsilon > 0$ such that with high probability the SA-$(1, \lambda)$-EA with parameter s finds the optimum of every (static or dynamic) monotone function efficiently if it starts at distance εn from the optimum. Hence, all positive and negative results for ONEMAX from [19] generalize to every single function in the class of dynamic monotone functions. This class falls into more general frameworks of partially ordered functions that are easy to optimize under certain generic assumptions [5,21].

To summarize, small success rates (large values of s) are problematic, but only if the fitness landscape is too easy. Based on this insight, and on the afore-mentioned fact that ONEMAX is the easiest function for the $(1 + 1)$-EA, Hevia Fajardo and Sudholt conjectured that ONEMAX is the most problematic situation for the SA-$(1, \lambda)$-EA: "given that for large values of s the algorithm gets stuck on easy parts of the optimisation and that ONEMAX is the easiest function with a unique optimum for the $(1 + 1)$-EA, we conjecture that any s that is efficient on ONEMAX would also be a good choice for any other problem." In the terminology above, the conjecture says that the threshold s_0 below which the SA-$(1, \lambda)$-EA is efficient for all dynamic monotone functions, is the same as the threshold s_0' below which the SA-$(1, \lambda)$-EA is efficient on ONEMAX. Note that the exact value of s_0' is not known theoretically except for the bounds $1 \leq s_0' \leq 18$, but that empirically $s_0' \approx 3.4$ [19]. If the conjecture was true, then experiments on ONEMAX could provide parameter control settings for the SA-$(1, \lambda)$-EA that work in much more general settings.

However, in this paper we disprove the conjecture. Moreover, our result makes it more transparent in which sense ONEMAX is the easiest benchmark for the $(1 + 1)$-EA or the $(1, \lambda)$-EA, and in which sense it is not. It is the easiest in the

---

[4] A function $f : \{0, 1\}^n \to \mathbb{R}$ is *monotone* if flipping a zero-bit into a one-bit always improves the fitness. In the dynamic monotone setting, selection may be based on a different function in each generation, but it must always be a monotone function. The formal definition is not relevant for this paper, but can be found in [23].

sense that for no other function with a unique global optimum, the distance to the optimum decreases faster than for ONEMAX [13,38]. But it is *not* the easiest function in the sense that it is the easiest to make a fitness improvement, i.e., to find a successful step. Rephrased, other functions make it easier to find a fitness improvement than ONEMAX. For the problems of the $(1 : s + 1)$-rule described above, the latter variant prevails, since the $(1 : s + 1)$-rule adjusts its population size based on the success probability of finding a fitness improvement.

1.1 Our Result

We are far from being able to determine the precise efficiency threshold s_0' even in the simple setting of ONEMAX, and the upper and lower bound $1 \leq s_0' \leq 18$ are far apart. Therefore, it is no option to just compute and compare the thresholds for different functions. Instead, we will identify a setting in which we can indirectly compare the efficiency thresholds for ONEMAX and for some other function, without computing either of the thresholds explicitly. For this reason, we only study the following, rather specific setting that makes the proof feasible.

We show that there are $\varepsilon > 0$ and $s > 0$ such that with high probability the SA-$(1, \lambda)$-EA with parameter s (and suitably chosen other parameters), started at any search point at distance exactly εn from the optimum

– finds the optimum of ONEMAX in $O(n)$ generations;
– does not find the optimum of DYNAMIC BINVAL in polynomial time.

The definition of the DYNAMIC BINVAL function can be found in Sect. 2.2. The key ingredient to the proof is showing that at distance εn from the optimum, DYNAMIC BINVAL makes it easier to find a fitness improvement than ONEMAX (Lemma 7). Since *easy* fitness landscapes translate into poor choices of the population size of the SA-$(1, \lambda)$-EA, and thus to *large* runtimes, we are able to find a value of s that separates the two functions: for this s and for distance εn from the optimum, the algorithm will have drift *away* from the optimum for DYNAMIC BINVAL (leading to exponential runtime), but drift *towards* the optimum for ONEMAX. Since the fitness landscape for ONEMAX only gets harder closer to the optimum, we then show that the drift remains positive all the way to the optimum for ONEMAX. A sketch with more detail can be found in Sect. 3.1.

A limitation of our approach is that we start with a search point at distance εn from the optimum, instead of a uniformly random search point in $\{0,1\}^n$. This simplifies the calculations substantially, and it disproves the strong "local" interpretation of the conjecture in [19] that an s that works for ONEMAX in some specific part of the search space also works in the same part for all other dynamic monotone functions. Our choice leaves open whether some weaker version of the conjecture in [19] might still be true. But since our argument refutes the intuitive foundation of the conjecture, we do not think that this limitation is severe.

Another limitation is our use of a dynamic monotone function instead of a static one. We show that ONEMAX is not the easiest function in the class of dynamic monotone functions, but it could still be easiest in the smaller class of static monotone functions. We have made this choice for technical simplicity. We believe that our results for DYNAMIC BINVAL could also be obtained with very similar arguments for a static HOTTOPIC function as introduced in [28]. However, DYNAMIC BINVAL is simpler than HOTTOPIC functions, and the dynamic setting allows us to avoid some technical difficulties. We thus restrict ourselves to experiments for this hypothesis (Sect. 4), and find that ONEMAX indeed has a harder fitness landscape (in terms of improvement probability) than other static monotone or even linear functions, and consistently (but counter-intuitive) the SA-$(1, \lambda)$-EA chooses a higher population size for ONEMAX. For some values of s, this leads to positive drift and efficient runtime on ONEMAX, while the same algorithm has negative drift and fails on other functions.

Finally, apart from the success ratio s, the SA-$(1, \lambda)$-EA also comes with other parameters. For the *mutation rate* we use the standard choice $1/n$, and any c/n for a constant $0 < c < 1$ would also work. The *update strength* $F > 1$ is the factor by which λ is reduced in case of success, see Sect. 2.1 for details. A slight mismatch with [19] is that we choose $F = 1 + o(1)$, while [19] focused on constant F. Again, this simplifies the analysis, but the restriction does not seem crucial for the conceptual understanding that we gain in this paper.

2   Preliminaries and Definitions

Our search space is always $\{0, 1\}^n$. Throughout the paper we will assume that $s > 0$ is independent of n while $n \to \infty$, but $F = 1 + o(1)$ will depend on n. We say that an event $\mathcal{E} = \mathcal{E}(n)$ holds *with high probability* or *whp* if $\Pr[\mathcal{E}] \to 1$ for $n \to \infty$. We will write $x = a \pm b$ as shortcut for $x \in [a - b, a + b]$. Throughout the paper we will measure drift *towards* the optimum, so a positive drift always points towards the optimum, and a negative drift points away from the optimum.

2.1   The Algorithm: SA-$(1, \lambda)$-EA

The $(1, \lambda)$-EA is the algorithm that generates λ offspring in each generation, and picks the fittest one as the unique parent for the next generation. All offspring are generated by standard bit mutation, where each of the n bits of the parent is flipped independently with probability $1/n$. The performance of the $(1, \lambda)$-EA for static population size λ is well-understood [1,35].

We will consider the self-adjusting $(1, \lambda)$-EA with $(1 : s + 1)$-success rule to control the population size λ, with success rate s and update strength F, and we denote this algorithm by SA-$(1, \lambda)$-EA. It is given by the following pseudocode. The key difference from the standard $(1, \lambda)$-EA is that the population size λ is updated at each step: whenever a fitness improvement is found, the population is reduced to λ/F and otherwise the population is increased to $\lambda F^{1/s}$. Note that the parameter λ may take non-integral values during the execution of the algorithm, and the number of children is the integer $\lfloor \lambda \rceil$ closest to λ.

One way to think about the SA-$(1, \lambda)$-EA is that for each search point x it implicitly has a *target population size* $\lambda^* = \lambda^*(x)$ such that, up to rounding,

---

Algorithm 1. SA-$(1, \lambda)$-EA with success rate s, update strength F, mutation rate c/n, initial start point $x^{\text{init}} \in \{0,1\}^n$ and initial population size $\lambda^{\text{init}} = 1$ for maximizing a fitness function $f : \{0,1\}^n \to \mathbb{R}$.

Initialization: Set $x^0 = x^{\text{init}}$ and $\lambda^0 := 1$
Optimization: for $t = 0, 1, \ldots$ do
  Mutation: for $j \in \{1, \ldots, \lfloor \lambda^t \rceil\}$ do
    $y^{t,j} \leftarrow$ mutate(x^t) by flipping each bit of x^t independently with prob. $1/n$
  Selection: Choose $y^t = \arg\max\{f(y^{t,1}), \ldots, f(y^{t,\lfloor\lambda\rceil})\}$, breaking ties randomly
  Update:
  if $f(y^t) > f(x^t)$ **then** $\lambda^{t+1} \leftarrow \max\{1, \lambda^t/F\}$;   **else** $\lambda^{t+1} \leftarrow F^{1/s}\lambda^t$;
  $x^{t+1} \leftarrow y^t$;

---

the probability to have success (the fittest of λ^* offspring is strictly fitter than the parent) equals the *target success rate* $s^* = 1/(s+1)$. The $(1:s+1)$-rule ensures that there is a drift towards λ^*: whenever $\lfloor \lambda \rceil > \lambda^*$, then λ decreases in expectation, and it increases for $\lfloor \lambda \rceil < \lambda^*$, both on a logarithmic scale. We refer the reader to [23, Section 2.1] for a more detailed discussion.

For the results of this paper, we will specify $s > 0$ as a suitable constant, the initial population size is $\lambda^{\text{init}} = 1$, the initial search point has exactly εn zero-bits for a given ε, and the update strength is $F = 1 + \eta$ for some $\eta \in \omega(\log n/n) \cap o(1/\log n)$. We often omit the index t if it is clear from the context.

2.2 The Benchmarks: OneMax and Dynamic BinVal

The first benchmark, the OneMax function, counts the number of one-bits

$$\text{Om}(x) = \text{OneMax}(x) = \sum_{i=1}^{n} x_i.$$

of $x \in \{0,1\}^n$. We also define the ZeroMax function $Z(x) := n - \text{Om}(x)$ as the number of zero-bits in x. Throughout the paper, we will denote $Z^t := Z(x^t)$, and we will frequently use the scaling $\varepsilon = Z/n$.

Our other benchmark is a *dynamic* function [32]. That means that in each generation t, we choose a different function f^t and use f^t in the selection update step of Algorithm 1. We choose Dynamic BinVal or DBv [30,31], which is the binary value function BinVal, applied to a randomly selected permutation of the positions of the input string. This function has been used to model dynamic environments [30,31] and uncertain objectives [27]. In detail, BinVal is the function that interprets a bit string as an integer representation and returns its value, so $\text{BinVal}(x) = \sum_{i=1}^{n} 2^{i-1} \cdot x_i$. For the dynamic version, for each generation t we draw uniformly at random a permutation π^t of the set $\{1, \ldots, n\}$. The DBv function for generation t is then defined as

$$\text{DBv}^t(x) = \sum_{i=1}^{n} 2^{i-1} \cdot x_{\pi^t(i)}.$$

2.3 Tools

We will use drift analysis [29] to analyze two random quantities: The distance $Z^t = Z(x^t)$ of the current search point from the optimum, and the population size λ^t (or rather, $\log \lambda^t$). We use the following drift theorems to transfer results on the drift into expected hitting times. Due to space constraints we refer the reader to the cited sources for their statements.

Theorem 1 (Tail Bound for Additive Drift [26]).

Theorem 2 (Negative Drift Theorem [26,33]).

To switch between differences and exponentials, we will frequently make use of the following estimates, taken from Lemma 1.4.2 – Lemma 1.4.8 in [8].

Lemma 3. *1. For all $r \geq 1$ and $0 \leq s \leq r$,*

$$(1 - 1/r)^r \leq 1/e \leq (1 - 1/r)^{r-1} \quad and \quad (1 - s/r)^r \leq e^{-s} \leq (1 - s/r)^{r-s}.$$

2. For all $0 \leq x \leq 1$, it holds that: $1 - e^{-x} \geq x/2$.
3. For all $0 \leq x \leq 1$ and all $y \geq 1$, it holds that $\frac{xy}{1+xy} \leq 1 - (1-x)^y \leq xy$.

3 Main Proof

We start this section by defining some helpful notation. Afterwards, we give an informal sketch of the main ideas, and provide details afterwards.

Definition 4. *Consider the SA-$(1,\lambda)$-EA optimizing a dynamic function $f = f^t$, and let $Z^t = Z(x^t)$. For all times t and all $i \in \mathbb{Z}$, we define*

$$p_i^{f,t} := \Pr[Z^t - Z^{t+1} = i \mid x^t, \lambda^t] \quad and \quad \Delta_i^{f,t} := i \cdot p_i^{f,t}.$$

We will often drop the superscripts f and t when the function and the time are clear from context. We also define $p_{\geq i} := \sum_{j=i}^{\infty} p_j$ and $\Delta_{\geq i} := \sum_{j=i}^{\infty} \Delta_j$; both $p_{\leq i}$ and $\Delta_{\leq i}$ are defined analogously. Finally, we write

$$\Delta^{f,t} := \mathbf{E}[Z^t - Z^{t+1} \mid x^t, \lambda^t] = \sum_{i=-\infty}^{\infty} \Delta_i^{f,t}.$$

Note that $i > 0$ and $\Delta > 0$ corresponds to steps/drift *towards* the optimum and $i < 0$ and $\Delta < 0$ *away from* the optimum.

Definition 5 (Improvement Probability, Equilibrium Population Size).
Let $x \in \{0,1\}^n$ and f be a strictly monotone function. Let y be obtained from x by flipping every bit independently with probability $1/n$. We define

$$p_{\mathrm{imp}}^f(x) := \Pr[f(y) > f(x)] \quad and \quad q_{\mathrm{imp}}^f(x,\lambda) := 1 - (1 - p_{\mathrm{imp}}^f(x))^{\lambda},$$

as the probability that respectively a single offspring or any offspring improves the fitness of x. We also define the equilibrium population size *as*

$$\lambda^{*,f}(x,s) := \log_{(1-p^f_{\mathrm{imp}}(x))}\left(\tfrac{s}{1+s}\right). \tag{1}$$

As usual, we drop the superscript when f is clear from context. As the two functions we consider are symmetric (i.e. all bits play the same role) p_{imp} only depends on $Z(x) = \varepsilon n$ so in slight abuse of notation we sometimes write $p_{\mathrm{imp}}(\varepsilon n)$ instead of $p_{\mathrm{imp}}(x)$, and sometimes drop the parameters by writing just p_{imp} when they are clear from context. Similarly we sometimes write q_{imp} and λ^*.

Remark 6. For all x, s and f, $\lambda^{*,f}(x,s)$ is chosen to satisfy $q^f_{\mathrm{imp}}(x, \lambda^{*,f}(x,s)) = \frac{1}{s+1}$. Note that the equilibrium population size λ^* need not be an integer, and rounding λ to the next integer can change the success probability by a constant factor. Thus we must account for these effects. Fortunately, as we will show, the effect of changing the function f from ONEMAX to DBv is much larger.

3.1 Sketch of Proof

We have three quantities that depend on each other: the target population size λ^*, the target success rate $1/(s + 1)$ and the distance $\varepsilon := Z/n$ of the starting point from the optimum. Essentially, choosing any two of them determines the third one. In the proof we will choose λ^* and s to be large, and ε to be small. As ε is small, it is very unlikely to flip more than one zero-bit and the positive contribution to the drift is dominated by the term Δ_1 (Lemma 8 (ii)). For ONEMAX we are also able to give a tight estimate of $\Delta_{\leq-1}$: for λ large enough we can guarantee that $|\Delta_{\leq-1}| \approx (1 - e^{-1})^\lambda$ (Lemma 8 (iii)).

The key to the proof is that under the above assumptions, the improvement probability p_{imp} for DBv is by a constant factor larger than for ONEMAX (Lemma 7). Since it is unlikely to flip more than one zero-bit, the main way to improve the fitness for ONEMAX is by flipping a single zero-bit and no one-bits. Likewise, DBv also improves the fitness in this situation. However, DBv may also improve the fitness if it flips, for example, exactly one zero-bit and one one-bit. This improves the fitness if the zero-bit has higher weight, which happens with probability $1/2$. This already makes $p^{\mathrm{DBv}}_{\mathrm{imp}}$ by a constant factor larger than $p^{\mathrm{OM}}_{\mathrm{imp}}$. (There are actually even more ways to improve the fitness for DBv.) As a consequence, for the same values of s and ε, the target population size λ^* for DBv is by a constant factor smaller than for ONEMAX (Lemma 7 (iii)).

This enables us to (mentally) fix some large λ, choose ε such that the drift for ONEMAX at $Z = \varepsilon n$ is slightly positive (towards the optimum) and choose the s that satisfies $\lambda^{*,\mathrm{OM}}(\varepsilon n, s) = \lambda$. Here, 'slightly positive' means that $\Delta_1 \approx 4|\Delta_{\leq-1}|$. This may seem like a big difference, but in terms of λ it is not. Changing λ only affects Δ_1 mildly. But adding just a single child (increasing λ by one) reduces $\Delta_{\leq-1}$ by a factor of $\approx 1-1/e$, which is the probability that the additional child is not a copy of the parent. So our choice of λ^*, ε and s ensures positive drift for ONEMAX as long as $\lambda^t \geq \lambda^* - 1$, but not for a much wider range.

However, as we show, λ^t stays concentrated in this small range due to our choice of $F = 1 + o(1)$. This already would yield progress for ONEMAX in a small range around $Z = \varepsilon n$. To extend this to all values $Z \leq \varepsilon n$, we consider the potential function $G^t = Z^t - K \log_F(\lambda^t)$, and show that this potential function has drift towards zero (Corollary 12) whenever $Z > 0$, similar to [19,23]. For DBV, we show that λ^t stays below $\lambda^{*,\mathrm{DBV}}(\varepsilon n, s) + 1$, which is much smaller than $\lambda^{*,\mathrm{OM}}(\varepsilon n, s)$, and that these values of λ^t give drift away from the optimum, Lemma 13. Hence, the algorithm is not able to cross the region $Z = \varepsilon n$ for DBV.

3.2 Proof Details

In the remainder of this section we give more detailed sketches of the proof. In particular, we derive some relations between λ, s, ε to find a suitable such triple. Those relations only hold if the number of bits n is large enough. For instance, we wish to start at a distance εn from the optimum, meaning we need εn to be an integer. We tacitly assume that εn is a positive integer. In particular, this implies $\varepsilon \geq 1/n$.[5]

The purpose of the next key lemma is twofold. On the one hand it gives useful bounds and estimates of the probabilities of improvement; on the other hand it compares those probabilities of improvement for ONEMAX and DYNAMIC BIN-VAL. In particular, the success probability for ONEMAX is substantially smaller than for DYNAMIC BINVAL, meaning that DYNAMIC BINVAL is easier than ONEMAX with respect to fitness improvements.

Lemma 7. *Let f be any dynamic monotone function and $1/n \leq \varepsilon \leq 1$. Then $p^f_{\mathrm{imp}}(\varepsilon n) \leq \varepsilon$. More specifically for OM and DBV, for all $n \geq 10$ we have*

$$p^{\mathrm{OM}}_{\mathrm{imp}}(\varepsilon n) = e^{-1}\varepsilon \pm 2\varepsilon^2 \qquad and \qquad p^{\mathrm{DBV}}_{\mathrm{imp}}(\varepsilon n) = (1 - e^{-1})\varepsilon \pm 11\varepsilon^2.$$

In particular, there exists $c > 0$ such that the following holds.

(i) For every $\delta \leq 1$, $\lambda \in \mathbb{N}$, $\lambda \leq c\delta/\varepsilon$, and every dynamic monotone function f,

$$q^f_{\mathrm{imp}}(\varepsilon n, \lambda) = (1 \pm \delta)\lambda p^f_{\mathrm{imp}}(\varepsilon n).$$

(ii) For every $s \geq 1$, every constant $0 < \varepsilon \leq c$, and every dynamic monotone function f there exists a constant $\varepsilon' > 0$ such that

$$\lambda^{*,f}((\varepsilon - \varepsilon')n, s) - \lambda^{*,f}((\varepsilon + \varepsilon')n, s) \leq 1/4.$$

(iii) For every $1/n \leq \varepsilon \leq c$ and $s > 0$ we have

$$0.5\lambda^{*,\mathrm{OM}}(\varepsilon n, s) \leq \lambda^{*,\mathrm{DBV}}(\varepsilon n, s) \leq 0.6\lambda^{*,\mathrm{OM}}(\varepsilon n, s).$$

---

[5] In Lemma 7 ((ii)) and Lemma 13 we consider a constant $\varepsilon > 0$, introduce an $\varepsilon' = \varepsilon'(\varepsilon)$ and look at all states in the range $(\varepsilon \pm \varepsilon')n$. Again, we implicitly assume that $(\varepsilon - \varepsilon')n$ and $(\varepsilon + \varepsilon')n$ are integers, since we use those in the calculations.

Proof. (*Sketch*) Any monotone f needs to flip a zero-bit to improve, so $p_{\text{imp}}^{f}(\varepsilon n) \leq \varepsilon$. For ONEMAX this is only an improvement if no one-bit flips, which happens in an $\frac{1}{e}$-fraction of these cases. For DYNAMIC BINVAL, given a single zero-bit flip we improve despite any number of flipped one-bits if they all have lower weight than the zero-bit, the latter occurring with probability $\frac{1}{(1-\varepsilon)n+1} \sum_{i=1}^{(1-\varepsilon)n+1}(1 - 1/n)^{i-1} \approx 1 - e^{-1}$. Flipping at least two zero-bits, which may or may not improve the fitness for ONEMAX or DYNAMIC BINVAL, has probability at most $\frac{\varepsilon^2}{2}$. Now (i) follows from the definition and Lemma 3, (ii) from the definition and the observation that, since $p_{\text{imp}}^{f} \leq \varepsilon$, we can set $\log(1 - p_{\text{imp}}^{f}) = -(1 \pm C\varepsilon)p_{\text{imp}}^{f}$ for some C possibly large but absolute constant. This gives the result for small changes of ε. The ratio $\frac{\lambda^{*,\text{DBV}}(\varepsilon n,s)}{\lambda^{*,\text{OM}}(\varepsilon n,s)}$ tends to ≈ 0.58 as $\varepsilon \to 0$, yielding (iii).

Next are estimates of the drift of Z. In particular, the first statement is one way of stating that ONEMAX is the easiest function with respect to minimizing the distance from the optimum. We only apply it to $f = $ DBV, but believe the result is interesting enough to be mentioned.

Lemma 8. *There exists a constant $c > 0$ such that the following holds for every dynamic monotone function f. Let $\delta > 0$, there exists λ_0 such that the following holds.*

(i) *For all $\lambda \geq 1$, all $x \in \{0,1\}^n$ and all $i \in \mathbb{Z}^+$ we have*

$$\Delta_{\geq i}^{f}(x,\lambda) \leq \Delta_{\geq i}^{\text{OM}}(x,\lambda) \quad and \quad |\Delta_{\leq -i}^{f}(x,\lambda)| \geq |\Delta_{\leq -i}^{\text{OM}}(x,\lambda)|.$$

(ii) *For every integer $\lambda \geq \lambda_0$ and all $1/n \leq \varepsilon \leq c\delta/\lambda$ we have*

$$\Delta_{\geq 2}^{f}(\varepsilon n,\lambda) \leq \delta \Delta_1^{\text{OM}}(\varepsilon n,\lambda).$$

(iii) *For every integer $\lambda \geq \lambda_0$ and all $1/n \leq \varepsilon \leq c\delta/\lambda$ we have*

$$|\Delta_{\leq -1}^{\text{OM}}(\varepsilon n,\lambda)| = (1 \pm \delta)(1 - e^{-1})^\lambda.$$

Note in the right hand side of (ii), the drift is with respect to ONEMAX, not f.

Proof. (*Sketch*) For ONEMAX, $Z^{t+1} \geq Z^t - j$ only occurs if $Z(y) \geq Z(x) - j$ holds for all offspring y of x, in which case the *selected* offspring also satisfies this for any other f. This guarantees $p_{\leq j}^{\text{OM}} \leq p_{\leq j}^{f}$ and $p_{\geq j}^{\text{OM}} \geq p_{\geq j}^{f}$ for all $j \in \mathbb{Z}$. Since the probability of flipping at least i zero-bits is at most $\frac{\varepsilon^i}{i!}$, implying $p_{\geq i}^{\text{OM}} \leq \lambda \varepsilon^i/i!$, so that $\Delta_{\geq 2} \leq 2\lambda\varepsilon^2$. A single offspring flipping exactly a zero-bit and no one-bit implies that Z^t decreases, so $\Delta_1 = p_1 \geq \varepsilon(1-1/n)^{n-1} \geq e^{-1}\varepsilon$. Choosing $c = 1/(2e)$ ensures that $\Delta_{\geq 2} \leq \delta \Delta_1$, and the corresponding bound for $\Delta_{\geq 2}^{\text{DBV}}$ follows from (i). For (iii), note if every child flips at least a one-bit and no zero-bit, then Z^t must increase: $|\Delta_{\leq -1}| \geq \left((1 - (1-1/n)^{(1-\varepsilon)n}) \cdot (1 - 1/n)^{\varepsilon n}\right)^\lambda \geq (1 - \delta)(1 - e^{-1})^\lambda$, for c small and $\varepsilon < c\delta/\lambda$. On the other hand, we can upper-bound $|\Delta_{\leq -1}| = p_{\leq -1} + \sum_{i=2}^{\infty} p_{\leq -i}$ term-wise: $p_{\leq -1} \leq (1 - e^{-1})^\lambda$, and $p_{\leq -i} \leq \left(\binom{(1-\varepsilon)n}{i}n^{-i}\right)^\lambda \leq (i!)^{-\lambda} \leq 2^{(1-i)\lambda}$. As $p_{\leq -i}$ is the faster decaying term, the statement follows for large enough λ_0.

Repeated applications of Lemma 7 and 8 allow us to derive the core lemma of our proof — finding a suitable triple $(\lambda^*, \varepsilon, s)$ (proof omitted):

Lemma 9. *For every $\delta > 0$ there exists $\lambda_0 \geq 1$ such that the following holds. For every integer $\lambda \geq \lambda_0$, there exist constants $\tilde{\varepsilon}, \tilde{s}$ depending only on λ such that $\lambda = \lambda^{*,\mathrm{OM}}(\tilde{\varepsilon}n, \tilde{s})$ and*

$$\Delta_{\geq 1}^{\mathrm{OM}}(\tilde{\varepsilon}n, \lambda) = (4 \pm \delta)|\Delta_{\leq -1}^{\mathrm{OM}}(\tilde{\varepsilon}n, \lambda)| = (1 \pm \delta)/(\tilde{s}+1).$$

Additionally $\tilde{\varepsilon}(\lambda) = o_\lambda(1/\lambda)$ and $\tilde{s}(\lambda) = \omega_\lambda(1)$.[6] In particular, for every $\delta > 0$, a sufficiently large λ_0 guarantees that one may apply Lemmas 7 and 8.

Naturally, the lemma above implies that for parameters $\lambda, \tilde{s}$ and at distance $\tilde{\varepsilon}n$ from the optimum, the drift of Z for ONEMAX is roughly $\Delta^{\mathrm{OM}} = \Delta_{\geq 1}^{\mathrm{OM}} + \Delta_{\leq -1}^{\mathrm{OM}} \approx \frac{3}{4}\Delta_{\geq 1}^{\mathrm{OM}} > 0$. Moreover, we want to show that the SA-$(1,\lambda)$-EA not only passes this point, but continues all the way to the optimum. To this end, we define a more general potential function already used in [19] and [23].

Definition 10. *We define $h(\lambda) := -K \log_F \lambda$, with $K = 1/2$. We also define $g(x, \lambda) := Z(x) + h(\lambda)$.*

Similarly to $Z^t = Z(x^t)$, we write $H^t = h(\lambda^t)$ and $G^t = Z^t + H^t = g(x^t, \lambda^t)$.

Lemma 11. *Let $f = \mathrm{OM}$. At all times t such that $\lambda^t \geq F$ we have*

$$\mathbf{E}\left[H^t - H^{t+1} \mid x^t, \lambda^t\right] = \frac{K}{s}(1 - (s+1)q_{\mathrm{imp}}(x^t, \lfloor\lambda^t\rfloor)).$$

We omit the proof, which is a straightforward calculation. With these choices, the drift of G^t is positive for *all* $\varepsilon \leq \tilde{\varepsilon}$ and $\lambda^t \geq \lambda - 1$. The next corollary and Lemma (proofs omitted) follow by assembling the estimates from Lemma 7 (ii), Lemma 11, Lemma 8 (iii) and Lemma 9.

Corollary 12. *Let $f = \mathrm{OM}$. There exists $\lambda_0 \geq 1$ such that the following holds for all $\lambda \geq \lambda_0$. Let $s = \tilde{s} = \tilde{s}(\lambda)$, $\varepsilon = \tilde{\varepsilon} = \tilde{\varepsilon}(\lambda)$ be as in Lemma 9. There exist $\rho(\lambda), \varepsilon'(\lambda)$ such that if $1 \leq Z^t \leq (\varepsilon + \varepsilon')n$ and $\lambda^t \geq \lambda - 1$, then*

$$\mathbf{E}[G^t - G^{t+1} \mid Z^t, \lambda^t] \geq \rho.$$

We have just shown that when within distance at most $\tilde{\varepsilon}n$ from the optimum, ONEMAX has drift *towards* the optimum. We now turn to DYNAMIC BINVAL: the following lemma states that at distance $\tilde{\varepsilon}n$ from the optimum, DYNAMIC BINVAL has drift *away* from the optimum.

Lemma 13. *Let $f = \mathrm{DBV}$. There exists $\lambda_0 \geq 1$ such that for all $\lambda \geq \lambda_0$ there are $\nu, \varepsilon' > 0$ such that the following holds. Let $\tilde{s} = \tilde{s}(\lambda), \tilde{\varepsilon} = \tilde{\varepsilon}(\lambda)$ as in Lemma 9. If $Z^t = (\tilde{\varepsilon} \pm \varepsilon')n$ and $\lambda^t \leq \lambda^{*,\mathrm{DBV}}(\tilde{\varepsilon}n, \tilde{s}) + 1$ we have*

$$\mathbf{E}[Z^t - Z^{t+1} \mid x^t, \lambda^t] \leq -\nu.$$

---

[6] The subscript indicates dependency on λ, i.e., for all $c, C > 0$ there exists λ_0 such that for all $\lambda \geq \lambda_0$ we have $\tilde{\varepsilon}(\lambda) \leq c/\lambda$ and $\tilde{s}(\lambda) \geq C$.

Corollary 12 and Lemma 13 respectively give drift towards the optimum for ONEMAX and away from the optimum for DYNAMIC BINVAL.

Lemma 14. *Consider the SA-$(1, \lambda)$-EA on* ONEMAX *or* DYNAMIC BINVAL. *With probability $1 - n^{-\omega(1)}$, either the optimum is found or $\lambda^t \leq n^2$ holds for super-polynomially many steps, and in all of these steps $|Z^t - Z^{t+1}| \leq \log n$.*

Proof. (*Sketch*) Only a non-improving step at $\lambda^t \geq n^2/F^{1/s}$ would yield $\lambda > n^2$. Even one step away from the optimum, a fixed offspring finds an improvement with probability at least $1/n \cdot (1 - 1/n)^{n-1} \geq 1/(en)$, so a non-improving generation occurs with probability at most $(1 - 1/(en))^{\lambda^t} \leq e^{-(\lambda^t - 1)/(en)} = e^{-\Omega(n)}$. The bound of $e^{-\Omega(n)}$ holds even after a union-bound over superpoly-nomially many steps. Note that $|Z^t - Z^{t+1}|$ is at most the maximal number of bits flipped by any offspring. The probability of flipping at least $k \geq 1$ bits is at most $2/(k!)$. The statement then follows from $k = \log n$ and a union bound.

Now we are ready to state the main result of this paper.

Theorem 15. *Let $F = 1 + \eta$ for some $\eta = \omega(\log n/n) \cap o(1/\log n)$. There exist constants s, ε such that with high probability the SA-$(1, \lambda)$-EA with success ratio s, update strength F and mutation rate $1/n$ starting with $Z^0 = \varepsilon n$,*

a) *finds the optimum of* ONEMAX *in $O(n)$ generations;*
b) *does not find the optimum of* DYNAMIC BINVAL *in a polynomial number of generations.*

Proof (Sketch). We start by choosing λ large and take $\varepsilon = \tilde{\varepsilon}(\lambda)$ and $s = \tilde{s}(\lambda)$. For a), we can show that λ^t quickly reaches a value of at least $\lambda - 7/8$ and then stays above $\lambda - 1$. Afterwards, the potential G^t, which is $O(n)$, has a constant drift towards zero, so $Z^t = 0$ after $O(n)$ steps by Theorem 1. For b) the potential Z^t has a constant drift away from the optimum, so by Theorem 2, it stays above $(\varepsilon - \varepsilon')n$ for a superpolynomial number of steps. Details omitted.

4   Simulations

This section aims to provide empirical support to our theoretical results. Namely, we show that there exist parameters s and F such that ONEMAX is optimized efficiently by the SA-$(1, \lambda)$-EA, while DYNAMIC BINVAL is not. Moreover, in the simulations we find that the claim also extends to non-dynamic functions, such as BINVAL and BINARY[7]. In all experiments, we set $n = 1000$, the update strength $F = 1.5$, and the mutation rate to be $1/n$. Then we start the SA-$(1, \lambda)$-EA with the zero string and an initial offspring size of $\lambda^{\text{init}} = 1$. The algorithm terminates when the optimum is found or after $500n$ generations. The code for the simulations can be found at https://github.com/zuxu/OneLambdaEA.

---

[7] Defined as BINVAL$(x) = \sum_{i=1}^{n} 2^{i-1} x_i$ and BINARY$(x) = \sum_{i=1}^{\lfloor n/2 \rfloor} x_i n + \sum_{i=\lfloor n/2 \rfloor + 1}^{n} x_i$.

We first show that the improve-
ment probability p_{imp} of ONEMAX
is the lowest among all considered
monotone functions (Fig. 1), while
it is highest for DYNAMIC BINVAL
and BINARY (partly covered by the
violet line). Hence the fitness landscape looks hardest for ONEMAX
with respect to fitness improvements.
Therefore, to maintain a target success probability of $1/(1+s)$, more offspring are needed for ONEMAX, and
the SA-$(1, \lambda)$-EA chooses a slightly
higher λ (first panel of Figs. 2, partly
covered by the green line). This contributes positively to the drift towards
the optimum, so that the drift for
ONEMAX is higher than for the other
functions (second panel).

Figure 2 summarizes our main
result. The SA-$(1, \lambda)$-EA gets stuck
on all considered monotone functions
except ONEMAX when $s = 3$ and
the number of one-bits in the search
point is in $(0.55n, 0.65n)$. Although
the algorithm spends a bit more gen-

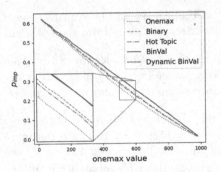

Fig. 1. The probability of fitness improvement with a single offspring for search
points with different ONEMAX values. Each
data point in the figure is estimated by
first sampling 1000 search points of the
corresponding ONEMAX value, then sampling 100 offspring for each of the sampled
search points, and calculating the frequency
of an offspring fitter than its parent. The
parameters of HOTTOPIC [28] are $L = 100$,
$\alpha = 0.25$, $\beta = 0.05$, and $\varepsilon = 0.05$. (Same
for Fig. 2.)

erations on ONEMAX between $0.5n$ and $0.8n$ compared to other parts, the optimum is found rather efficiently (not explicit in the figure due to the normalization). The reason is, all functions except ONEMAX have negative drifts within
the interval $(0.58n, 0.8n)$. The drift of ONEMAX here is also small compared to
the other regions, but remains positive. We note that the optimum is also found
for HOTTOPIC [28] despite its negative drift at $0.7n$, probably because the drift
is so weak that it can be overcome by random fluctuations. In [22], we show that
if s is decreased to 2, the optimum is found for all functions.

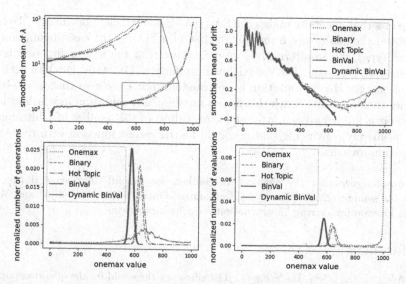

Fig. 2. Smoothed average of λ, smoothed average drift, average number of generations, and average number of evaluations of the self-adjusting $(1, \lambda)$-EA with $s = 3$, $F = 1.5$, and $c = 1$ in 100 runs at each OneMax value when optimizing monotone functions with $n = 1000$. The average of λ is shown in log scale. The average of λ and the average drift are smoothed over a window of size 15. The number of generations/evaluations is normalized such that its sum over all OneMax values is 1.

5   Conclusion

The key insight of our work is that there are two types of "easiness" of a benchmark function, which need to be separated carefully[8]. The first type relates to the question of how much progress an elitist hillclimber can make on the function. In this sense, it is well-known that OneMax is indeed the easiest benchmark among all functions with unique global optimum. However, a second type of easiness is how likely it is that a mutation gives an improving step. Here OneMax is not the easiest function.

Once those concepts are mentally separated, it is not hard to see OneMax is not the easiest function of second type. We have shown that Dynamic BinVal is easier (Lemma 7), but we conjecture this holds for many more functions, including static ones. This is backed up by experimental data, but we lack more systematic understanding of which functions are hard or easy in this aspect.

We have also shown that the second type of easiness is *relevant*. In particular, the SA-$(1, \lambda)$-EA relies on an empirical sample of the second type of easiness (aka the improvement probability) to choose the population size. Since the SA-$(1, \lambda)$-EA may make bad choices for too easy settings (of second type) if the parameter s is set too high, it is important to understand how easy a fitness landscape can get. These easiest fitness landscapes will determine the

---

[8] We note that there are other types of "easiness", e.g. with respect to a fixed budget.

range of s that generally makes the SA-$(1, \lambda)$-EA an efficient optimizer. We have disproved the conjecture from [19,20] that ONEMAX is the easiest function (of second type). As an alternative, we conjecture that the easiest function is the 'adversarial' DYNAMIC BINVAL, defined similarly to DYNAMIC BINVAL with the exception that the permutation is not random, but chosen so that any 0-bit is heavier than all 1-bits. With this fitness function, any mutation in which at least one 0-bit is flipped gives a fitter child, regardless of the number of 1-bit flips, so it is intuitively convincing that it should be the easiest function with respect to fitness improvement.

Acknowledgements. We thank Dirk Sudholt for helpful discussions during the Dagstuhl seminar 22081 "Theory of Randomized Optimization Heuristics" and Mario Hevia Fajardo for sharing his simulation code for comparison.

References

1. Antipov, D., Doerr, B., Yang, Q.: The efficiency threshold for the offspring population size of the (μ, λ) EA. In: Genetic and Evolutionary Computation Conference (GECCO), pp. 1461–1469 (2019)
2. Auger, A.: Benchmarking the $(1+1)$ evolution strategy with one-fifth success rule on the BBOB-2009 function testbed. In: Genetic and Evolutionary Computation Conference (GECCO), pp. 2447–2452 (2009)
3. Badkobeh, G., Lehre, P.K., Sudholt, D.: Unbiased black-box complexity of parallel search. In: Bartz-Beielstein, T., Branke, J., Filipič, B., Smith, J. (eds.) PPSN 2014. LNCS, vol. 8672, pp. 892–901. Springer, Cham (2014). https://doi.org/10.1007/978-3-319-10762-2_88
4. Böttcher, S., Doerr, B., Neumann, F.: Optimal fixed and adaptive mutation rates for the LeadingOnes problem. In: Schaefer, R., Cotta, C., Kołodziej, J., Rudolph, G. (eds.) PPSN 2010. LNCS, vol. 6238, pp. 1–10. Springer, Heidelberg (2010). https://doi.org/10.1007/978-3-642-15844-5_1
5. Colin, S., Doerr, B., Férey, G.: Monotonic functions in EC: anything but monotone! In: Genetic and Evolutionary Computation Conference (GECCO), pp. 753–760 (2014)
6. Corus, D., He, J., Jansen, T., Oliveto, P.S., Sudholt, D., Zarges, C.: On easiest functions for mutation operators in bio-inspired optimisation. Algorithmica **78**(2), 714–740 (2017)
7. Devroye, L.: The compound random search. Ph.D. dissertation, Purdue Univ., West Lafayette, IN (1972)
8. Doerr, B.: Probabilistic tools for the analysis of randomized optimization heuristics. In: Theory of Evolutionary Computation. NCS, pp. 1–87. Springer, Cham (2020). https://doi.org/10.1007/978-3-030-29414-4_1
9. Doerr, B., Doerr, C.: Theory of parameter control for discrete black-box optimization: provable performance gains through dynamic parameter choices. In: Theory of Evolutionary Computation. NCS, pp. 271–321. Springer, Cham (2020). https://doi.org/10.1007/978-3-030-29414-4_6
10. Doerr, B., Doerr, C., Ebel, F.: From black-box complexity to designing new genetic algorithms. Theor. Comput. Sci. **567**, 87–104 (2015)
11. Doerr, B., Doerr, C., Lengler, J.: Self-adjusting mutation rates with provably optimal success rules. Algorithmica **83**(10), 3108–3147 (2021)

12. Doerr, B., Doerr, C., Yang, J.: Optimal parameter choices via precise black-box analysis. Theor. Comput. Sci. **801**, 1–34 (2020)
13. Doerr, B., Johannsen, D., Winzen, C.: Multiplicative drift analysis. Algorithmica **64**, 673–697 (2012)
14. Doerr, B., Witt, C., Yang, J.: Runtime analysis for self-adaptive mutation rates. Algorithmica **83**(4), 1012–1053 (2021)
15. Doerr, C., Wagner, M.: Simple on-the-fly parameter selection mechanisms for two classical discrete black-box optimization benchmark problems. In: Genetic and Evolutionary Computation Conference (GECCO), pp. 943–950 (2018)
16. Droste, S.: A rigorous analysis of the compact genetic algorithm for linear functions. Nat. Comput. **5**(3), 257–283 (2006)
17. Eiben, A.E., Hinterding, R., Michalewicz, Z.: Parameter control in evolutionary algorithms. IEEE Trans. Evol. Comput. **3**, 124–141 (1999)
18. He, J., Chen, T., Yao, X.: On the easiest and hardest fitness functions. IEEE Trans. Evol. Comput. **19**(2), 295–305 (2014)
19. Hevia Fajardo, M.A., Sudholt, D.: Self-adjusting population sizes for non-elitist evolutionary algorithms: why success rates matter. arXiv preprint arXiv:2104.05624 (2021)
20. Hevia Fajardo, M.A., Sudholt, D.: Self-adjusting population sizes for non-elitist evolutionary algorithms: why success rates matter. In: Genetic and Evolutionary Computation Conference (GECCO), pp. 1151–1159 (2021)
21. Jansen, T.: On the brittleness of evolutionary algorithms. In: Stephens, C.R., Toussaint, M., Whitley, D., Stadler, P.F. (eds.) FOGA 2007. LNCS, vol. 4436, pp. 54–69. Springer, Heidelberg (2007). https://doi.org/10.1007/978-3-540-73482-6_4
22. Kaufmann, M., Larcher, M., Lengler, J., Zou, X.: Onemax is not the easiest function for fitness improvements. arXiv preprint arXiv:2204.07017 (2022)
23. Kaufmann, M., Larcher, M., Lengler, J., Zou, X.: Self-adjusting population sizes for the $(1, \lambda)$-EA on monotone functions (2022). https://arxiv.org/abs/2204.00531
24. Kaufmann, M., Larcher, M., Lengler, J., Zou, X.: Self-adjusting population sizes for the $(1, \lambda)$-EA on monotone functions. In: Rudolph, G., Kononova, A.V., Aguirre, H., Kerschke, P., Ochoa, G., Tusar, T. (eds.) Parallel Problem Solving from Nature (PPSN) XVII. PPSN 2022. Lecture Notes in Computer Science, vol. 13399, pp. 569–585. Springer, Cham (2022). https://doi.org/10.1007/978-3-031-14721-0_40
25. Kern, S., Müller, S.D., Hansen, N., Büche, D., Ocenasek, J., Koumoutsakos, P.: Learning probability distributions in continuous evolutionary algorithms-a comparative review. Nat. Comput. **3**(1), 77–112 (2004)
26. Kötzing, T.: Concentration of first hitting times under additive drift. Algorithmica **75**(3), 490–506 (2016)
27. Lehre, P., Qin, X.: More precise runtime analyses of non-elitist evolutionary algorithms in uncertain environments. Algorithmica 1–46 (2022). https://doi.org/10.1007/s00453-022-01044-5
28. Lengler, J.: A general dichotomy of evolutionary algorithms on monotone functions. IEEE Trans. Evol. Comput. **24**(6), 995–1009 (2019)
29. Lengler, J.: Drift analysis. In: Theory of Evolutionary Computation. NCS, pp. 89–131. Springer, Cham (2020). https://doi.org/10.1007/978-3-030-29414-4_2
30. Lengler, J., Meier, J.: Large population sizes and crossover help in dynamic environments. In: Bäck, T., et al. (eds.) PPSN 2020. LNCS, vol. 12269, pp. 610–622. Springer, Cham (2020). https://doi.org/10.1007/978-3-030-58112-1_42
31. Lengler, J., Riedi, S.: Runtime analysis of the $(\mu + 1)$-EA on the dynamic BinVal function. In: Zarges, C., Verel, S. (eds.) EvoCOP 2021. LNCS, vol. 12692, pp. 84–99. Springer, Cham (2021). https://doi.org/10.1007/978-3-030-72904-2_6

32. Lengler, J., Schaller, U.: The $(1 + 1)$-EA on noisy linear functions with random positive weights. In: Symposium Series on Computational Intelligence (SSCI), pp. 712–719. IEEE (2018)
33. Oliveto, P., Witt, C.: On the analysis of the simple genetic algorithm. Theor. Comput. Sci. **545**, 2–19 (2014)
34. Rechenberg, I.: Evolutionsstrategien. In: Schneider, B., Ranft, U. (eds.) Simulationsmethoden in der Medizin und Biologie. Medizinische Informatik und Statistik, vol. 8, pp. 83–114. Springer, Berlin (1978). https://doi.org/10.1007/978-3-642-81283-5_8
35. Rowe, J.E., Sudholt, D.: The choice of the offspring population size in the $(1, \lambda)$ evolutionary algorithm. Theor. Comput. Sci. **545**, 20–38 (2014)
36. Schumer, M., Steiglitz, K.: Adaptive step size random search. IEEE Trans. Autom. Control **13**(3), 270–276 (1968)
37. Sudholt, D.: A new method for lower bounds on the running time of evolutionary algorithms. IEEE Trans. Evol. Comput. **17**(3), 418–435 (2012)
38. Witt, C.: Tight bounds on the optimization time of a randomized search heuristic on linear functions. Comb. Probab. Comput. **22**(2), 294–318 (2013)

The Cost of Randomness in Evolutionary Algorithms: Crossover can Save Random Bits

Carlo Kneissl and Dirk Sudholt[(✉)]

Chair of Algorithms for Intelligent Systems University of Passau, Passau, Germany
dirk.sudholt@uni-passau.de

Abstract. Evolutionary algorithms make countless random decisions during selection, mutation and crossover operations. These random decisions require a steady stream of random numbers.

We analyze the expected number of random bits used throughout a run of an evolutionary algorithm and refer to this as the cost of randomness. We give general bounds on the cost of randomness for mutation-based evolutionary algorithms using 1-bit flips or standard mutations using either a naive or a common, more efficient implementation that uses $\Theta(\log n)$ random bits per mutation.

Uniform crossover is a potentially wasteful operator as the number of random bits used equals the Hamming distance of the two parents, which can be up to n. However, we show for a (2+1) Genetic Algorithm that is known to optimize the test function ONEMAX in roughly $(e/2)n \ln n$ expected evaluations, twice as fast as the fastest mutation-based evolutionary algorithms, that the total cost of randomness during all crossover operations on ONEMAX is only $\Theta(n)$. Consequently, the use of crossover can reduce the cost of randomness below that of the fastest evolutionary algorithms that only use standard mutations.

Keywords: Evolutionary algorithms · crossover · population diversity · runtime analysis · theory

1 Introduction

Evolutionary algorithms (EAs) are popular metaheuristics inspired by the principle of natural evolution that have found countless applications for optimization and design problems [15]. EAs are popular in practice because they are easy to use, they can provide solutions of acceptable quality in scenarios where computationally expensive exact approaches fail, and they can be applied in a black-box scenario, when the problem is not well understood and evaluating candidate solutions is the only way of learning about the problem in hand. The black-box scenario also applies when there is not enough expertise, time or money to design a problem-specific algorithm.

Despite countless successful applications in practice, the reasons behind the success of EAs are not well understood. It is often not clear when and why

© The Author(s), under exclusive license to Springer Nature Switzerland AG 2023
L. Pérez Cáceres and T. Stützle (Eds.): EvoCOP 2023, LNCS 13987, pp. 179–194, 2023.
https://doi.org/10.1007/978-3-031-30035-6_12

EAs perform well and when they don't. The performance of EAs depends crucially, and often unpredictably, on algorithmic design choices like the choice of operators and the choice of parameters such as the population size or the mutation strength [14]. Over the past 20 years, the runtime analysis of evolutionary algorithms has emerged as a fruitful research area (see, e. g., [13,14,20,26]) providing rigorous bounds on the performance of evolutionary algorithms on interesting problems. Problems range from simple pseudo-Boolean test problems like $\textsc{OneMax}(x) := \sum_{i=1}^{n} x_i$ (counting the number of ones in a bit string x) and $\textsc{LeadingOnes}(x) := \sum_{i=1}^{n} \prod_{j=1}^{i} x_i$ (the length of the longest prefix of ones in x) to various problems from combinatorial optimization [14,26]. These analyses have led to a better understanding of EAs' working principles (e. g. the advantages of population diversity [7] and the robustness of populations in stochastic environments [18,29]) and have inspired novel, probably more effective evolutionary and genetic algorithms[1] (e. g. choosing mutation rates from a heavy-tailed distribution to enable large changes [12], changing the order of crossover and mutation and amplifying the probability of finding an improving mutation [9], parent selection preferring worse search points [3] or choosing mutation rates adaptively based on information obtained from the current run [11]).

Existing runtime analyses focus on the expected number of function evaluations until a target (e. g. some global optimum) is reached, also called the optimization time. This performance measure is motivated by the fact that function evaluations are often expensive and may dominate the execution time of an EA.

We complement this line of research by asking a more specific question: *how many random bits* does an evolutionary algorithm consume during the course of a run? Randomness is crucial for the success of EAs and most operators from mutation to crossover and selection make heavy use of randomness to promote exploration of the search space and to create truly novel solutions. A steady stream of randomness is required to implement these operators. Random numbers can be created from physical resources such as atmospheric noise or from pseudo-random number generators (PRNGs). PRNGs create the illusion of randomness through a deterministic sequence of numbers. Established PRNGs such as the Mersenne Twister [25] produce sequences of numbers that are hard to distinguish from perfect randomness and can be produced more efficiently than hardware-implemented instructions [30].

We consider the expected number of random bits drawn by an EA until a global optimum is reached, and refer to this as the *cost of randomness*. By studying the cost of randomness we aim to provide a new perspective and obtain a better understanding of the working principles of EAs. As generating random bits is inexpensive [30], we do not expect random number generation to play a significant role in the execution time of an EA—unless randomness is used naively (see Sect. 3 for a discussion and analysis of naive implementations of mutations). The motivation for our work is to obtain a deeper understanding of the working principles of EAs and the various resources used throughout a run,

---

[1] We use the term "genetic algorithm" (GA) for EAs that use crossover.

in addition to established performance measures such as the number of function evaluations.

We show that studying the cost of randomness indeed gives new insights into search dynamics. We provide a detailed analysis of the cost of randomness incurred during uniform crossover operations on a run of a (2+1) GA on the test function $\text{ONEMAX}(x) := \sum_{i=1}^{n} x_i$. It has been shown in [33] that the (2+1) GA is twice as fast as the fastest evolutionary algorithm using only mutation in terms of function evaluations (up to small-order terms). Since uniform crossover is an additional, potentially wasteful operator in terms of random bits, it is not clear whether the (2+1) GA will retain this advantage when considering the cost of randomness as performance measure. However, we show in a thorough analysis of population dynamics that the cumulative population diversity during the run is so small that all crossover operations only need a total of $\Theta(n)$ random bits throughout the whole run. This means that crossover still provides at least a factor 2 speedup over all mutation-based EAs when using the cost of randomness as performance measure.

1.1   Our Contributions

We first review costs for sampling from probability distributions typically found in EAs in Sect. 2. In Sect. 3 we give general statements on the cost of randomness for EAs that only use mutation (and no crossover), including Randomized Local Search (RLS) and the well-known (1+1) EA, considering a naive and a more sophisticated implementation of standard mutations (i.e. flip each bit independently with a mutation rate $p \in (0,1)$). These analyses are problem-independent as the cost of randomness is essentially proportional to the expected number of function evaluations. The additional factor is $\Theta(\log n)$ for 1-bit flips and $\Theta(\log(1/p))$ for standard mutation with mutation rate p using the more efficient implementation.

We observe more interesting effects when using crossover as an additional operator. For a uniform crossover of two search points x and y, we will argue that the number of random bits used equals the Hamming distance $H(x,y)$ of x and y. Hence, here the cost of randomness depends on the search dynamics and captures the cumulative population diversity across all generations applying crossover. We give a general upper bound for all $(\mu+\lambda)$ Genetic Algorithms in Sect. 4 and study the aforementioned (2+1) GA on ONEMAX in more detail in Sect. 5. We show that, although the first generation already uses $\Theta(n)$ random bits in expectation, the cost of randomness in all crossover operations during the run is only $\Theta(n)$, and hence we obtain a proof that crossover in the (2+1) GA can reduce the cost of randomness beyond that of mutation-only algorithms.

In this extended abstract, most proofs have been omitted or reduced to proof sketches due to space restrictions.

1.2 Related Work

Most runtime analyses use the number of function evaluations before a global optimum is reached as the sole performance measure. Jansen and Zarges [21] proposed a refined cost model based on empirical data about the computational cost of individual operators. Additional performance measures include studying the parallel time (number of generations) in parallel evolutionary algorithms [23] and the communication effort (number of migrated individuals) in island models [24].

Doerr, Fouz and Witt [10] studied the effect of partially removing randomness in quasi-random EAs. They proposed algorithms in which the number of flipping bits is chosen randomly, but the choice which bits to flip is made deterministically. One algorithm was shown to be twice as fast on ONEMAX as the (1+1) EA, in expectation.

2 Preliminaries and Analysis Tools

We focus on the most common search space $\{0,1\}^n$ and the aim is to maximize a pseudo-Boolean fitness function $f\colon \{0,1\}^n \to \mathbb{R}$. We consider black-box algorithms that query search points based on the quality of search points evaluated previously. For a black-box algorithm $\mathcal{A}$ and a fitness function f we denote by $T_{\mathcal{A},f}$ the random number of function evaluations until a global optimum of f is found for the first time (optimization time).

For measuring the cost of randomness, we assume that algorithms have access to a stream of random bits that each take values 0 and 1 with probability $1/2$. Many PRNGs create a *word* of random bits, that is, a vector $r_1 r_2 \ldots r_w$ of random bits r_i for a specified word size w. It is straightforward to convert such a word to a stream as the stream could simply return $r_1, r_2, \ldots, r_w$ in this order as required and generate a new word of random numbers once all w random bits are exhausted. The reason for considering a stream of random bits is that it makes our considerations independent from the word size w, and it prevents wasting randomness. Consider, for instance, a uniform random selection between two search points. In most implementations, a word of random numbers is drawn for a decision that requires just one bit of randomness. This wastes $w - 1$ random bits as a stream of random bits would only use one random bit.

Definition 1 (Cost of randomness). *We define $R_{\mathcal{A},f}$ as the number of random bits used during a run of a black-box algorithm $\mathcal{A}$ on a function f.*

We first summarize bounds on the minimum (expected) number of random bits required for typical distributions used in EAs. The first two statements are quite trivial; the last two follow from adapting results from the full version of [2].

Theorem 1. *The minimum expected number of random bits needed to sample from the following distributions is bounded as follows.*

1. *Sampling a search point in $\{0,1\}^n$ from the uniform distribution requires n random bits, which is optimal.*

Algorithm 1. Randomised Local Search (RLS)

Let x be a search point drawn uniformly at random.
repeat
    Choose an index i uniformly at random from $\{1,\ldots,n\}$. Create an offspring y from x by copying x and flipping bit i.
    If $f(y) \geq f(x)$ **then** y replaces x.
until false

Algorithm 2. (1+1) EA

Let x be a search point drawn uniformly at random.
repeat
    Create an offspring y from x by copying x and flipping each bit independently with a mutation rate p.
    If $f(y) \geq f(x)$ **then** y replaces x.
until false

---

2. *The minimum expected number of random bits required for sampling from a uniform distribution over n elements, $T_{\mathrm{Unif}}(n)$, is $T_{\mathrm{Unif}}(n) = \log n$ if n is a power of 2, and $\log n \leq T_{\mathrm{Unif}}(n) \leq 2\lceil \log n \rceil$ otherwise.*
3. *The minimum expected number of random bits required for sampling from a Bernoulli distribution $\mathrm{Ber}(p)$ with parameter p is $T_{\mathrm{Ber}}(p) \leq 2$.*
4. *The minimum expected number of random bits required for sampling from a geometric distribution $\mathrm{Geo}(p)$ with $p \in (0,1/2)$ is $T_{\mathrm{Geo}}(p) = \Theta(\log(1/p))$.*

The following lemma gives an exact formula for the cost of randomness incurred by some operator, if the number of random bits used is either a deterministic value $r(n)$ or given by a sequence of iid random variables with expectation $r(n)$. The proof for the latter case uses Wald's equation [16].

Lemma 1. *Consider an operator in a black-box algorithm $\mathcal{A}$ that uses X_t random bits in step t. Assume that the sequence $(X_t)_{t \in \mathbb{N}}$ is either deterministic with $X_t = r(n)$ for all $t \in \mathbb{N}$ and some function $0 < r(n) < \infty$ or a sequence of iid random variables with $0 < \mathrm{E}(X_t) = r(n) < \infty$. Then the expected total number of random bits used by the operator throughout a run of $\mathcal{A}$ on a function f is*

$$r(n) \cdot \mathrm{E}(T_{\mathcal{A},f}).$$

3 Cost of Randomness in Mutation-Based EAs

We first give general bounds on the expected number of random bits used by simple EAs that only use mutation (and no crossover). The possibly simplest such algorithm is Randomized Local Search (RLS) (Algorithm 1). It creates an initial search point uniformly at random. In every step, it creates an offspring by cloning the current search point and flipping one bit chosen uniformly at random. If the offspring is no worse than the parent, it replaces its parent.

Theorem 2. *Consider RLS on an arbitrary fitness function f, then*

$$E(R_{\mathrm{RLS},f}) = n + T_{\mathrm{Unif}}(n) \cdot E(T_{\mathrm{RLS},f}).$$

This is $n + \log(n) \cdot E(T_{\mathrm{RLS},f})$ if n is a power of 2.

Proof. If $E(T_{\mathrm{RLS},f}) = \infty$, the claim is trivial. For $E(T_{\mathrm{RLS},f}) < \infty$ the initialization requires n random bits and the uniform random choice of a bit to flip requires $T_{\mathrm{Unif}}(n)$ random bits in every iteration. Then the statement follows from Lemma 1 and Theorem 1.

Standard mutations flip each bit independently with a *mutation rate p*, enabling EAs to escape from local optima. A naive implementation is to perform n Bernoulli trials with parameter p to decide which bits to flip. Replacing the mutation operator in RLS by standard mutation yields the well known (1+1) EA (see Algorithm 2).

Theorem 3. *Consider the (1+1) EA with mutation rate $p \in (0,1)$ on an arbitrary fitness function f. Using the naive implementation of standard mutations,*

$$E\big(R_{(1+1)\mathrm{EA},f}\big) = n + n \cdot T_{\mathrm{Ber}}(p) \cdot E(T_{(1+1)\mathrm{EA},f}) \leq n + 2n \cdot E(T_{(1+1)\mathrm{EA},f}).$$

Proof. The statement follows directly from the cost of n bits for the initialisation and combining Theorem 1 with Lemma 1.

It is common to use a more sophisticated "geometric" implementation of standard mutations, see [19,31]. The idea is to determine the position of the next bit that should be flipped by sampling from a geometric distribution with parameter p, to flip this bit, and then to iterate this process until the end of the search point is reached. If the mutation rate p is small, we save on samples since in total only a few bits will be flipped. The procedure keeps track of a global variable m that indicates the position of the next bit to be flipped. When reaching the end of a bit string, m is reduced by n and used in the following standard mutation in order to fully exploit the available randomness.

Theorem 4. *Consider the (1+1) EA with mutation rate $p \in (0,1/2)$ on an arbitrary fitness function f. Using the geometric implementation of standard mutations,*

$$E\big(R_{(1+1)\ \mathrm{EA},f}\big) = n + T_{\mathrm{Geo}}(p) \cdot (1 + pn \cdot E(T_{\mathcal{A},f}))$$
$$= n + \Theta(\log(1/p)) \cdot (1 + pn \cdot E(T_{\mathcal{A},f})).$$

Proof. The initialization requires n random bits. After initialization and after every bit that is flipped, the geometric implementation draws a fresh geometric random variable. The number of geometric samples drawn at time t equals the number of flipping bits at time t, denoted as F_t. Then $F_1, F_2, \ldots$ are iid random variables with expectation pn. By Lemma 1, and including the initial geometric distribution, the expected number of random variables needed in all mutations is $(1 + pn \cdot E(T_{\mathcal{A},f}))$. Multiplying with a factor of $T_{\mathrm{Geo}}(p) = \Theta(\log(1/p))$ (cf. Theorem 1) yields the claimed result on the cost of randomness.

Note that the proofs of Theorems 2, 3 and 4 are independent from the population model and any selection mechanisms. Hence, they immediately generalize to much larger classes of algorithms with minor modifications, such as adding the cost of randomness for probabilistic selection mechanisms.

We briefly discuss some implications to well known results, for the geometric implementation of standard mutations. It is well known (see, e.g., Theorem 8 in [32]) that the expected number of function evaluations of the (1+1) EA with mutation rate χ/n, $\chi > 0$ constant on ONEMAX is at most $\frac{e^\chi n \ln(n)}{\chi} + O(n)$. The expected optimization time of the (1+1) EA with mutation rate p on LEADINGONES was shown in [1] to be $\frac{1}{2p^2} \cdot ((1-p)^{-n+1} - (1-p))$.

Corollary 1. *For the expected cost of randomness for the (1+1) EA with mutation rate $p = \chi/n$, $\chi > 0$ constant, we have*

$$\mathrm{E}\big(R_{(1+1)\text{ EA,ONEMAX}}\big) \leq T_{\text{Geo}}(\chi/n) \cdot (e^\chi n \ln(n) + O(n))$$
$$= \Theta(\log(n/\chi)) \cdot (e^\chi n \ln(n) + O(n)) \text{ and}$$

$$\mathrm{E}\big(R_{(1+1)\text{ EA,LEADINGONES}}\big) = n + T_{\text{Geo}}(p)\left(1 + \frac{n}{2p} \cdot ((1-p)^{-n+1} - (1-p))\right).$$

For $\chi \to 0$ the term $e^\chi n \ln(n)$ converges to $n \ln(n)$. This makes sense as for very small mutation rates, the (1+1) EA behaves like RLS in steps in which at least one bit is flipped, as the conditional probability of flipping more than one bit is very small. The expected optimization time of RLS on ONEMAX is at most $n \ln(n) + O(n)$. Likewise, for $p \to 0$ the term $\frac{n}{2p} \cdot ((1-p)^{-n+1} - (1-p))$ converges to $n^2/2$, which equals the expected optimization time of RLS on LEADINGONES when starting with a leading zero. Note that every non-idling step of the (1+1) EA has a cost of randomness of at least $T_{\text{Geo}}(p) = \Theta(\log(1/p))$, which diverges to ∞ as $p \to 0$. And, of course, the expected optimization time will increase due to idle steps when p is chosen very small, leading to a trade-off between expected optimization times and the cost of randomness.

4 Cost of Randomness with Crossover

Now we turn to genetic algorithms that use crossover and mutation. We focus on uniform crossover in which every bit in the offspring is chosen from either parent uniformly at random.

The cost of randomness in one uniform crossover is given by the Hamming distance between both parents:

Lemma 2. *The number of random bits needed to implement a uniform crossover of x and y equals the Hamming distance $H(x,y)$.*

Proof. For bits that agree in x and y, the result is deterministic. All $H(x,y)$ disagreeing bits are chosen uniformly at random.

Since $H(x,y) \leq n$ for all $x,y \in \{0,1\}^n$, we get the following upper bound.

Algorithm 3. (2+1) GA with a diversity-preserving mechanism

Initialize population P by drawing two search points uniformly at random
for $t := 1, 2, \ldots$ **do**
    Sample b from $\mathrm{Ber}(p_c)$
    if $b = 1$ **then**
        Select parents x_1 and x_2 from P independently and uniformly at random
        $y :=$ uniform crossover(x_1, x_2)
    else
        Choose y uniformly at random from P.
    Let z be the result of a standard mutation applied to y.
    Set P to the 2 best individuals from $P \cup \{z\}$; break ties by rejecting z if $z \in P$.

Theorem 5. *The expected cost of randomness, excluding the cost for selection, in a $(\mu+\lambda)$ GA using uniform crossover with crossover probability p_c and standard mutation with mutation rate p on any fitness function f is at most*

$$\mu n + T_{\mathrm{Geo}}(p) + (pn \cdot T_{\mathrm{Geo}}(p) + T_{\mathrm{Ber}}(p_c) + p_c n) \cdot \mathrm{E}\big(T_{(\mu+\lambda)\ \mathrm{GA}, f}\big).$$

5   Detailed Analysis for the (2+1) Genetic Algorithm

We take a closer look at the (2+1) GA from [33] as one of the simplest algorithms that use crossover, see Algorithm 3. With probability p_c it picks two parents uniformly at random and with replacement, then applies uniform crossover to them and mutates the result using standard mutation. With the remaining probability $1 - p_c$ it only chooses one parent uniformly at random and mutates it. In the replacement selection, the two best search points from parents and the offspring survive. The algorithm uses a diversity preserving mechanism in the tie-breaking rule: if there are multiple search points of the same fitness, the offspring is rejected if it equals one of its parents. This simplifies the original tie-breaking rule from [33] that computes the number of duplicates. The simplification is taken from [27] and it does not affect the results from [33].

For the (2+1) GA from [33] on ONEMAX with any constant crossover probability $p_c \in (0, 1)$ and mutation rate χ/n, $\chi > 0$ constant, Theorem 4 in [33] showed that

$$\mathrm{E}\big(T_{(2+1)\ \mathrm{GA}, \mathrm{ONEMAX}}\big) \leq \frac{e^\chi n \ln n}{\chi \cdot (1 + \chi)} + O(n). \tag{1}$$

Parent selection is a uniform choice of two search points. If crossover is used, two parents are selected and otherwise one parent is selected, thus at most 2 random bits are required per generation for parent selection. By Lemma 1, the total cost of randomness for parent selection is at most $2\mathrm{E}\big(T_{(2+1)\ \mathrm{GA}, \mathrm{ONEMAX}}\big) = O(n \log n)$. By Theorem 5, we get:

Corollary 2. $\mathrm{E}\big(R_{(2+1)\ \mathrm{GA}, \mathrm{ONEMAX}}\big) = O(n^2 \log n)$.

Since the (1+1) EA only requires $O(n \log^2 n)$ expected random bits by Corollary 1, the upper bound from Corollary 2 is by a factor of order $n/\log n$ larger. This observation might suggest that crossover operations can be very costly in terms of randomness. Of course, Corollary 2 only gives an upper bound that may not be tight. We show that the bound is far from tight by proving that the cumulative population diversity during a run is typically small. This result may be of independent interest.

We start with a bit of notation. For a population $P = \{x_1, x_2\}$ of size 2 we abbreviate the Hamming distance between its members as $H(P) = H(x_1, x_2)$ and refer to $H(P)$ as the Hamming distance of P. We call a population P with $H(P) = 0$ *monomorphic* as in [28], borrowing a term from population genetics. For a monomorphic population P we define $f(P)$ as the fitness of its (identical) members. The diversity mechanism implies that a non-monomorphic population with parents of equal fitness cannot become monomorphic, unless the fitness increases. Starting from a monomorphic population, the next monomorphic population is either identical to the current one or has a better fitness.

Lemma 3. *Let P_{t_1} be a monomorphic population and P_{t_2} be the next population reached that is also monomorphic. Then either $t_2 = t_1 + 1$ and $P_{t_2} = P_{t_1}$ or $t_2 > t_1 + 1$ and $f(P_{t_2}) > f(P_{t_1})$.*

We define a set of non-monomorphic populations and a subset where parents have different fitness; this is beneficial for reaching monomorphic populations.

Definition 2. *Let $\mathcal{P}$ be the set of all non-optimal non-monomorphic populations, that is, all sets $\{x_1, x_2\}$ with $f(x_1) < n$ and $f(x_2) < n$ as well as $x_1 \neq x_2$. Let $\mathcal{P}_{\neq} \subset \mathcal{P}$ be the set of all populations $\{x_1, x_2\} \in \mathcal{P}$ with $f(x_1) \neq f(x_2)$.*

Now we define the expected cost of randomness during crossover throughout a phase that starts with a non-monomorphic population and ends with a monomorphic (or optimal) population.

Definition 3. *For a population $P_t \in \mathcal{P}$, $T(P_t)$ denotes the cost of randomness during crossover until a monomorphic (or an optimal) population is reached for the next time:*

$$T(P_t) = \inf \left\{ \sum_{i=t+1}^{t'} H(P_i) \mid P_t, t' > t, P_{t'} \notin \mathcal{P} \right\}.$$

We now define expected costs of randomness, parameterised by a maximum Hamming distance of k in the initial population.

Definition 4. *For $1 \leq k \leq n$, let $T(k)$ be the maximum expected cost of randomness during crossover before reaching a monomorphic population, starting from a population with Hamming distance at most k,*

$$T(k) := \max\{\mathrm{E}(T(P_t)) \mid P_t \in \mathcal{P}, H(P_t) \leq k\}.$$

Let $T_{\neq}(k)$ denote the same quantity, but restricting the initial population to $\mathcal{P}_{\neq}$:

$$T_{\neq}(k) := \max\{\mathrm{E}(T(P_t)) \mid P_t \in \mathcal{P}_{\neq}, H(P_t) \leq k\}.$$

Note that, by definition, the following inequalities hold, since the maximum is taken over larger sets of possible populations P_t.

$$T(1) \leq T(2) \leq \cdots \leq T(n) \tag{2}$$
$$T_{\neq}(1) \leq T_{\neq}(2) \leq \cdots \leq T_{\neq}(n) \tag{3}$$
$$\forall k \in \{1, \ldots, n\}: T_{\neq}(k) \leq T(k) \tag{4}$$

The following lemma states that the Hamming distance can only increase by the number of flipping bits (it cannot increase through uniform crossover).

Lemma 4. *Consider a population P with Hamming distance $H(P) = k$. If F denotes the random number of bits flipped in the next mutation step, the Hamming distance of the next population is at most $k + F$.*

Next, we identify beneficial events $\mathcal{E}_1$ and $\mathcal{E}_2$ that, together, create a monomorphic population of higher fitness.

Lemma 5. *Consider one generation of the (2+1) GA with mutation rate χ/n and current population $P \in \mathcal{P}$. Let F be the random number of bits flipped in the mutation step. There is an event $\mathcal{E}_1 \subseteq (F = 0)$ for reaching a population from $\mathcal{P}_{\neq}$ with Hamming distance at most $H(P)$ in the next generation, with*

$$P(\mathcal{E}_1) \geq \eta_1 := p_c \cdot \frac{1}{8} \cdot \left(1 - \frac{\chi}{n}\right)^n.$$

For every current population from $\mathcal{P}_{\neq}$, there is an event $\mathcal{E}_2 \subseteq (F = 0)$ for reaching a monomorphic population in the next generation, with

$$P(\mathcal{E}_2) \geq \eta_2 := \frac{1}{4} \cdot \left(1 - \frac{\chi}{n}\right)^n.$$

Proof (Sketch of proof for Lemma 5). Following [33], for populations with two different and equally fit search points, $\mathcal{E}_1$ can be achieved by a uniform crossover on different parents (probability $p_c/2$), creating a surplus of ones on the differing bits (probability $1/4$) and not flipping any bits during mutation (prob. $(1 - \chi/n)^n$). Event $\mathcal{E}_2$ can be achieved by cloning the fitter search point.

The following lemma is at the heart of our analysis. It gives closed forms for $T(k)$ and $T_{\neq}(k)$ by solving a system of cross-coupled recurrence equations.

Lemma 6. *For all $k \in \{1, \ldots, n\}$,*

$$T_{\neq}(k) \leq \frac{1 + \eta_1 - \eta_2}{\eta_1 \eta_2} \cdot k + \frac{(1 - \eta_2)(1 + \eta_1)(1 + \chi)}{\eta_1^2 \eta_2^2}$$

$$T(k) \leq \frac{1 + \eta_1}{\eta_1 \eta_2} \cdot k + \frac{(1 - \eta_1 \eta_2)(1 + \eta_1)(1 + \chi)}{\eta_1^2 \eta_2^2}$$

Both upper bounds are linear functions in k. If the next population has a maximum Hamming distance of $k = 0$, a monomorphic population has been reached and the remaining time for reaching a monomorphic population is 0. The following lemma gives a closed form for the expectation of $f(X)$ where f is a function that is linear for positive arguments and 0 otherwise.

Lemma 7. *Let* $f\colon \mathbb{N}_0 \to \mathbb{N}_0$ *be a function defined as* $f(k) := a \cdot k + b$ *if* $k > 0$ *for* $a, b \in \mathbb{R}$ *and* $f(0) := 0$. *Let* X *be a random variable defined on* $\mathbb{N}_0$, *then*

$$\mathrm{E}(f(X)) = a \cdot \mathrm{E}(X) + \mathrm{P}(X > 0) \cdot b.$$

Proof. The statement follows from the weak linearity of expectation, $\mathrm{E}(a \cdot X + b) = a \cdot \mathrm{E}(X) + b$, and the fact that the additive term b is only present when $X > 0$.

Now we prove Lemma 6, solving a system of cross-coupled recurrences.

Proof (Proof sketch for Lemma 6). We use induction over decreasing k.

For every population considered in $T_{\neq}(n)$, the Hamming distance is at most n, hence the cost of randomness increases by at most n in the next generation. With probability at least η_2, a monomorphic population is reached and no further costs are incurred. With the remaining probability $1 - \eta_2$, some other population is reached. We then estimate the remaining costs by $T(n)$ as this expression is maximal by Eqs. (2)–(4) and thus it is an upper bound applying to all populations from $\mathcal{P}$. Thus, we have shown the recurrence:

$$T_{\neq}(n) \le n + (1 - \eta_2) \cdot T(n). \tag{5}$$

For every population considered in $T(n)$, the Hamming distance is at most n. With probability at least η_1 we reach a population from $\mathcal{P}_{\neq}$ and then the remaining cost is at most $T_{\neq}(n)$. With the remaining probability $1 - \eta_1$ we reach some unspecified population and bound the remaining cost by $T(n)$. Hence,

$$T(n) \le n + \eta_1 \cdot T_{\neq}(n) + (1 - \eta_1) \cdot T(n).$$

Plugging in (5),

$$\begin{aligned} T(n) &\le n + \eta_1 \cdot (n + (1 - \eta_2) \cdot T(n)) + (1 - \eta_1) \cdot T(n) \\ &= (1 + \eta_1)n + (1 - \eta_1\eta_2)T(n). \end{aligned}$$

Subtracting $(1 - \eta_1\eta_2)T(n)$ on both sides and dividing by $\eta_1\eta_2$,

$$T(n) \le \frac{1 + \eta_1}{\eta_1\eta_2} \cdot n \tag{6}$$

which implies the claimed bound for $T(n)$. Plugging (6) into (5),

$$T_{\neq}(n) \le n + (1 - \eta_2) \cdot \frac{1 + \eta_1}{\eta_1\eta_2} \cdot n = \frac{\eta_1\eta_2 + (1 - \eta_2)(1 + \eta_1)}{\eta_1\eta_2} \cdot n = \frac{1 + \eta_1 - \eta_2}{\eta_1\eta_2} \cdot n$$

which implies the claimed bound for $T_{\neq}(n)$.

Now assume that the claimed bounds for $T(k')$ and $T_{\neq}(k')$ hold for all $k' > k$. We first show the claimed bound for $T_{\neq}(k)$, setting up an appropriate recurrence equation. Since the Hamming distance is at most k, the cost of randomness in the next generation is at most k. Afterwards, a population is reached with a random Hamming distance that we denote by X. Conditional on $X = x$, the remaining cost of randomness is bounded by either $T_{\neq}(x)$ or $T(x)$. Since by (4), for all x, $T_{\neq}(x) \leq T(x)$, we bound these terms for all x using the function T. This would yield the following recurrence:

$$T_{\neq}(k) \leq k + \mathrm{E}(T(X)).$$

Assume for a moment that the bound on $T(x)$ from the statement was already proven for all $x \in \{1,\ldots,n\}$. Note that the bound for T from the statement is a linear function, that is, we may write $T(x) \leq a \cdot x + b$ for two constants $a, b > 0$ for all $x \in \{1,\ldots,n\}$. For $x = 0$ we have $T(0) = 0$ as the argument 0 implies a Hamming distance of 0 and thus a monomorphic population. By Lemma 7, we would then get $\mathrm{E}(T(X)) \leq a \cdot \mathrm{E}(X) + \mathrm{P}(X > 0) \cdot b$.

To avoid a circular argument and to be able to use our induction hypothesis, we will replace X with a random variable X' that stochastically dominates X and only takes on values in $\{0, k+1, \ldots, n\}$ for which we have proven upper bounds. The inequalities (2) show that this approach is pessimistic. Then Lemma 7 will yield an upper bound of $k + a \cdot \mathrm{E}(X') + \mathrm{P}(X' > 0) \cdot b$.

Let F be the random number of flipping bits during the mutation step and let $F_{\geq 1} := (F \mid F \geq 1)$ be the number of flipping bits conditional on at least one bit flipping, then we define X' as $X' := 0$ if $X = 0$ and $X' := \min\{k + F_{\geq 1}, n\}$ if $X > 0$. As $F_{\geq 1} \geq 1$, X' only takes on values in $\{0, k+1, k+2, \ldots, n\}$ and we argue that $X' \geq X$ by distinguishing three cases.

Since $\mathcal{E}_2 \subseteq (F = 0)$ by Lemma 5, $\overline{\mathcal{E}_2}$ is the disjoint union of two events: $(F \geq 1)$ and $(F = 0) \setminus \mathcal{E}_2$. The event space can be partitioned into three disjoint events: $\mathcal{E}_2$, $(F \geq 1)$, and $(F = 0) \setminus \mathcal{E}_2$. We show that in all three events, $X' \geq X$.

Note that $(X \mid \mathcal{E}_2) = 0$ and thus, conditional on $\mathcal{E}_2$, $X' \geq X = 0$. Conditional on $F \geq 1$, the number of flipping bits is $F_{\geq 1} = (F \mid F \geq 1)$. Then the next generation's Hamming distance is bounded by $k + F_{\geq 1}$ by Lemma 4 (and bounded trivially by n), thus it is bounded by X'. Conditioning on $(F = 0) \setminus \mathcal{E}_2$, no bits are flipped and by Lemma 4, the Hamming distance cannot increase. Therefore $X \leq k$ and $X' > k \geq X$ under this condition. Thus, in all cases $X' \geq X$.

Using $\mathrm{P}(X' > 0) = \mathrm{P}(X > 0) \leq 1 - \eta_2$ and bounding the conditional expectation $\mathrm{E}(F_{\geq 1}) = \mathrm{E}(F \mid F \geq 1) \leq 1 + \mathrm{E}(F) = 1 + \chi$, where the inequality follows from Lemma 7.3 in [8], the expectation of X' is bounded by

$$\mathrm{E}(X') \leq \mathrm{P}(X > 0) \cdot (k + 1 + \mathrm{E}(F)) \leq (1 - \eta_2)(k + 1 + \chi).$$

Along with Lemma 7, this implies

$$\begin{aligned} T_{\neq}(k) &\leq k + a \cdot \mathrm{E}(X') + \mathrm{P}(X > 0) \cdot b \\ &\leq k + a \cdot (1 - \eta_2)(k + 1 + \chi) + (1 - \eta_2)b \\ &= (1 + a(1 - \eta_2)) \cdot k + (1 - \eta_2)(a(1 + \chi) + b). \end{aligned}$$

It is easy to verify that the coefficients from this linear function match the ones from the claimed bound for $T_{\neq}$.

The bound on $T(k)$ is shown in a similar way. The cost incurred by the first crossover is at most k. With probability at least η_1, a population in $\mathcal{P}_{\neq}(k)$ is reached, whose Hamming distance is at most k. Hence the remaining expected cost of randomness is bounded by $T_{\neq}(k)$. Otherwise, some other population is reached with a Hamming distance Y that is stochastically dominated by $Y' := \min\{k + F_{\geq 1}, n\}$. The stochastic domination of Y' follows as above for X' since $\mathcal{E}_1 \subseteq (F = 0)$ and thus the previous arguments apply. Setting up and solving recurrence equations as for $T_{\neq}$ completes the proof.

Theorem 6. *The expected cost of randomness during crossover operations in a run of the (2+1) GA with mutation rate χ/n, $\chi > 0$ constant and crossover probability $p_c = \Omega(1)$ is in $\Theta(n)$.*

Proof. A lower bound of $\Omega(n)$ follows as the expected Hamming distance of the two initial search points is $n/2$ and with probability at least $p_c/2 - 2^{-n+1} = \Omega(1)$, a uniform crossover between these is performed in the first generation. The term -2^{-n+1} bounds the probability of any initial individual being optimal.

Starting with an arbitrary population, the expected cost of randomness until a monomorphic population is reached for the first time is at most $T(n) = O(n)$ by Lemma 6.

Starting from a monomorphic population, denote by N_{run} the number of sampled random bits used for crossover until a monomorphic population of higher fitness is reached. We claim that $E(N_{\text{run}}) = O(1)$. If the next population is monomorphic, no costs are incurred. Hence, we consider the first non-monomorphic population reached. By Lemma 4, a generation creating a non-monomorphic population from a monomorphic one requires $F \geq 1$ flipping bits. Hence, as in the proof of Lemma 6 we are working under the condition $F \geq 1$. The Hamming distance Z of the non-monomorphic population has an expectation of at most $E(Z) \leq E(F \mid F \geq 1) \leq 1 + \chi$ and using the weak linearity of expectation, the expected cost of randomness during crossover is $E(T(Z)) \leq a \cdot E(Z) + b \leq a \cdot (1 + \chi) + b = O(1)$. As by Lemma 3 the next monomorphic population has a strictly higher fitness, $E(N_{\text{run}}) = E(T(Z)) = O(1)$.

Since the fitness strictly increases after an expected cost of $E(N_{\text{run}}) = O(1)$, it suffices to repeat the above argument at most n times.

Now we can put everything together to give a refined bound on the cost of randomness in the (2+1) GA on ONEMAX.

Theorem 7. *For the (2+1) GA on ONEMAX with $p_c \in (0,1)$ and mutation rate χ/n, $\chi > 0$ constant,*

$$E(R_{(2+1)\text{GA,ONEMAX}}) = (\chi \cdot T_{\text{Geo}}(\chi/n) + 4) \cdot E(T_{(2+1)\text{GA,ONEMAX}}) + O(n)$$

$$= T_{\text{Geo}}(\chi/n) \cdot \left(\frac{e^\chi n \ln n}{1 + \chi} + O(n)\right).$$

Proof. The cost of randomness for initialization, mutation and decisions whether to apply crossover or not is bounded by

$$2n + T_{\text{Geo}}(\chi/n) + (pn \cdot T_{\text{Geo}}(\chi/n) + 2) \cdot E(T_{(2+1)\text{ GA,ONEMAX}})$$

as in Theorem 5. The cost of randomness for all crossover operations is $O(n)$ by Theorem 6. The cost for parent selection is at most $2\mathrm{E}\big(T_{(2+1)\ \mathrm{GA,ONEMAX}}\big)$ as discussed at the start of this section. Using $pn = \chi$ and absorbing $2n + T_{\mathrm{Geo}}(\chi/n)$ in the $O(n)$ term implies the first bound. The second bound follows from plugging in (1), canceling χ and using $4\mathrm{E}\big(T_{(2+1)\ \mathrm{GA,ONEMAX}}\big) + O(n) = \Theta(n\log n) = T_{\mathrm{Geo}}(\chi/n) \cdot \Theta(n)$ since $T_{\mathrm{Geo}}(\chi/n) = \Theta(\log n)$.

This upper bound strictly increases with χ, and the term $\frac{e^{\chi}n\ln n}{1+\chi}$ converges to $n\ln n$ as $\chi \to 0$. For the default value $\chi = 1$, the dominant term in the upper bound is half the dominant term in the cost of randomness for the $(1+1)$ EA on ONEMAX from Corollary 1. Hence, the use of crossover decreases the cost of randomness by a factor of 2 (bar small-order terms). So here we obtain the rather surprising result that using an additional operator that draws on random bits and that can be potentially wasteful actually reduces the cost of randomness.

6   Conclusions and Future Work

Studying the cost of randomness in EAs can give new and refined insights into the performance of EAs and evolutionary operators and deepen our understanding of the working principles of EAs. For 1-bit flips and standard mutation with the geometric implementation and mutation rates of $n^{-\Theta(1)}$, the cost of randomness is essentially larger than the expected number of function evaluations by a term of $\Theta(\log n)$.

Analyzing uniform crossover is harder as the use of random bits depends on the Hamming distances between parents. In the $(2+1)$ GA crossover only uses a total of $\Theta(n)$ expected random bits, which is by a $\Theta(\log^2 n)$ factor smaller than the cost of randomness during mutation. Here the cost of randomness reflects the cumulative population diversity during a run (in all generations using crossover), which we believe to be of independent interest.

In the $(2+1)$ GA the cumulative diversity is very small as crossover quickly generates improvements from any available diversity, and a better monomorphic population is reached after having used only $O(1)$ expected random bits during crossover. This implies that the overhead of using crossover is negligible, and the factor-2 speedup over all mutation-only EAs proven in [33] for the number of function evaluations carries over to the cost of randomness. We conclude that here using an additional operator that consumes additional random bits actually reduces the cost of randomness; in other words, crossover is beneficial for reducing the number of function evaluations as well as the cost of randomness on ONEMAX.

Future work may include studying the cost of randomness in other algorithms analyzed theoretically such as the standard $(2+1)$ GA with uniform random tie-breaking [4,5,28], heavy-tailed mutation operators [12], the $(1+(\lambda,\lambda))$ GA [9], ant colony optimizers, estimation-of-distribution algorithms [17] and artificial immune systems [6,22].

Acknowledgment. The authors thank participants of Dagstuhl seminar 22081 "Theory of Randomized Optimization Heuristics" for fruitful discussions.

References

1. Böttcher, S., Doerr, B., Neumann, F.: Optimal fixed and adaptive mutation rates for the LeadingOnes problem. In: Schaefer, R., Cotta, C., Kołodziej, J., Rudolph, G. (eds.) PPSN 2010. LNCS, vol. 6238, pp. 1–10. Springer, Heidelberg (2010). https://doi.org/10.1007/978-3-642-15844-5_1
2. Bringmann, K., Friedrich, T.: Exact and efficient generation of geometric random variates and random graphs. In: Fomin, F.V., Freivalds, R., Kwiatkowska, M., Peleg, D. (eds.) ICALP 2013. LNCS, vol. 7965, pp. 267–278. Springer, Heidelberg (2013). https://doi.org/10.1007/978-3-642-39206-1_23
3. Corus, D., Lissovoi, A., Oliveto, P.S., Witt, C.: On steady-state evolutionary algorithms and selective pressure: why inverse rank-based allocation of reproductive trials is best. ACM Trans. Evol. Learn. Optim. **1**(1), 2:1-2:38 (2021)
4. Corus, D., Oliveto, P.S.: Standard steady state genetic algorithms can hill climb faster than mutation-only evolutionary algorithms. IEEE Trans. Evol. Comput. **22**(5), 720–732 (2018)
5. Corus, D., Oliveto, P.S.: On the benefits of populations for the exploitation speed of standard steady-state genetic algorithms. Algorithmica **82**(12), 3676–3706 (2020)
6. Corus, D., Oliveto, P.S., Yazdani, D.: Fast immune system-inspired hypermutation operators for combinatorial optimization. IEEE Trans. Evol. Comput. **25**(5), 956–970 (2021)
7. Dang, D.-C., et al.: Escaping local optima using crossover with emergent diversity. IEEE Trans. Evol. Comput. **22**(3), 484–497 (2018)
8. Doerr, B.: Probabilistic tools for the analysis of randomized optimization heuristics. In: Doerr and Neumann [13], pp. 1–87
9. Doerr, B., Doerr, C., Ebel, F.: From black-box complexity to designing new genetic algorithms. Theoret. Comput. Sci. **567**, 87–104 (2015)
10. Doerr, B., Fouz, M., Witt, C.: Sharp bounds by probability-generating functions and variable drift. In: Proceedings of the 13th Annual Genetic and Evolutionary Computation Conference (GECCO 2011), pp. 2083–2090. ACM Press (2011)
11. Doerr, B., Gießen, C., Witt, C., Yang, J.: The $(1+\lambda)$ evolutionary algorithm with self-adjusting mutation rate. Algorithmica **81**(2), 593–631 (2019)
12. Doerr, B., Phuoc Le, H., Makhmara, R., Nguyen, T.D.: Fast genetic algorithms. In: Proceedings of the Genetic and Evolutionary Computation Conference (GECCO 2017), pp. 777–784. ACM (2017)
13. Doerr, B., Neumann, F. (eds.): Theory of Evolutionary Computation - Recent Developments in Discrete Optimization. Natural Computing Series, Springer, Cham (2020). https://doi.org/10.1007/978-3-030-29414-4
14. Doerr, B., Neumann, F.: A survey on recent progress in the theory of evolutionary algorithms for discrete optimization. ACM Trans. Evol. Learn. Optim. **1**(4), 1–43 (2021)
15. Eiben, A.E., Smith, J.E.: Introduction to Evolutionary Computing, 1st edn. Springer, Cham (2015). https://doi.org/10.1007/978-3-662-44874-8
16. Feller, W.: An Introduction to Probability Theory and Its Applications, vol. 2. Wiley, New York (1971)

17. Friedrich, T., Kötzing, T., Krejca, M.S., Sutton, A.M.: The benefit of recombination in noisy evolutionary search. In: Elbassioni, K., Makino, K. (eds.) ISAAC 2015. LNCS, vol. 9472, pp. 140–150. Springer, Heidelberg (2015). https://doi.org/10.1007/978-3-662-48971-0_13

18. Gießen, C., Kötzing, T.: Robustness of populations in stochastic environments. Algorithmica 75(3), 462–489 (2016)

19. Grefenstette, J.: Efficient implementation of algorithms. In: Handbook of Evolutionary Computation, pp. E2.1:1–E2.1:6. IOP Publishing Ltd., 1st edn. (1997)

20. Thomas Jansen. Analyzing Evolutionary Algorithms - The Computer Science Perspective. Springer, 2013

21. Jansen, T., Zarges, C.: Analysis of evolutionary algorithms: from computational complexity analysis to algorithm engineering. In: Proceedings of the 11th Workshop on Foundations of Genetic Algorithms (FOGA 2011), pp. 1–14. ACM (2011)

22. Jansen, T., Zarges, C.: Analyzing different variants of immune inspired somatic contiguous hypermutations. Theoret. Comput. Sci. 412(6), 517–533 (2011)

23. Lässig, J., Sudholt, D.: General upper bounds on the running time of parallel evolutionary algorithms. Evol. Comput. 22(3), 405–437 (2014)

24. Mambrini, A., Sudholt, D.: Design and analysis of schemes for adapting migration intervals in parallel evolutionary algorithms. Evol. Comput. 23(4), 559–582 (2015)

25. Matsumoto, M., Nishimura, T.: Mersenne twister: a 623-dimensionally equidistributed uniform pseudo-random number generator. ACM Trans. Model. Comput. Simul. 8(1), 3–30 (1998)

26. Neumann, F., Witt, C.: Bioinspired Computation in Combinatorial Optimization - Algorithms and Their Computational Complexity. NCS, 1st edn. Springer, Cham (2010). https://doi.org/10.1007/978-3-642-16544-3

27. Nguyen, P.T.H., Sudholt, D.: Memetic algorithms outperform evolutionary algorithms in multimodal optimisation. Artif. Intell. 287, 103345 (2020)

28. Oliveto, P.S., Sudholt, D., Witt, C.: Tight bounds on the expected runtime of a standard steady state genetic algorithm. Algorithmica 84(6), 1603–1658 (2022)

29. Qian, C., Bian, C., Yang, Yu., Tang, K., Yao, X.: Analysis of noisy evolutionary optimization when sampling fails. Algorithmica 83(4), 940–975 (2021)

30. Route, M.: Radio-flaring ultracool dwarf population synthesis. Astrophys. J. 845(1), 66 (2017)

31. Rudolph, G., Ziegenhirt, J.: Computation time of evolutionary operators. In: Handbook of Evolutionary Computation, pp. E2.2:1–E2.2:4. IOP Publishing Ltd. 1st edn. (1997)

32. Sudholt, D.: A new method for lower bounds on the running time of evolutionary algorithms. IEEE Trans. Evol. Comput. 17(3), 418–435 (2013)

33. Sudholt, D.: How crossover speeds up building-block assembly in genetic algorithms. Evol. Comput. 25(2), 237–274 (2017)

Multi-objectivization Relaxes Multi-funnel Structures in Single-objective NK-landscapes

Shoichiro Tanaka[(✉)], Keiki Takadama, and Hiroyuki Sato

Graduate School of Informatics and Engineering, The University of
Electro-Communications, Tokyo, Japan
{stanaka,h.sato}@uec.ac.jp, keiki@inf.uec.ac.jp

Abstract. This paper investigated the impacts of multi-objectivization on solving combinatorial single-objective NK-landscape problems with multiple funnel structures. Multi-objectivization re-formulates a single-objective target problem into a multi-objective problem with a helper problem to suppress the difficulty of the target problem. This paper analyzed the connectivity of two funnels involving global optima in the target and the helper NK-landscape problems via the Pareto local optimal solutions in the multi-objectivized problem. Experimental results showed that multi-objectivization connects the two funnels with global optima of the target and the helper problems as a single bridging domain consisting of the Pareto local optimal solutions. Also, this paper proposed an algorithm named the multi-objectivized local search (MOLS) that searched for the global optimum of the target problem from the global optimum of an artificially generated helper problem via the Pareto local optimal solutions. Experimental results showed that the proposed MOLS achieved a higher success rate of the target single-objective optimization than iterative local search algorithms on target NK-landscape problems with multiple funnels.

Keywords: Multi-objectivization · Combinatorial optimization · Landscape analysis · Local optima networks · Local search

1 Introduction

The optimization performance of a search algorithm is determined by its behavior on the landscape of the target optimization problem. There are two main approaches to tackling optimization problems with rugged and complex landscapes. One is to develop search algorithms that work robustly on optimization problems with such landscapes. The other is to transform optimization problems with such landscapes into problems with simple landscapes. This paper focuses on the latter and multi-objectivization as one of the approaches [7]. Multi-objectivization re-formulates an objective function (*target function*) of a single-objective target optimization problem into a multi-objective one. There are two types of multi-objectivization methods: those based on decomposition

L. Pérez Cáceres and T. Stützle (Eds.): EvoCOP 2023, LNCS 13987, pp. 195–210, 2023.
https://doi.org/10.1007/978-3-031-30035-6_13

[7] and those based on additional objectives [5]. We limit the scope of this paper to the latter one, the multi-objectivization, by adding a new objective function (*helper function*) for solving the single-objective target problem. Previous works have reported that helper functions are effective for solving target problems with several difficult landscapes such as plateaus [2] and multimodalities [3,5]. Multi-objectivization impacts the problem's difficulties since the fundamental change in the landscape [4].

Among several landscape indicators, it has been known that local optima cause premature convergence of local search and harm global search methods such as evolutionary algorithms. On the other hand, recent landscape analysis research has shown that local optima distribution and connectivity affect the problem difficulties rather than the number of local optima [13]. A cluster of local optima densely connected internally and sparsely connected externally is called *funnel* and has recently attracted attention as a new indicator of the landscape. In an optimization problem with multiple funnels, a search algorithm may converge to a funnel including no global optimum. Even global search methods face difficulty in escaping from such a funnel. Indeed, in several optimization problem classes, funnels have been found to make the problem difficult to solve [9,10,14]. To the best of our knowledge, the impact of multi-objectivization on the multi-funnel structure has yet to be studied.

In this paper, we investigate the impact of multi-objectivization on a single-objective target problem with a multi-funnel structure. We analyze the change in the landscape caused by the addition of helper problems. Each target and the helper problem is NK-landscape [6], a combinatorial optimization problem with adjustable local optima and funnels. For combinations of the target and the helper NK-landscape problems with different problem parameters, we conduct a landscape analysis that observes the connectivity of their optima. In addition, we propose a search algorithm named the multi-objectivized local search (MOLS) that searches for the target problem from the global optimum of an artificially generated helper problem. We conduct an algorithm benchmark that compares the search performances of the proposed MOLS and the conventional iterated local search (ILS). Contributions of this paper are (1) to reveal how multi-objectivization relaxes multi-funnel structures in single-objective NK-landscape problems and (2) the proposal of MOLS that achieves better search performance on the target problem with multiple funnels than the conventional ILS.

2   Single-objective Landscape

2.1   Single-objective Optimization

Suppose we have a single-objective optimization problem with a pseudo-Boolean function $f : \{0,1\}^N \to \mathbb{R}$, which must be maximized. The search space is $X = \{0,1\}^N$, and the variable vector is $x \in X$ which is a bit string of bit length N. The purpose of the single objective optimization is to search for the set of solutions $G := \{x \in X \mid \nexists x' \in X; f(x) < f(x')\}$ with the highest function value in the search space X. The solutions in G are called global optima. The problems treated in this paper have only one global optimum g. In the search

---

Algorithm 1: Local Search (best-improvement)

Input: initial solution x, objective function f
Procedure LS(x, f):
 | while x is updated **do**
 | | $x \leftarrow \arg\max_{x' \in \mathcal{N}(x)} f(x)$
 | **return** x

---

space X, local optima often interrupt convergence to a global optimum. Let $\mathcal{N}_\epsilon(x) = \{x' \in X \mid x' \neq x, d(x, x') \leq \epsilon\}$ be the ϵ-neighborhood of solution x, where $d(x, x')$ is the distance between the solutions x and x'. The set of local optima L of the single objective function f can be defined as

$$L := \{x \in X \mid \nexists x' \in \mathcal{N}_{\epsilon=1}(x); f(x) < f(x')\}. \tag{1}$$

Note that $G \subseteq L$ from the above definitions.

2.2 Basin of Attraction

In this paper, the local search is the method of updating the search point to a solution with the highest function value among the neighborhoods of the search point. The pseudo-code of local search is shown in Algorithm 1. The Basin of attraction can be defined as the set of solutions that converge to the same local optimum $l \in L$ via local search. The convergence from a solution x to a local optimum $l \in L$ by local search is denoted as LS(x) = l. The basin $B(l)$ of the local optimum l is defined as

$$B(l) := \{x \in X \mid \text{LS}(x) = l\}. \tag{2}$$

The search space X can be divided into a finite number of basins $B(l)$ ($l \in L$) when no plateau exists. For local search, the initial solution needs to exist in the optimal basin $B(g)$ to converge to a global optimum $g \in G$. It is difficult to find global optimum $g \in G$ by local search in an optimization problem with a tiny optimal basin $B(g)$.

2.3 Local Optima Networks

The funnel is defined using a local optima networks (LONs) model [11], which represents the connectivity between local optima based on given search operators. LON is a weighted directed graph $G = (L, A)$, with the set of local optima L as nodes and the set of transitions between local optima A as arcs. This paper refers to directed edges as arcs. In this section, we describe a variant of the LON, the monotonic LON [12] and its arc, the perturbation arc [15].

- **Monotonic LON:** Monotonic LON (MLON) is a subgraph of LON. Nodes are local optima, and arcs are only ameliorate transitions between local optima by given search operators.

Algorithm 2: Iterated Local Search

Input: initial solution x, objective function f
1 **Procedure** ILS(x, f):
2 $\quad A \leftarrow \emptyset$
3 $\quad l \leftarrow$ LS (x)
4 $\quad$ **while** terminal criteria are not fulfilled **do**
5 $\quad\quad x' \leftarrow$ Perturbation (l)
6 $\quad\quad$ **if** $x' \notin A$ **then**
7 $\quad\quad\quad l' \leftarrow$ LS (x')
8 $\quad\quad\quad A \leftarrow A \cup \{x'\}$
9 $\quad\quad$ **if** $f(l') > f(l)$ **then**
10 $\quad\quad\quad l \leftarrow l'$
11 $\quad\quad\quad A \leftarrow \emptyset$

12 $\quad$ **return** x

- **Perturbation Arcs:** The perturbation arc represents the transition based on the ILS operator. ILS is the trajectory-based algorithm and repeats local search and perturbation. The perturbation is a random bit-flip up to a fixed number of bits α for the solution x. The pseudo-code for iterative local search is shown in Algorithm 2. When a local optimum $l \in L$ is updated to a local optimum $l' \in L$, l and l' are connected by an arc $a(l, l')$. The arc weight $w(l, l')$ is the fraction of solutions in the α-bit neighborhood $\mathcal{N}_\alpha(l)$ that converge to l' by the local search. When the perturbation arc $a(l, l')$ satisfies $f(l') > f(l)$, it is called *monotonic*.

2.4 Funnel

A monotonic LON with monotonic perturbation arcs has no cycles in the graph. Nodes in MLON involves a set of sink nodes $\hat{L} = \{l \in L \mid \sum_{l' \neq l} w(l, l') = 0\}$ that outgoing degree is zero. That is, each of sink nodes $\hat{L}$ has no transitions that improve the function value. When the local optimum l repeats the most probable transition, it converges to the sink point $\hat{l} \in \hat{L}$. This sequence of solutions is called a monotonic sequence [12] and is denoted Path$(l) = \hat{l}$. A set of local optima that converges to the same sink point $\hat{l}$ via transitions is called funnel $F(\hat{l})$ in this paper. The funnel $F(\hat{l})$ is defined as

$$F(\hat{l}) := \{l \in L \mid \text{Path}(l) = \hat{l}\}. \tag{3}$$

Compared to Eq. (2), a funnel is a basin of attraction at the level of local optima [12]. In single objective optimization, it has been reported that the number and depth of funnels affect the search [9, 14].

3 Multi-objective Landscape

3.1 Multi-objective Optimization

Suppose that we have a multi-objective optimization problem with M pseudo-Boolean functions $f_m : [0,1]^N \to \mathbb{R}$ $(m = 1, \ldots, M)$, which must be maximized. The search space is $X = \{0,1\}^N$, and the variable vector is $x \in X$ which is a bit string of bit length N. The purpose of multi-objective optimization is to search for a set of Pareto optimal solutions with an optimal trade-off relationship of multiple function values. For two solutions $x, x' \in X$, x dominates x' $(x \succ x')$ if

$$\forall m \in \{1, \ldots, M\}; f_m(x) \geq f_m(x') \text{ and } \exists m \in \{1, \ldots, M\}; f_m(x) > f_m(x'). \quad (4)$$

A solution $x \in X$ is said to be non-dominated in a solution set $S \subseteq X$ if no solution dominates x in S. The set of non-dominated solutions between neighborhoods is the Pareto local optimal solution set $PL := \{x \in \mathcal{X} \mid \nexists x' \in \mathcal{N}_{\epsilon=1}(x); x' \succ x\}$. The non-dominated solution set to the whole search space X is the Pareto optimal solution set $P := \{x \in X \mid \nexists x' \in X; x' \succ x\}$. Note that $P \subseteq PL$ from the above definitions.

A local optimum l of the m-th objective function f_m have higher $f_m(l)$ than its neighbors. Thus, its neighbors do not dominate the l. Let L_m be the local optima of the function f_m, the following relation holds: $\forall m \in \{1, \ldots, M\}; L_m \subseteq PL$. This paper defines $PL\text{-}domain$ as a sub-domain consisting of a set of Pareto local optimal solutions in the search space. The search space X can be divided into one or more PL-domains and non PL-domains. Since the Pareto optimal solutions and the local optimum of each objective function are Pareto local optimal solutions, each of them is involved in one of the PL-domains. For the sake of intuitive understanding, Fig. 1 shows a two-dimensional continuous $x_1 - x_2$ search space for a two-objective optimization problem $f(x_1, x_2) = (f_1(x_1, x_2), f_2(x_1, x_2))$. Note that this paper deals with a discrete space, not the one shown in this figure. The red circles are the local optima for f_1, and the blue circle is the local optimum for f_2. The dark gray areas are PL-domains, and the green area is the Pareto set. We see that both local and Pareto optimal solutions exist in a PL-domain in this problem.

3.2 Pareto Local Optimal Solutions Networks

The Pareto local optimal solution network (PLOS-net) [8] models the connectivity of Pareto local optimal solutions in combinatorial search space. PLOS-net is an unweighted undirected graph $G = (PL, E)$ with the Pareto local optimal solution set PL as nodes and the transitions between the solutions E as edges. When Pareto local optimal solutions $pl, pl' \in PL$ are adjacent to each other $pl \in \mathcal{N}(pl')$, pl and pl' are connected by an edge $E(pl, pl')$. PLOS-net has one or more components corresponding to PL-domains. This paper focuses on the NK-landscape problem [6] and the MNK-landscape problem [1].

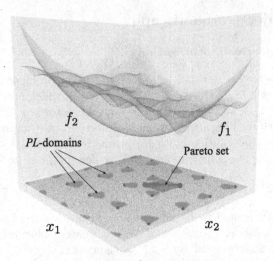

Fig. 1. Visual example using a two-dimensional continuous search space for a two-objective minimization problem. The red circles are the local optima of f_1, the blue one is the global optimum of f_2, the green is the Pareto set of $\boldsymbol{f} = (f_1, f_2)$, and the gray are PL-domains, sets of the Pareto local optimal solutions. (Color figure online)

4   NK- And MNK-landscape Problems

4.1   NK-landscape Problem

Solution is a bit string $\boldsymbol{x} = (x_1, x_2, \ldots, x_N) \in \{0,1\}^N$ of bit length N. The search space size $|X|$ is 2^N. The objective function of the NK-landscape problem that must be maximized is a pseudo-Boolean function and is defined as

$$f^{NK}(\boldsymbol{x}) = \frac{1}{N} \sum_{j=1}^{N} g_j(x_j, x_{j,1}, x_{j,2}, \ldots, x_{j,K}), \tag{5}$$

where g_j $(j = 1, \ldots, N)$ are the sub-functions, $x_{j,1}, x_{j,2}, \ldots, x_{j,K}$ are the randomly selected co-variables for the j-th sub-function g_j and K is the number of co-variables. In this paper, all instances of NK-landscape problems have only one global optimum $\boldsymbol{g}$ and no plateau. The NK-landscape problem framework is tunable from a smooth unimodal landscape to a rugged multimodal landscape as K increases from 0 to N. In general, as K increases, the difficulty increases. The function value $f^{NK}(\boldsymbol{x})$ is calculated by variable relationships determined by co-variables and a random number table. By changing the variable relationships randomly determined by co-variables or the table, we can generate different instances with the same problem parameters N and K.

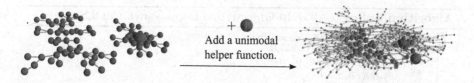

A target function with bi-funnel structure,   A multi-objectivized function.

Fig. 2. Change in the landscape with the addition of a unimodal helper objective. Red circles are the local optima of target objective f_{target}, and the blue one is the global optimum of helper f_{helper}, green ones are the Pareto set of $f = (f_{\text{target}}, f_{\text{helper}})$, and gray ones are PL-domain, the set of Pareto local optimal solutions. (Color figure online)

4.2 MNK-landscape Problem

The MNK-landscape problem is an extension of the NK-landscape one and involves M objective functions with the same solution space X. The objective functions of the MNK-landscape problem are defined as

$$f^{MNK}(x) = (f_1^{NK}(x), \ldots, f_M^{NK}(x)), \tag{6}$$

where

$$f_i^{NK}(x) = \frac{1}{N} \sum_{j=1}^{N} g_{i,j}(x_j, x_{i,j,1}, \ldots, x_{i,j,K_i}) \quad (i = 1, \ldots, M). \tag{7}$$

Each objective function f_i^{NK} is an NK-landscape problem and consists of N sub-functions $g_{i,j}$ ($j = 1, \ldots, N$), and its each sub-function $g_{i,j}$ takes variable x_j and K_i co-variables. In this paper, NK-landscapes, which constitutes MNK-landscape, are uncorrelated.

5 Proposed Method: Multi-Objectivized Local Search

5.1 Motivation

As the variable relationships K increase, local optima generally increase. As a result, the size of the optimal basin $B(g)$ and funnel $F(g)$, including a global optimum g, becomes small. This is one of the intuitive reasons why optimization problems become more difficult to solve. In contrast, a study [8] using PLOS-net suggests that the PL-domain appears as one large domain independent of variable relationship strength.

Figure 2 shows the change in the landscape when a multimodal instance of a single objective target NK-landscape f_{target} is multi-objectivized with a unimodal instance of an additional helper one f_{helper}. The red, blue, and gray graphs

---

Algorithm 3: Proposed Multi-Objectivized Local Search (MOLS)

Input: objective function f_{target}, helper function f_{helper}
1   **Procedure** MOLS(f_{target}, f_{helper}):
2      $x \leftarrow$ a global optimum of f_{helper}
3      $f \leftarrow (f_{target}, f_{helper})$
4      $A \leftarrow \{x\}$
5   **while** terminal criteria are not fulfilled **do**
6          $\mathcal{N}^+ \leftarrow \emptyset$
7   **foreach** $x' \in \mathcal{N}_{\epsilon=1}(x)$ **do**
8   **if** $x \not\succ x' \wedge x' \notin A$ **then**
9                  $\mathcal{N}^+ \leftarrow \mathcal{N}^+ \cup \{x'\}$
10          $x \leftarrow \arg\max_{x' \in \mathcal{N}^+} f_{target}(x)$
11          $A \leftarrow A \cup \{x\}$
12      $x \leftarrow \arg\max_{x \in A} f_{target}(x)$
13   **return** x

---

represent the LON of the target f_{target}, the helper f_{helper}, and the PLOS-net of $f = (f_{target}, f_{helper})$, respectively, while the green nodes represent the Pareto set. From the left side of Fig. 2, we see that the red LON of the target f_{target} has two red components, funnels. From the right side of Fig. 2, the target bi-funnel structure in red is transformed into a landscape with one huge PL-domain in gray by adding the helper function f_{helper}. If the search space has a multi-funnel structure, search algorithms may converge to a funnel with no global optimum. On the other hand, if the PL-domain can connect the multiple funnels, the helper function enables transitions between the multiple funnels via Pareto local optimal solutions. Multi-objectivization aggregates the multiple funnels into a PL-domain and suppresses the search difficulty due to the multi-funnel structures.

5.2 Algorithm

In this paper, we propose an algorithm that searches for a global optimum of a target function f_{target} using a multi-objective problem consisting of the target function f_{target} and an additional helper function f_{helper}.

Algorithm 3 shows the pseudo-code of the proposed search algorithm named the multi-objectivized local search (MOLS). The input is the target function f_{target} with N variables and a helper function f_{helper}, which is a randomly generated unimodal instance of NK-landscape problem with N variables and $K = 0$ co-variables. It is not always appropriate to choose such a unimodal function that is independent of the target function for the helper function. However, there are two reasons for this design. First, in a black-box scenario, providing helper functions that are not independent of the target function, i.e., correlated in some respect, is not easy. Therefore, we decided to generate independent helper functions as a first step. Second, the reason for using unimodal helper functions is that it is easy to find a global optimum. This algorithm expects transitions between global optima and requires knowledge of the global optimum of the helper function. At line 2, we set an initial solution x. The proposed MOLS does not start with a randomly generated initial solution such as conventional search algorithms. The proposed MOLS starts with the global optimum g_{helper} of the helper function f_{helper}. The computational complexity of searching for the global solution g_{helper} of the helper function f_{helper} is $\mathcal{O}(N)$. Since the initial solution of MOLS is a global optimum $g_{helper} \in G_{helper}$ of the helper function f_{helper} and a Pareto local optimal solution at the same time, MOLS searches only within the PL-domain where g_{helper} exists. At line 4, the initial solution joins to the archive A, and the repetition started from line 5 searches the neighborhoods of the current solution x. At line 6, the set of candidate neighborhoods $\mathcal{N}^+$ is prepared as the empty set. In the repetition started from line 7, we collect candidate neighborhoods $\mathcal{N}^+$ from the neighborhoods $\mathcal{N}_{\epsilon=1}(x)$ of the current solution x. The neighborhoods $\mathcal{N}_{\epsilon=1}(x)$ not dominated by the current solution x and not in the archive A joins to the candidate neighborhoods $\mathcal{N}^+$. In the repetition started from line 10, we take the best solution in the candidate neighborhoods $\mathcal{N}^+$ in the viewpoint of the target objective function value f_{target} as the next focal solution x. At line 11, we put it to the archive A. We then repeat the above process.

In this way, the proposed MOLS searches the global optimum g_{target} of the target objective function f_{target} from the global optimum g_{helper} of the helper objective function f_{helper}. If global optima of both target and helper functions exist in the same PL-domain, MOLS can reach the global optimum g_{target} of the target function f_{target} in principle.

6 Experimental Setup

We conducted two experiments in this paper. One is (1) the landscape analysis, and the other is (2) the algorithm benchmarks.

6.1 Landscape Analysis

In the landscape analysis, we quantitatively investigated the landscape changes caused by multi-objectivization.

We generated independent 10 instances for each NK-landscape problem setting with $N = 18, K \in \{0, 2, 4, 6, 8, 10\}$, for a total of 60 instances. The 60 instances of NK-landscape problems were quantitatively evaluated using the metrics shown in Table 1. We obtained the funnel using MLONs with $\alpha = 2$.

As combinations in the 60 instances of single-objective NK-landscape problems generated above, we generated $\binom{60}{M=2} = 1,770$ instances of MNK-landscape problem setting with $M = 2, N = 18$, and $K_1, K_2 \in \{0, 2, 4, 6, 8, 10\}$. The 1,770 instances of MNK-landscape problems are quantitatively evaluated using the metrics shown in Table 2. We obtained the PL-domain using the PLOS-net [8]. Graph features are obtained for the only component corresponding to the optimal PL-domain.

Table 1. Metrics for single-objective landscapes

Metrics	Description
$\mathcal{B}$	The number of the basin of attraction.
$\mathcal{F}$	The number of the funnels.
$\mathcal{B}^*/\mathcal{X}$	The relative size of the optimal basin to the search space $|B(g)|/|X|$
$\mathcal{F}^*/\mathcal{L}$	The relative size of the optimal funnel to the search space $|F(g)|/|L|$

Table 2. Metrics for multi-objective landscapes

Metrics	Description
$\mathcal{PL}/\mathcal{X}$	The ratio of the Pareto local optimal solutions PL to the search space.
$\mathcal{D}$	The number of PL-domains.
$\%\mathcal{D}^*$	The existence probability of optimal PL-domain D^*.
$\mathcal{D}^*/\mathcal{X}$	The relative size of D^* to the search space X.
$\mathcal{D}^*/\mathcal{PL}$	The relative size of D^* to PL.
$path^*$	The length of shortest path $path^*$ between global optima in D^*.
$\mathcal{P}/path^*$	The fraction of the Pareto optimal solutions in $path^*$.
deg_D^*	The average degree of D^*.
deg_path^*	The average degree of $path^*$.

Table 3. Metrics of search performance

Metrics	Description
$\mathcal{SR}$	The fraction of trials in which a global optimum was found in $1,000$ trials.
$\mathcal{ST}$	The number of function evaluations when the global optimum is found.
$\mathcal{CT}$	The number of function evaluations at the end of the search

6.2 Algorithm Benchmarks

In the algorithm benchmark, we quantitatively assessed the optimization performance of the target function.

We compared three search algorithms: the proposed MOLS and the two conventional ILSs with different perturbation strengths $\alpha \in \{2, 3\}$, which gives perturbations of more than 2-bits and less than α-bits. We considered that ILS has converged if the focal solution has not been updated after performing a local search from all possible solutions obtained by perturbation. ILS terminated when it either converged or reached the maximum number of function evaluations $FE_{max} = N^2 \times 100$. The proposed MOLS terminated if there was no unexplored Pareto local optimal solution in the neighborhood of the focal solution or reached the maximum number of function evaluations $FE_{max} = N^2 \times 100$.

We used independent 30 instances for each NK-landscape problem setting with $N = 18, K \in \{2, 4, 6, 8, 10\}$, for a total of 150 instances. For each instance, each search algorithm was executed 1,000 times independently. We used three metrics shown in Table 3. Note that ST, the number of function evaluations until the global solution was found, was the average only among the successful runs.

7 Experimental Results and Discussions

7.1 Landscape Analysis

The metrics values in Table 2 for the 1,770 instances of the MNK-landscape are illustrated in Fig. 3. The average value for all instances corresponding to each parameter combination $\boldsymbol{K}$ is plotted. In each figure, the vertical axis is the metric value, the horizontal axis is the number of co-variables K_1 of the function , f_1, and the six plots are respectively different numbers of co-variables K_2. Since the results for $\boldsymbol{K} = (K_1, K_2)$ and $\boldsymbol{K} = (K_2, K_1)$ are identical, the results are plotted only when $K_1 \leq K_2$.

Figure 3(a) shows that the Pareto local optimal solution set PL increases in the increase of K_2, which is the maximum number of co-variables between two K values. Figure 3(b) shows how many PL-domains the Pareto local optimal solutions are divided into. As with the number of Pareto local optimal solutions, PL-domains are affected by K_2 but do not increase monotonically. Figure 3(c) shows the existing probability of the optimal PL-domain D^* that contains global optima of both target and helper problems. Except for the combination $\boldsymbol{K} = (0, 2)$, we see the optimal PL-domain exists in more than 85% of instances. Figures 3(d) and 3(e) respectively show the size of the optimal PL-domain and the percentage of the Pareto local optimal solutions within the optimal PL-domain among all Pareto local optimal solutions. The optimal PL-domain and the number of the Pareto local optimal solutions become larger as K_2 is higher. These results indicate that the higher K_1 and K_2 are, the more Pareto local optimal solutions are generated, resulting in more significant optimal PL-domains. Increasing the size of the optimal PL-domain makes reaching the domain easier from outside. Still, it also makes searching for global optimum within the domain more challenging.

Since the optimal PL-domain involves two global optima of the target and the helper problems, the shortest path $path^*$ between them connecting the Pareto local optimal solutions exists. Figure 3(f) shows the length of $path^*$. The higher K_2 is, the longer $path^*$ is. That is, the number of transition steps between two global optima of the target and the helper problems via Pareto local optimal solutions increases with K_2 even in the same optimal PL-domain.

Figure 3(g) shows the ratio of Pareto optimal solutions in $path^*$. The higher K_2 is, the smaller the ratio in the shortest path. That is, the Pareto local optimal solutions connecting the two global optima is not necessarily the Pareto optimal solution.

Figurse 3(h) and 3(i) respectively show the average degrees in the optimal PL-domain D^* and $path^*$. As K_2 increases, the degree in D^* decreases, which means that the connection becomes sparse. The average degree in $path^*$ tends to be similar to that in D^*. However, we see the tendency that the average degree

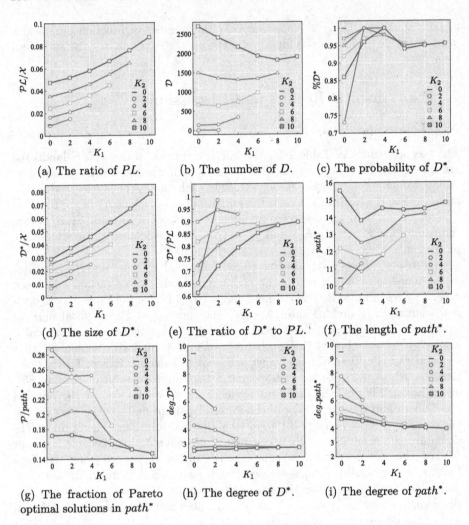

(a) The ratio of PL.

(b) The number of D.

(c) The probability of D^*.

(d) The size of D^*.

(e) The ratio of D^* to PL.

(f) The length of $path^*$.

(g) The fraction of Pareto optimal solutions in $path^*$

(h) The degree of D^*.

(i) The degree of $path^*$.

Fig. 3. The metrics of multi-objective landscapes by combining NK-landscapes with different difficulties. The color and shape of markers indicate the maximum K of the NK-landscape in the MNK-landscape.

in $path^*$ is higher than the one D^*. This result suggests that $path^*$ is highly centered in the optimal PL-domain. Graph features would be useful to find the appropriate trajectory in D^*.

Figure 4 finally compares the number and size of basins, funnels for 60 instances of NK-landscape, and optimal PL-domains for 600 instances of NK-landscape with $K_1 = 0$. We see the number of basins and PL-domains increase exponentially with increasing K or K_2. This result indicates that a single local optimum forms a single PL-domain. We see the funnel increase but more slowly than the number of PL-domain. Local optimum forms a funnel rather than a

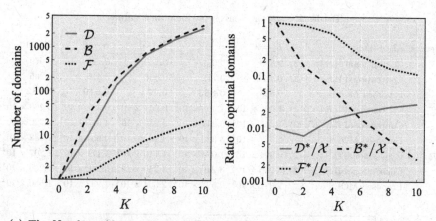

(a) The Number of basins $\mathcal{B}$, funnels $\mathcal{F}$, and PL-domains $\mathcal{D}$.

(b) The relative size of the optimal basin $\mathcal{B}^*/\mathcal{X}$, funnels $\mathcal{F}^*/\mathcal{L}$, and PL-domains $\mathcal{D}^*/\mathcal{X}$ to search space.

Fig. 4. Comparison of single-objective landscapes and multi-objectivized landscapes with unimodal helper functions.

completely random distribution. As the number of basins and funnels increases, the size of the optimal ones becomes relatively small. In contrast, the optimal PL-domain is slightly affected by K and grows in size as K increases. Small D^* makes an exploration within D^* easier, and the larger D^* makes finding D^* in search space easier. While conventional optimization algorithms search the optimal funnel or basin from the search space, the proposed MOLS explores within the optimal PL-domain from the known global optimum of the helper function. MOLS is considered most effective in small and dense connected D^*.

7.2 Algorithm Benchmarks

Table 4 and Fig. 5 respectively show the average search performance values and the boxplots of search performance values with their distribution. From the results on $\mathcal{SR}$, the success rate of finding global optima, we see that the proposed MOLS performs better than the conventional ILSs for all K except for $K = 2$. This performance deterioration in $K = (0, 2)$ instances is caused by the low probability of D^* organized, as shown in Fig. 3(c). This is because the proposed MOLS is an algorithm that searches within D^*, and if D^* does not exist, the search fails to reach the global optimum of the target problem. From Fig. 4, we see that instances of the NK-landscape with $K = 2$ have a large funnel of nearly 90% of the search space X. If the optimal funnel can be easily searched, as in the case of NK-landscape with $K = 2$, there is a risk that multi-objectivization makes the target function more difficult.

On the other hand, when K is high, the success rate is only about 30%, even though D^* is more than 85% likely to exist. This result indicates that the proposed MOLS cannot explore the entire D^*. One reason the search performance

Table 4. Average search performance values

Metrics	Algorithms	$K = 2$	$K = 4$	$K = 6$	$K = 8$	$K = 10$
$\mathcal{SR}$	Conventional ILS ($\alpha = 2$)	0.835	0.561	0.289	0.115	0.079
	Conventional ILS ($\alpha = 3$)	**0.915**	0.772	0.586	0.410	0.265
	Proposed MOLS	0.825	**0.883**	**0.771**	**0.519**	**0.282**
$\mathcal{ST}$	Conventional ILS ($\alpha = 2$)	1.306×10^3	1.408×10^3	0.941×10^3	0.381×10^3	0.260×10^3
	Conventional ILS ($\alpha = 3$)	1.356×10^3	3.247×10^3	4.236×10^3	4.315×10^3	2.722×10^3
	Proposed MOLS	2.357×10^3	3.322×10^3	4.904×10^3	4.654×10^3	2.832×10^3
$\mathcal{CR}$	Conventional ILS ($\alpha = 2$)	0.991×10^4	1.074×10^4	1.119×10^4	1.171×10^4	1.207×10^4
	Conventional ILS ($\alpha = 3$)	3.240×10^4	3.240×10^4	3.240×10^4	3.240×10^4	3.240×10^4
	Proposed MOLS	2.372×10^4	2.592×10^4	2.354×10^4	1.992×10^4	1.536×10^4

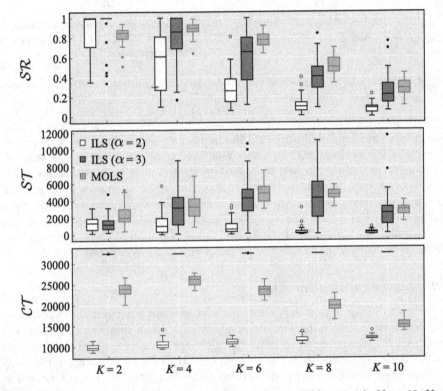

Fig. 5. Search performances on target NK-landscape problems with $N = 18, K \in \{2, 4, 6, 8, 10\}$.

of the proposed MOLS deteriorates as K increases would be due to the connectivity of D^*. From Fig. 3(h), we can expect that the higher K of the target function is, the more sparse the D^* becomes. Due to the nature of the proposed MOLS, which cannot transition back to a solution once searched, it is likely to converge on the terminal part of D^*.

From the results on $\mathcal{ST}$, the number of function evaluations when the global optimum is found, we see the proposed MOLS shows the highest values for all K. The range of $\mathcal{ST}$ for MOLS was within the range of $\mathcal{ST}$ for ILS with $\alpha = 3$ except when $K = 2$, and the proposed MOLS shows smaller variances of $\mathcal{ST}$ and more stable than ILSs. From the results on $\mathcal{CT}$, the number of function evaluations at the end of the search, ILS with $\alpha = 3$ shows the highest values for all K. This is because the number of perturbations required to determine convergence increases as the ILS parameter α increases. In contrast, the termination condition of the proposed MOLS is whether or not a new Pareto local optimal solution exists in the neighborhood. Although it does not directly refer to the objective function value of the target function, the search can be terminated before the upper bound of function evaluation. The faster convergence with increasing K may be due to the sparse connection of D^*. MOLS needs a mechanism that prevents it from falling into a dead end on a sparsely connected domain.

8   Conclusions

This paper investigated the effect of multi-objectivization on single-objective NK-landscape problems with multi-funnel structures. First, we used instances of NK-landscape problems with different K and conducted the landscape analysis to quantitatively observe the landscape changes of the target problem by adding helper problems. Experimental results showed that the global optima of the target and the helper problems exist within one domain consisting of the Pareto local optimal solutions in many problem cases. The size of the bridging domain grew as the target function's global multimodality increased. This trend contrasted with existing concepts such as basin of attraction and funnels. On the other hand, the connections within the bridging domain became sparse and inversely proportional to its size. In addition, we proposed a search algorithm named MOLS that searched for the global optimum of the target problem from the global optimum of an artificially generated helper problem via Pareto local optimal solutions. The proposed MOLS showed a better success rate of finding the target problem's global optimum than the conventional ILS in NK-landscape with multiple funnels. This result suggests that the multi-objectivization relaxed the problem difficulty of the target function due to the multiple funnel structure.

The results in this paper are limited to relatively small-scale NK-landscape problems. Also, the proposed MOLS is intuitive but consists of simple operators, and there is room for improvement in the update and termination phases. Experimental results suggested that graph features such as degree would be utilized in the search. In future work, we will improve the proposed search algorithm and analyze landscapes with a broad range of problems to examine the effectiveness of multi-objectivization.

References

1. Aguirre, H.E., Tanaka, K.: Insights on properties of multiobjective MNK-landscapes. In: Proceedings of the 2004 Congress on Evolutionary Computation (IEEE Cat. No. 04TH8753), vol. 1, pp. 196–203. IEEE (2004)
2. Brockhoff, D., Friedrich, T., Hebbinghaus, N., Klein, C., Neumann, F., Zitzler, E.: Do additional objectives make a problem harder? In: Proceedings of the 9th Annual Conference on Genetic and Evolutionary Computation, pp. 765–772 (2007)
3. Deb, K., Saha, A.: Finding multiple solutions for multimodal optimization problems using a multi-objective evolutionary approach. In: Proceedings of the 12th Annual Conference on Genetic and Evolutionary Computation, pp. 447–454 (2010)
4. Garza-Fabre, M., Toscano-Pulido, G., Rodriguez-Tello, E.: Multi-objectivization, fitness landscape transformation and search performance: a case of study on the HP model for protein structure prediction. Eur. J. Oper. Res. **243**(2), 405–422 (2015)
5. Jensen, M.T.: Helper-objectives: using multi-objective evolutionary algorithms for single-objective optimisation. J. Math. Modell. Algorithms **3**(4), 323–347 (2004)
6. Kauffman, S., Levin, S.: Towards a general theory of adaptive walks on rugged landscapes. J. Theoret. Biol. **128**(1), 11–45 (1987)
7. Knowles, J.D., Watson, R.A., Corne, D.W.: Reducing local optima in single-objective problems by multi-objectivization. In: Zitzler, E., Thiele, L., Deb, K., Coello Coello, C.A., Corne, D. (eds.) EMO 2001. LNCS, vol. 1993, pp. 269–283. Springer, Heidelberg (2001). https://doi.org/10.1007/3-540-44719-9_19
8. Liefooghe, A., Derbel, B., Verel, S., López-Ibáñez, M., Aguirre, H., Tanaka, K.: On pareto local optimal solutions networks. In: Auger, A., Fonseca, C.M., Lourenço, N., Machado, P., Paquete, L., Whitley, D. (eds.) PPSN 2018. LNCS, vol. 11102, pp. 232–244. Springer, Cham (2018). https://doi.org/10.1007/978-3-319-99259-4_19
9. McMenemy, P., Veerapen, N., Ochoa, G.: How perturbation strength shapes the global structure of TSP fitness landscapes. In: Liefooghe, A., López-Ibáñez, M. (eds.) EvoCOP 2018. LNCS, vol. 10782, pp. 34–49. Springer, Cham (2018). https://doi.org/10.1007/978-3-319-77449-7_3
10. Ochoa, G., Herrmann, S.: Perturbation strength and the global structure of QAP fitness landscapes. In: Auger, A., Fonseca, C.M., Lourenço, N., Machado, P., Paquete, L., Whitley, D. (eds.) PPSN 2018. LNCS, vol. 11102, pp. 245–256. Springer, Cham (2018). https://doi.org/10.1007/978-3-319-99259-4_20
11. Ochoa, G., Tomassini, M., Vérel, S., Darabos, C.: A study of NK landscapes' basins and local optima networks. In: Proceedings of the 10th Annual Conference on Genetic and Evolutionary Computation, pp. 555–562 (2008)
12. Ochoa, G., Veerapen, N., Daolio, F., Tomassini, M.: Understanding phase transitions with local optima networks: number partitioning as a case study. In: Hu, B., López-Ibáñez, M. (eds.) EvoCOP 2017. LNCS, vol. 10197, pp. 233–248. Springer, Cham (2017). https://doi.org/10.1007/978-3-319-55453-2_16
13. Thomson, S.L., Daolio, F., Ochoa, G.: Comparing communities of optima with funnels in combinatorial fitness landscapes. In: Proceedings of the Genetic and Evolutionary Computation Conference, pp. 377–384 (2017)
14. Thomson, S.L., Ochoa, G.: On funnel depths and acceptance criteria in stochastic local search. In: Proceedings of the Genetic and Evolutionary Computation Conference, pp. 287–295 (2022)
15. Vérel, S., Daolio, F., Ochoa, G., Tomassini, M.: Local optima networks with escape edges. In: Hao, J.-K., Legrand, P., Collet, P., Monmarché, N., Lutton, E., Schoenauer, M. (eds.) EA 2011. LNCS, vol. 7401, pp. 49–60. Springer, Heidelberg (2012). https://doi.org/10.1007/978-3-642-35533-2_5

Decision/Objective Space Trajectory Networks for Multi-objective Combinatorial Optimisation

Gabriela Ochoa[1]([📧]) [ID], Arnaud Liefooghe[2] [ID], Yuri Lavinas[3] [ID],
and Claus Aranha[3] [ID]

[1] University of Stirling, Stirling FK9 4LA, UK
gabriela.ochoa@stir.ac.uk
[2] Univ. Lille, CNRS, Inria, Centrale Lille, UMR 9189 CRIStAL, 59000 Lille, France
arnaud.liefooghe@univ-lille.fr
[3] University of Tsukuba, 1-1-1 Tennodai, Tsukuba 305-8577, Japan
lavinas.yuri.xp@alumni.tsukuba.ac.jp, caranha@cs.tsukuba.ac.jp

Abstract. This paper adapts a graph-based analysis and visualisation tool, search trajectory networks (STNs) to multi-objective combinatorial optimisation. We formally define multi-objective STNs and apply them to study the dynamics of two state-of-the-art multi-objective evolutionary algorithms: MOEA/D and NSGA2. In terms of benchmark, we consider two- and three-objective ρmnk-landscapes for constructing multi-objective multi-modal landscapes with objective correlation. We find that STN metrics and visualisation offer valuable insights into both problem structure and algorithm performance. Most previous visual tools in multi-objective optimisation consider the objective space only. Instead, our newly proposed tool asses algorithm behaviour in the decision and objective spaces simultaneously.

Keywords: algorithm analysis · search trajectory networks · STNs · combinatorial optimisation · visualisation · multi-objective optimisation

1 Introduction

Understanding the behaviour of search and optimisation algorithms remains a challenge to which visualisation techniques can contribute. The performance of multi-objective optimisation algorithms is usually visualised in the objective space, where the Pareto front (or an approximation of it) for two or three objectives is shown in a standard scatter plot; an idea that has been extended for more than three objectives using dimensionality reduction [21]. Incorporating the design space into the visualisation, however, can help to improve our understanding. Only a small number of approaches point in this direction, and most of them are tailored to continuous optimisation, such as cost landscapes [7], gradient field heatmaps [11], and the plot of landscapes with optimal trade-offs [20]. In the combinatorial domain, local optima networks [16,19] have been adapted to multi-objective optimisation [6,14]. These insightful visual approaches, however, concentrate on the fitness landscape structure, rather than on the algorithms dynamic behaviour.

The goal of this article is to adapt search trajectory networks (STNs) to multi-objective combinatorial optimisation. STNs were originally proposed for single-objective optimisation [17,18] as a graph-based tool to visualise and analyse the dynamics of any type of metaheuristic: evolutionary, swarm-based or single-point, on both continuous and discrete search spaces. STNs were later extended to multi-objective optimisation [12], but so far have been applied to continuous benchmark problems only. The extension of STNs from single- to multi-objective optimisation relies on the notion of *decomposition* [23], where the multi-objective problem is transformed into multiple single-objective scalar sub-problems. The idea is then to aggregate the STN of each these sub-problems to construct the multi-objective STN. One limitation of the approach proposed in [12] is that it considers a small number of decomposition vectors (5 to be precise), which restricts the granularity and expressing power of the modelling tool. In this paper, our contributions can be summarised as follows:

(1) We apply multi-objective STNs to combinatorial benchmarks, where both the landscape ruggedness and the correlation among objectives can be tuned.
(2) We offer a more formal definition of multi-objective STNs.
(3) We improve the granularity and accuracy of the modelling tool by increasing the number of decomposition vectors.
(4) We propose a 2D graph layout that conveys the design and objective spaces simultaneously in a single plot — this applies to two-objective problems only.

The paper is organised as follows. Section 2 introduces the necessary background on multi-objective optimisation. Section 3 formally defines the multi-objective STNs, together with the related metrics and visualisation techniques. Section 4 gives the experimental setup. Section 5 presents the experimental results of our analysis for both small and large multi-objective landscapes. At last, Sect. 6 concludes the paper and discusses further research.

2   Multi-objective Combinatorial Optimisation

This section provides definitions for multi-objective combinatorial optimisation, and presents two well-established multi-objective evolutionary algorithms.

2.1   Definitions

We assume an m-dimensional objective function vector $f \colon X \mapsto Z$ is to be maximised, such that every solution from the (discrete) solution space $x \in X$ maps to a vector in the objective space $z \in Z$, with $z = f(x)$ and $Z \subseteq \mathbb{R}^m$. Given two objective vectors $z, z' \in Z$, we say that z is dominated by z' if $z_i \leqslant z_i'$ for all $i \in \{1, \ldots, m\}$, and there is a $j \in \{1, \ldots, m\}$ such that $z_j < z_j'$. Similarly, a solution $x \in X$ is dominated by $x' \in X$ if $f(x)$ is dominated by $f(x')$. An objective vector $z^\star \in Z$ is non-dominated if there is no $z \in Z$ such that $z^\star$ is dominated by z. A solution $x^\star \in X$ is Pareto optimal if $f(x)$ is non-dominated. The set of Pareto optimal solutions is the Pareto set (PS), and its mapping in the objective space is the Pareto front (PF). Evolutionary multi-objective optimisation (EMO) algorithms aim at identifying a PS approximation that is to be presented to the decision maker for further consideration [2,4].

2.2   Multi-objective Evolutionary Algorithms

We consider two state-of-the-art EMO algorithms that are described below.

MOEA/D is a decomposition-based EMO algorithm that seek a high-quality solution in multiple regions of the objective space by decomposing the original (multi-objective) problem into a number of scalar (single-objective) sub-problems [23]. Let μ be the population size. A set $(\lambda^1, \ldots, \lambda^i, \ldots, \lambda^\mu)$ of uniformly-distributed weighting coefficient vectors defines the scalar sub-problems, and a population $P = (x^1, \ldots, x^i, \ldots, x^\mu)$ is maintained such that each solution x^i maps to the sub-problem defined by λ^i. Different scalarising functions can be used, and the weighted Chebyshev scalarising function [15] defined in the next section is a well-established example. A neighbourhood $\mathcal{B}(i)$ is additionally defined for each sub-problem $i \in \{1, \ldots, \mu\}$, by considering its T closest weighting coefficient vectors. At each iteration, the population evolves with respect to a given sub-problem. Two solutions are selected at random from $\mathcal{B}(i)$ and an offspring is produced by means of variation operators. Then, for each neighbouring sub-problem $j \in \mathcal{B}(i)$, the offspring is used to replace the current solution x^j if there is an improvement in terms of the scalarising function. The algorithm iterates over sub-problems until a stopping condition is satisfied.

NSGA2 is an elitist dominance-based EMO algorithm using Pareto dominance for selection [5]. At a given iteration t, the current population P_t is merged with its offspring Q_t, and is divided into non-dominated fronts $\{F_1, F_2, \ldots\}$ based on the non-dominated sorting procedure [9]. The front in which a given solution belongs to gives its rank within the population. Crowding distance is also calculated within each front. Selection is based on dominance ranking, and crowding distance is used as a tie breaker. Survival selection consists in filling the new population P_{t+1} with solutions having the best (smallest) ranks. In case a front F_i overfills the population size, the required number of solutions from F_i are chosen based on their crowding distance. Parent selection for reproduction consists of binary tournaments between randomly-chosen solutions, following the lexicographic order induced by ranks first, and crowding distance next.

3   Search Trajectory Networks

In order to define a graph-based model, we need to specify its nodes and edges. We start by giving these definitions for single-objective optimisation before describing how to construct the models for multiple objectives.

3.1 Definitions

Nodes are unique candidate solutions to the optimisation problem at each iteration, representing the status of the search process. In population-based algorithms, the best solution from the population (measured by the objective function) is typically chosen at each iteration as the representative solution. The set of nodes is denoted by N.

A Search Trajectory is given by a sequence of representative solutions (nodes) in the order in which they are encountered during the search process.

Edges are directed and connect two consecutive nodes in the search trajectory. Edges are weighted with the number of times a transition between two given nodes occurred during the process of sampling and constructing the STN. The set of edges is denoted by E.

Single-Objective STN Model. An STN is a directed graph $STN = G(N, E)$ with nodes N and edges E as defined above. For constructing a single-objective STN, multiple runs of the algorithm under study are performed, and explored solutions and their transitions are aggregated into a single graph model. Notice that some solutions and transitions may appear multiple times during the sampling process. However, the graph retains as nodes each unique solution, and as edges each unique transitions among encountered solutions. Counters are maintained as attributes of the graph, indicating the frequency of occurrence of each (unique) node and edge.

Decomposition-Based STN Sub-model. In multi-objective optimisation based on decomposition, the problem is decomposed into p scalar (single-objective) sub-problems that target different regions of the Pareto front [23]. A set of uniformly-generated weight vectors $\Lambda = (\lambda^1, \lambda^2, \ldots, \lambda^p)$ represents the scalar sub-problems defined by decomposition. For a given sub-problem $\lambda^j \in \Lambda$, the well-established Chebyshev scalarising function [15], to be minimised, is defined as follows:

$$g(x \mid \lambda^j) := \max_{i \in \{1, \ldots, m\}} \lambda_i^j \cdot \left| z_i^\star - f_i(x) \right| \tag{1}$$

such that $x \in X$ is a solution, $\lambda^j \in \mathbb{R}^m$ is a weighting coefficient vector and $z^\star \in \mathbb{R}^m$ is a reference point. The reference point is set to the best-known value for each objective.

In order to define the multi-objective STN, nodes are as described above, and edges separately follow the trajectories of each weight vector $\lambda^j \in \Lambda$, $j \in \{1, \ldots, p\}$. In other words, for a given sub-problem, the STN follows the trajectory of the solution with the best (lowest) Chebyshev scalar value for the corresponding weight vector. The trajectories for all weight vectors are then aggregated to construct a single graph model. Edges in the multi-objective STN are labelled by the vector whose transition they represent. The number of weight vectors p is a parameter of the modelling process. Section 5 reports the setting

Table 1. Description of STN metrics.

metric	description
nodes	number of unique solutions visited
pareto	number of solutions in the Pareto set
mean_pareto_in	average incoming degree to Pareto nodes
pareto_num_path	number of paths to Pareto nodes
pareto_mean_path	average shortest path to Pareto nodes

considered in our experiments. We offer below a more formal definition of the multi-objective STN model.

Multi-objective STN model (STN$_{MO}$). Assuming we have $p = |\Lambda|$ single-objective sub-problems (weight vectors), the multi-objective STN model is obtained by the graph union of the p single-objective STNs. More formally, let $STN_{v1} = G(N_{v1}, E_{v1}), STN_{v2} = G(N_{v2}, E_{v2}), \dots, STN_{vp} = G(N_{vp}, E_{vp})$ be the single-objective STNs for the sub-problems represented by vectors $(\lambda^1, \lambda^2, \dots, \lambda^p)$, respectively. We construct STN$_{MO}$ as the graph union of the STN_{vj} graphs, $j \in \{1, \dots, p\}$. Specifically, $STN_{MO} = G(N_{v1} \cup N_{v2} \cup \dots \cup N_{vp}, E_{v1} \cup E_{v2} \cup \dots \cup E_{vp})$. The union graph contains the nodes and edges that are traversed for at least one of the weight vectors. Node and edge attributes indicate which weight vector(s) visited them.

3.2 Network Metrics

We introduce five network metrics to describe the behaviour of the algorithms. These metrics, summarised in Table 1, were selected as they have been found to relate to search performance in single-objective problems [18]. The number of nodes expresses the algorithm exploratory power, the number of Pareto optimal solutions indicates effectiveness, the mean incoming degree to Pareto nodes is reflective of how many trajectories were successful, the number of paths as well as the average shortest path to Pareto nodes are indicative of the algorithm efficiency in reaching Pareto optimal solutions.

3.3 Network Visualisation

Visualising networks is a powerful and often beautiful way of appreciating their structure, which can offer insights and even reflect features not easily captured by network metrics. Node-edge diagrams are the most familiar form of network visualisation, they assign nodes to points in the two-dimensional Euclidean space and connect adjacent nodes by straight lines or curves. Nodes and edges can be decorated with visual properties such as size, colour and shape to highlight important features.

Our proposed multi-objective STN visualisations (see Fig. 2 for an example we will analyse later), use node colours and shapes to identify four relevant types of nodes: (1) start of trajectories, (2) end of trajectories that do not reach a Pareto optimal solution, (3) intermediate solutions in the trajectories, and (4) solutions in the Pareto set. The size of nodes and the thickness of edges are proportional to their sampling frequency.

A key aspect of network visualisation is the graph-layout, which accounts for the positions of nodes in the 2D Euclidean space. Graphs are mathematical objects, they do not have a unique visual representation. Many graph-layout algorithms have been proposed. *Force-directed* layout algorithms, such as Fruchterman-Reignold [8], are based on physical analogies defining attracting and repelling forces among edges. They strive to satisfy generally accepted aesthetic criteria such as an even distribution of nodes on the plane, minimising edge crossings, and keeping a similar length of edges. We use force-directed layouts for visualising the multi-objective STNs with two and three objectives (Figs. 2, 3, 7). For two objectives, we additionally introduced a layout that takes advantage of the objective space. The idea is to use the two objective values as the nodes x and y coordinates (Figs. 4, 5, and 8). These plots allow us to appreciate the progression of the search trajectories in the design and objective spaces simultaneously. Our graph visualisations were produced using the igraph and ggraph packages of the R programming language.

4   Experimental Setup

This section describes the experimental setup of our analysis, including the considered benchmark problems as well as the parameters used for the STNs and for the algorithms.

4.1   Benchmark Problems

In terms of benchmark, we consider ρmnk-landscapes [22] for constructing multi-objective multi-modal landscapes with objective correlation. They extend single-objective nk-landscapes [10] and multi-objective nk-landscapes with independent objectives [1]. Candidate solutions are binary strings of size n. The objective function vector $f = (f_1, \ldots, f_i, \ldots, f_m)$ is defined as $f \colon \{0,1\}^n \mapsto [0,1]^m$ such that each objective f_i is to be maximised. The objective value $f_i(x)$ of a solution $x = (x_1, \ldots, x_j, \ldots, x_n)$ is an average value of the individual contributions associated with each variable x_j. Given objective f_i, $i \in \{1, \ldots, m\}$, and variable x_j, $j \in \{1, \ldots, n\}$, a component function $f_{ij} \colon \{0,1\}^{k+1} \mapsto [0,1]$ assigns a real-valued contribution for every combination of x_j and its k variable interactions $\{x_{j_1}, \ldots, x_{j_k}\}$. These f_{ij}-values are uniformly distributed in $[0,1]$. Thus, the individual contribution of a variable x_j depends on its own value and on the values of $k < n$ variables other than x_j.

The variable interactions, i.e. the k variables that influence the contribution of x_j, are set uniformly at random among the $(n-1)$ variables other than x_j, following the random model from [10]. By increasing the number of variable interactions k, landscapes can be gradually tuned from smooth to rugged. In

Table 2. Benchmark parameters for small and large ρmnk-landscapes.

description	values
number of variables	$n = 16$ (small), $n = 128$ (large)
number of interactions	$k \in \{1, 4\}$
number of objectives	$m \in \{2, 3\}$
objective correlation	$\rho \in \{-0.4, 0.0, 0.4\}$

ρmnk-landscapes, f_{ij}-values additionally follow a multivariate uniform distribution of dimension m, defined by an $m \times m$ positive-definite symmetric covariance matrix (c_{pq}) such that $c_{pp} = 1$ and $c_{pq} = \rho$ for all $p, q \in \{1, \ldots, m\}$ with $p \neq q$, where $\rho > \frac{-1}{m-1}$ defines the correlation among the objectives; see [22] for details. The positive (resp. negative) correlation ρ decreases (resp. increases) the degree of conflict between the objective values.

Interestingly, ρmnk-landscapes exhibit different characteristics and degrees of difficulty for EMO algorithms [3,13]. The source code of the ρmnk-landscapes generator is available at the following URL: http://mocobench.sf.net.

4.2 Parameter Setting

We generate 12 small and 12 large ρmnk-landscapes with the parameter settings listed in Table 2. This allows us to investigate the differences between small and large instances, two and three objectives, conflicting, independent or correlated objectives, all this for relatively smooth to relatively rugged landscapes.

In terms of algorithms, we experiment with both MOEA/D and NSGA2 under the parameters from Table 3. Each algorithm is run independently 10 times on each instance. Algorithm performance is given in terms of hypervolume [24]. More particularly, we measure the relative hypervolume deviation with respect to the exact PF (for small instances) or best-known PF (for large instances). Let hv be the hypervolume covered by the population, the relative hypervolume deviation is $(hv^{\star} - hv)/hv^{\star}$, such that $hv^{\star}$ is the best-known hypervolume. A lower value is thus better. The hypervolume reference point is set to the origin.

Table 3. Algorithm parameters for MOEA/D and NSGA2.

description	values
population size	$\mu = 101$ ($m = 2$), $\mu = 231$ ($m = 3$)
neighbourhood size	$T = 10$ (MOEA/D)
variation	1-point crossover, bit-flip mutation with rate $1/n$
number of generations	$g = 20$ ($n = 16$), $g = 500$ ($n = 128$)

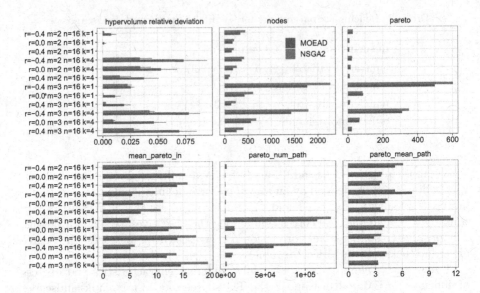

Fig. 1. Algorithm performance and STN metrics for *small* instances.

4.3 Reproducibility

For reproducibility purposes, relevant data and code are be available at: https://github.com/gabro8a/STNs-MOCO

5 Results

This section reports and comments the STNs obtained for small instances, and then for large instances. STN metrics are also discussed and related with algorithm performance.

5.1 Small Instances

We start with results for *small* instances with $n = 16$. In this case, the STN modelling used $p = 101$ decomposition vectors for instances with two objectives and $p = 231$ vectors for instances with three objectives; i.e. the same setting as the algorithms' population size. This give us the maximum possible modelling granularity (one vector per each individual member in the population), while still producing interpretable images.

Network Metrics. Algorithm performance for the 12 small instances is reported in Fig. 1 (top left), together with the five network metrics described in Table 1. For this set of instances, NSGA2 consistently outperforms MOEA/D, as indicated by the lower hypervolume relative deviation values. The higher STN metric values obtained by NSGA2 for nodes, pareto and pareto_num_path clearly support this trend. Another clear trend from the STN metrics is the decrease in values when we go from conflicting ($\rho = -0.4$) to positively correlated objectives ($\rho = 0.4$), which is observed for both values of $k \in \{1, 4\}$ and $m \in \{2, 3\}$. Finally, a salient observation from Fig. 1 is the large metric values observed for instances with $m = 3$ and conflicting objectives ($\rho = -0.4$). The values of nodes, pareto, pareto_num_path and pareto_mean_path are higher for $k = 1$ than for $k = 4$. This is consistent with previous findings: although there are more local optima for larger k values, the number of global optima (i.e. Pareto optimal solutions) has the opposite trend and decreases with increasing k [22].

Network Visualisation with a Force-Directed Layout. Figures 2 and 3 provide examples using a force-directed layout for $m = 2$ and $m = 3$ objectives, respectively. They report the multi-objective STN obtained for MOEA/D (top) and NSGA2 (bottom) for conflicting (left), independent (middle) and correlated objectives (right). The network visualisations confirm the trends observed in the metrics. Notably, the number of nodes in the networks consistently decreases when moving from negatively correlated objectives ($\rho = -0.4$, left) to positively correlated objectives ($\rho = 0.4$, right). We can also visually confirm the much denser STNs obtained for $m = 3$ objectives, as reported in Fig. 3.

Network Visualisation with the Objective-Space Layout. Figures 4 and 5 shows our proposed objective-space network layouts applied to the two studied levels of ruggedness $k \in \{1, 4\}$, respectively. Notice that this layout is only applicable for $m = 2$ objectives if we restrict ourselves to the 2D Euclidean space. We argue that this layout may be more useful to the multi-objective optimisation community (as compared to the force-directed layouts shown in Figs. 2 and 3) as they resemble the familiar Pareto front scatter plots. However, they offer additional insights, revealing not only the Pareto front when it is reached, but also the search progress towards it, thus giving indication of unsuccessful runs as well. Notice that in these plots, an additional graphical layer is shown in the form of blue diamonds. They correspond to the exact Pareto front and are not part of the STN nodes. They serve as a tool to appreciate if and when the STN trajectories reach the Pareto front. The objective-space layout, therefore, might be more suitable for appreciating the performance difference between algorithms.

With respect to NSGA2 outperforming MOEA/D (as indicated by the performance metric in Fig. 1), this can only be clearly appreciated for $k = 4$ and $m = 2$ (Fig. 5). Looking at the left plots for $\rho = -0.4$, we can confirm that the MOEA/D STN (top plot) has a larger number end nodes (orange triangles) that are also of larger size as those of the NSGA2 STN (bottom plot). Remember that the size of nodes is proportional to their sampling frequency. Therefore, this is a visual reflection that MOEA/D has a larger number of unsuccessful runs, that is, trajectories ending into sub-optimal solutions. The NSGA2 STN (bottom plot) reveals in this case a larger number of red nodes (Pareto solutions), which are of larger size. Remember that the super-imposed blue diamond scatter

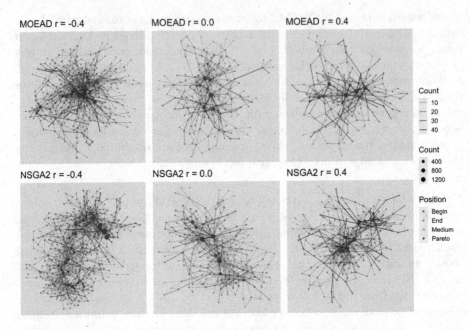

Fig. 2. STN visualisation with a force-directed layout for *small* instances with $k = 1$ and $m = 2$ objectives.

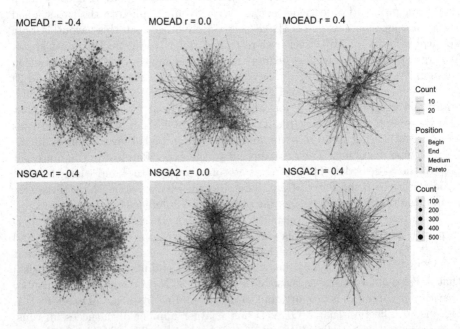

Fig. 3. STN visualisation with a force-directed layout for *small* instances with $k = 4$ and $m = 3$ objectives.

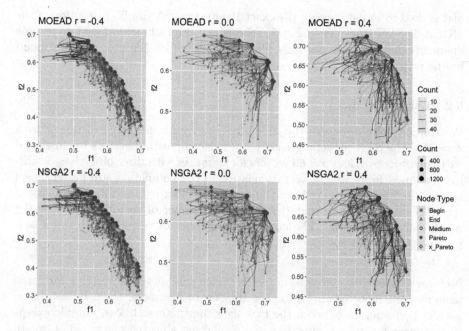

Fig. 4. STN visualisation with the objective-space layout for *small* instances with $k = 1$ and $m = 2$ objectives.

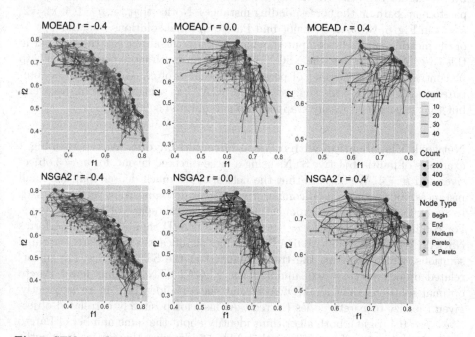

Fig. 5. STN visualisation with the objective-space layout for *small* instances with $k = 4$ and $m = 2$ objectives.

plot is used to visually locate the exact Pareto front. A careful inspection of the MOEA/D STN (top plot in Fig. 5) reveals one empty blue diamond, and some diamonds that are only partially filled with red nodes (Pareto solutions found by the trajectories).

5.2   Large Instances

We continue our discussion by analysing the results for *large* instances with $n = 128$. The STN modelling used $p = 51$ decomposition vectors for instances with two objectives and $p = 66$ vectors for instances with three objectives. In this case, we used fewer weight vectors relative to the population size for efficiency reasons, and for improving both the cosmetic rendering and interpretability of the STN images. Notice that the larger the number of vectors, the larger the number of nodes in the STN models. The number of vectors can be seen as a parameter to adjust the model granularity.

Network Metrics. Algorithm performance and network metrics for large instances are reported in Fig. 6. For this set of instances, there is less difference in performance between the two algorithms. Nevertheless, notable exceptions appear for $m = 3$ objectives and conflicting objectives ($\rho = -0.4$), where MOEA/D reaches significantly better hypervolume values. This is supported by the higher STN metric values obtained by MOEA/D for nodes, pareto and pareto_num_path on the corresponding instances. Notice that for $\rho = 0.4$, $m = 2$, $k = 4$ in Fig. 6, NSGA2 does not find Pareto optimal solutions, therefore, some of the metrics cannot be computed, which explains the absence of the blue bar in this case. We notice that the NSGA STNs contain much more nodes, which is to be contrasted by its number of pareto nodes that is often particularly low compared to MOEA/D. This suggests that NSGA2 has a higher rate of discovery, but that it gets more easily trapped into sub-optimal solutions.

Network Visualisation with a Force-Directed Layout. We report in Fig. 7 examples of multi-objective STNs using a force-directed layout, for $m = 3$ objectives and $k = 4$. We observe that the networks are much denser than for small instances, although we used comparatively fewer decomposition vectors. This is to be expected given the exponentially larger search space of large instances. For conflicting objectives ($\rho = -0.4$, left), MOEA/D significantly outperforms NSGA2. We observe that MOEA/D identifies significantly more Pareto optimal solutions, which confirms the trend observed in the STN metrics. For uncorrelated objectives ($\rho = 0.0$, middle), the NSGA2 STN contains fewer Pareto optimal solutions than for MOEA/D, but they are identified more frequently, given the size of pareto nodes (in red). At last, for positively correlated objectives ($\rho = 0.4$, right), both algorithms identify about the same number of Pareto optimal solutions, but we still see that NSGA2 identifies them more frequently, which supports the fact that NSGA2 is slightly better for this instance.

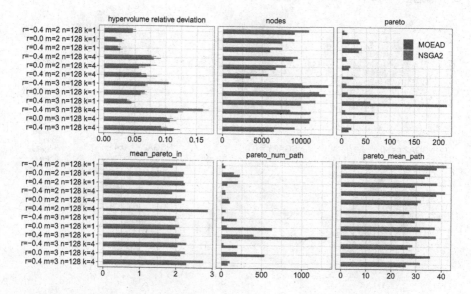

Fig. 6. Algorithm performance and STN metrics for *large* instances.

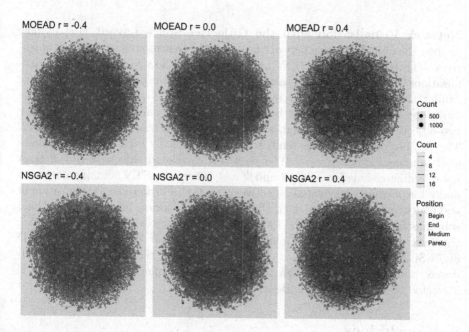

Fig. 7. STN visualisation with a force-directed layout for *large* instances with $k = 4$ and $m = 3$ objectives.

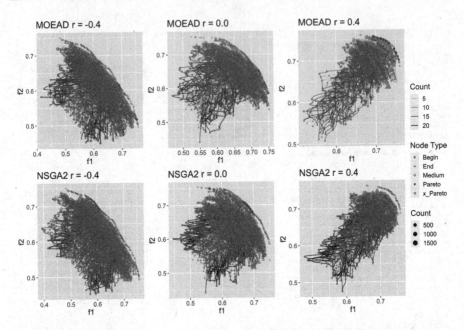

Fig. 8. STN visualisation with the objective-space layout for *large* instances with $k = 4$ and $m = 2$ objectives.

Network Visualisation with the Objective-Space Layout. Let us now analyse the objective-space network layout for large instances with $m = 2$ objectives. The multi-objective STNs are reported in Fig. 8 for $k = 4$. The visualisations for $k = 1$ are not shown due to space constraints, but they show similar trends. As anticipated by the analysis of STN metrics in Fig. 6, the multi-objective STNs obtained by the two algorithms are similar for the large two-objective instance, although solutions seem a bit more spread for NSGA2. The number of pareto nodes tends to be proportionally higher as we gradually shift from conflicting ($\rho = -0.4$) to positively correlated objectives ($\rho = 0.4$). A notable difference is for NSGA2 and $k = 4$, where the STN contains no pareto nodes for positively correlated objectives ($\rho = 0.4$). Furthermore, the position of end nodes (orange triangles) suggests that the trajectories end up in sub-optimal solutions farther away from the Pareto front for instances with $k = 4$, for which there are more local optima. Going back to the comparison between MOEA/D and NSGA2, the objective-space network layout of the STNs provide visual evidence confirming that, although NSGA2 seems to explore more solutions, it is attracted to lower quality solutions.

6   Conclusions

We argue that STNs are an accessible tool to analyse and visualise the behaviour of evolutionary multi-objective optimisation algorithms. Constructing STN models does not require any specific sampling techniques. Instead, data is collected from a set of runs of the studied algorithms, and then aggregated and processed to devise the models. Post-processing tools, however might be required to deal with large models. STNs provide insights into problem structure as well as into algorithm convergence behaviour and performance differences.

Future work could study additional multi-objective problems and algorithms, including real-world problems and 4+ objectives. The challenge we foresee here deals with the larger number of solutions attained by the trajectories. For this, we could thoroughly investigate coarser models including varying the number of decomposition vectors in the STN model, and of grouping multiple solutions within a single node, as has been done for single-objective STN models [17,18]. A number of repositories contain code and data to start with STN modelling and analysis for both single-objective[1] (including a web-based tool[2]), and multi-objective combinatorial[3] and continuous[4] problems. We should provide unified software tools to improve the usability of STN models.

Acknowledgements. We are deeply grateful to the SPECIES Society for funding a scholarship for Yuri Lavinas to visit the University of Stirling, Scotland, UK.

References

1. Aguirre, H.E., Tanaka, K.: Working principles, behavior, and performance of MOEAs on MNK-landscapes. Eur. J. Oper. Res. **181**(3), 1670–1690 (2007)
2. Coello Coello, C.A., Lamont, G.B., Van Veldhuizen, D.A.: Evolutionary Algorithms for Solving Multi-Objective Problems. Springer, New York (2007). https://doi.org/10.1007/978-0-387-36797-2
3. Daolio, F., Liefooghe, A., Verel, S., Aguirre, H., Tanaka, K.: Problem features versus algorithm performance on rugged multiobjective combinatorial fitness land-scapes. Evol. Comput. **25**(4), 555–585 (2017)
4. Deb, K.: Multi-Objective Optimization Using Evolutionary Algorithms. Wiley, Chichester (2001)
5. Deb, K., Pratap, A., Agarwal, S., Meyarivan, T.: A fast and elitist multi-objective genetic algorithm: NSGA-II. IEEE Trans. Evol. Comput. **6**(2), 182–197 (2002)
6. Fieldsend, J.E., Alyahya, K.: Visualising the landscape of multi-objective problems using local optima networks. In: Proceedings of the Genetic and Evolutionary Computation Conference Companion, GECCO 2019, pp. 1421–1429. Association for Computing Machinery, New York (2019)
7. Fonseca, C.M., Fleming, P.J.: On the performance assessment and comparison of stochastic multiobjective optimizers. In: Voigt, H.-M., Ebeling, W., Rechenberg, I., Schwefel, H.-P. (eds.) PPSN 1996. LNCS, vol. 1141, pp. 584–593. Springer, Heidelberg (1996). https://doi.org/10.1007/3-540-61723-X_1022
8. Fruchterman, T.M.J., Reingold, E.M.: Graph drawing by force-directed placement. Softw. Pract. Exper. **21**(11), 1129–1164 (1991)

---

[1] https://github.com/gabro8a/STNs.
[2] http://45.32.184.82.
[3] https://github.com/gabro8a/STNs-MOCO.
[4] https://github.com/gabro8a/STNs-MOEA.

9. Goldberg, D.E.: Genetic Algorithms in Search, Optimization and Machine Learning. Addison-Wesley, Boston (1989)
10. Kauffman, S.A.: The Origins of Order. Oxford University Press, Oxford (1993)
11. Kerschke, P., Grimme, C.: An expedition to multimodal multi-objective optimization landscapes. In: Trautmann, H., et al. (eds.) EMO 2017. LNCS, vol. 10173, pp. 329–343. Springer, Cham (2017). https://doi.org/10.1007/978-3-319-54157-0_23
12. Lavinas, Y., Aranha, C., Ochoa, G.: Search trajectories networks of multiobjective evolutionary algorithms. In: Jiménez Laredo, J.L., Hidalgo, J.I., Babaagba, K.O. (eds.) EvoApplications 2022. LNCS, vol. 13224, pp. 223–238. Springer, Cham (2022). https://doi.org/10.1007/978-3-031-02462-7_15
13. Liefooghe, A., Daolio, F., Verel, S., Derbel, B., Aguirre, H., Tanaka, K.: Landscape-aware performance prediction for evolutionary multi-objective optimization. IEEE Trans. Evol. Comput. **24**(6), 1063–1077 (2020)
14. Liefooghe, A., Derbel, B., Verel, S., López-Ibáñez, M., Aguirre, H., Tanaka, K.: On pareto local optimal solutions networks. In: Auger, A., Fonseca, C.M., Lourenço, N., Machado, P., Paquete, L., Whitley, D. (eds.) PPSN 2018. LNCS, vol. 11102, pp. 232–244. Springer, Cham (2018). https://doi.org/10.1007/978-3-319-99259-4_19
15. Miettinen, K.: Nonlinear Multiobjective Optimization. Kluwer Academic Publishers (1999)
16. Ochoa, G., Tomassini, M., Verel, S., Verel, C.: A study of NK landscapes? Basins and local optima networks. In: Genetic and Evolutionary Computation Conference. GECCO, pp. 555–562. ACM Press, New York (2008)
17. Ochoa, G., Malan, K.M., Blum, C.: Search trajectory networks of population-based algorithms in continuous spaces. In: Castillo, P.A., Jiménez Laredo, J.L., Fernández de Vega, F. (eds.) EvoApplications 2020. LNCS, vol. 12104, pp. 70–85. Springer, Cham (2020). https://doi.org/10.1007/978-3-030-43722-0_5
18. Ochoa, G., Malan, K.M., Blum, C.: Search trajectory networks: a tool for analysing and visualising the behaviour of metaheuristics. Appl. Soft Comput. **109**, 107492 (2021)
19. Ochoa, G., Veerapen, N., Daolio, F., Tomassini, M.: Understanding phase transitions with local optima networks: number partitioning as a case study. In: Hu, B., López-Ibáñez, M. (eds.) EvoCOP 2017. LNCS, vol. 10197, pp. 233–248. Springer, Cham (2017). https://doi.org/10.1007/978-3-319-55453-2_16
20. Schäpermeier, L., Grimme, C., Kerschke, P.: One PLOT to show them all: visualization of efficient sets in multi-objective landscapes. In: Bäck, T., et al. (eds.) PPSN 2020. LNCS, vol. 12270, pp. 154–167. Springer, Cham (2020). https://doi.org/10.1007/978-3-030-58115-2_11
21. Tušar, T., Filipič, B.: Visualization of Pareto front approximations in evolutionary multiobjective optimization: a critical review and the prosection method. IEEE Trans. Evol. Comput. **19**(2), 225–245 (2015)
22. Verel, S., Liefooghe, A., Jourdan, L., Dhaenens, C.: On the structure of multiobjective combinatorial search space: MNK-landscapes with correlated objectives. Eur. J. Oper. Res. **227**(2), 331–342 (2013)
23. Zhang, Q., Li, H.: MOEA/D: a multiobjective evolutionary algorithm based on decomposition. IEEE Trans. Evol. Comput. **11**(6), 712–731 (2007)
24. Zitzler, E., Thiele, L., Laumanns, M., Fonseca, C.M., Da Fonseca, V.G.: Performance assessment of multiobjective optimizers: an analysis and review. IEEE Trans. Evol. Comput. **7**(2), 117–132 (2003)

On the Effect of Solution Representation and Neighborhood Definition in AutoML Fitness Landscapes

Matheus C. Teixeira[ID] and Gisele L. Pappa[(✉)][ID]

Universidade Federal de Minas Gerais, Belo Horizonte, MG, Brazil
`glpappa@dcc.ufmg.br`

Abstract. The interest in AutoML search spaces has given rise to a plethora of studies conceived to better understand the characteristics of these spaces. Exploratory landscape analysis is among the most commonly investigated techniques. However, in contrast with other classical optimization problems, in AutoML defining the landscape may be as tough as characterizing it. This is because the concept of solution neighborhood is not clear, as the spaces have a high number of conditional hyperparameters and a somehow hierarchical structure. This paper looks at the impact of different solution representations and distance metrics on the definition of these spaces, and how they affect exploratory landscape analysis metrics. We conclude that these metrics are not able to deal with structured, complex spaces such as the AutoML ones, and problem-related metrics might be the way to leverage the landscape complexity.

Keywords: Fitness landscape analysis · Automated Machine Learning · distance measures

1 Introduction

Exploratory fitness landscape analysis has been an interesting tool used to better understand the characteristics of the search space of classical optimization algorithms [7]. More recently, some efforts have migrated from traditional optimization problems to those that deal with machine learning problems. In this direction, a few studies have looked at the loss landscape of neural networks, while others have focused on understanding both the search of Automated Machine Learning (AutoML) problems focusing on machine learning pipelines [18] and the architecture spaces of neural networks [9].

When compared to traditional optimization problems, these landscapes are harder to be defined and analyzed, as they have both categorical, discrete and continuous parameters, many of them conditional (i.e., one hyperparameter is only present if another hyperparameter is previously selected). The representation is somehow hierarchical, as changes in one hyperparameter may activate others. For example, suppose we consider different types of classification

L. Pérez Cáceres and T. Stützle (Eds.): EvoCOP 2023, LNCS 13987, pp. 227–243, 2023.
https://doi.org/10.1007/978-3-031-30035-6_15

algorithms. Each of them includes different sets of parameters, which will only become available when that algorithm is selected. From an optimization point of view, these problems can be defined as a mixed-integer nonlinear optimization problem [24].

This work focuses on analyzing the landscape of AutoML problems that generate machine learning pipelines [24]. A pipeline is defined as a sequence of pre-processing methods, algorithms and their hyperparameters that obtains accurate results in the target classification task. Many different methods have been used to solve this problem [3], but not much about the space itself is known. In these problems, the fitness landscape can be considered hard to define, given the loose definition that can be given to neighborhood.

A fitness landscape is defined by a tuple (S, f, N), where S is the set of all possible solutions (*i.e.* the search space), $f : S \rightarrow \mathbb{R}$ is a function that attributes a real-valued performance estimation for each solution in S, and $N(x)$ is a notion of neighborhood between solutions, usually defined as a distance metric $N(x) = \{y \in S | d(x, y) \leq \epsilon\}$ for a sufficiently small ϵ.

Although the concept of neighborhood is clear in most problems, there are cases where defining neighborhood is a challenge. In AutoML, these challenges are related to the conditional nature of the search space, and to whether changing one type of hyperparameter should have a larger impact on the fitness of the generated solutions than changing others. For example, using a mutation to change the number of trees of a Random Forest will probably affect less the fitness than changing the Random Forest by another classifier, say, Naive Bayes. This needs to be reflected in the distances between solutions and the definition of neighborhood.

Aiming to increase our understanding of the fitness landscapes of AutoML problems and the impacts of neighborhood definition, this paper analyses three different ways of representing the fitness landscapes of AutoML problems and studies how these definitions affect the shape of the space. For that, we define a simplified search space of AutoML pipelines that is fully enumerable, where continuous attributes are discretized. We generate the fitness landscape for 20 different datasets and extract metrics from traditional exploratory fitness landscape analysis. We observed that different representations yield different magnitudes of distances and change neighborhood, and that for different datasets the characteristics of the landscape vary drastically. Hence, it might be the case that the study of these landscapes requires metrics related to the problem domain and the underlying datasets being tackled by AutoML.

2   Related Work

There are a few studies in the literature that have looked at the landscape of AutoML problems in general, and they can be divided into two groups: those where the search space is made of machine learning pipelines and those where the space refers to neural network architectures. We start by looking at problems where candidate solutions are full pipelines. The authors in [2] were the first to perform the analysis of AutoML landscapes considering a subspace of TPOT (Tree-based Pipeline Optimization Tool), using metrics such as slope, roughness and neutrality. Their results suggest that many regions of high fitness exist in the space, but these are prone to overfitting. In this same direction, the authors in [12] looked at fitness landscape metrics to better understand the search space of a huge space of machine learning pipelines. They looked at Fitness Distance Correlation (FDC) and metrics of neutrality, and concluded FDC was a poor metric for performing the analyses.

In a similar fashion, but concerning the landscape of algorithm configuration problems, the authors in [13] evaluated them in terms of modality and convexity of parameter responses, and concluded that many of the parameter slices appear to be uni-modal and convex, both on instance sets and on individual instances. In a follow-up work [14], they tested the unimodality of the AutoML loss landscape considering the joint interaction of the hyperparameters and concluded that most landscapes have this property, but are not convex. They also observed that hyperparameters interact strongly in regions of configuration space farther from the optimal solutions. Finally, they empirically demonstrate that FDC has limitations in characterizing certain spaces.

Turning to analysis of spaces of neural network architectures, the authors in [9] analyzed the fitness landscape of NAS in the context of graph neural network architectures. They used FDC together with the dispersion metric, and also looked at the neutrality of the space. The analysis of neutrality indicated that the space was not neutral, but the authors highlighted the need to use more elaborate techniques for estimating neutrality.

Going further, the authors in [19] introduced the concept of fitness landscape footprint, given by an aggregation of eight general-purpose metrics to synthesize the landscape of a neural network architecture search problem. They looked at the classical image classification benchmarks and concluded the technique was able to characterize the relative difficulty of the problem, and the insights provided may be used to assess the expected performance of a search strategy in each dataset.

Note that all the works reviewed so far focus on using standard metrics of FLA. Lately, there have been a few works looking at these spaces with Local Optima Networks (LONs). For example, the work of [20] adapted LONs to analyze the global structure of parameter configuration spaces. For complex scenarios, they found a large number of sub-optimal funnels, while simpler problems had a single global funnel. With this same objective, the authors in [1] looked at parameter spaces for Particle Swarm Intelligence (PSO) and found that PSO's parameter landscapes are relatively simple at the macro level but a lot more complex at the micro level, making parameter tuning more difficult than they initially assumed. The authors in [18] looked at AutoML pipelines using LONs, and concluded their space was multi-modal and with a high number of local optima.

Finally, a few recent works have looked at the fitness landscapes of NAS problems from a point of view of both classical metrics of FLA and LONs.

The authors in [17] analyzed the NAS fitness landscape generating a fully enumerable space using FDC and LON. The results showed that the search space is easy and that most local optima are only one perturbation away from the global optimum. The authors in [10] have also looked at the fitness landscapes of NAS benchmarks. They concluded that the FL analyzed are multi-modal but have few local optima, making it not complicated for local search methods to escape these regions of the search space with simple perturbation operations.

3   AutoML Fitness Landscape

In order to generate different fitness landscapes, we use an AutoML search space that considers a trade-off between space size and solution effectiveness: the space has enough components to generate accurate solutions to real problems but it is simple enough so it can be fully enumerated. Given the hierarchical nature of the search space, which emerges due to a large number of conditional parameters, we represent the search space by a grammar, and its production rules are followed to generate feasible solutions. The grammar uses a subset of the original space defined in [12], and originally presented in [18]. It has 38 production rules, 92 terminals and 45 non-terminals[1]. In terms of preprocessing, it includes algorithms that deal with feature scaling and dimensionality reduction, such as Principal Component Analysis (PCA) and Select K-Best. It is also possible for a pipeline to use no preprocessing algorithms. In terms of classification methods, there are five possible options: Logistic Regression, Multilayer Perceptron, K-Nearest Neighbors (KNN), Random Forest, and Ada Boost. The number of hyperparameters varies from one classification algorithm to another, going from two (Ada Boost) to 7 (Random Forest). A few continuous parameters are left out of the grammar and assume their default values as defined in their Scikit-learn implementation [11].

3.1   Solution Representation and Neighborhood

Given the search space is defined by the grammar, the most intuitive way to represent the candidate solutions is by using the derivation tree extracted directly from the grammar. Figure 1 shows two examples of derivations trees (pipelines): the first performs no preprocessing and runs an Adaboost classifier, while the second performs feature selection before applying Adaboost. Note that the algorithms and hyperparameters are the leaf nodes of the tree.

Given a representation, we also need to define the concept of solution neighborhood, which is usually given by the distance between pairs of solutions. In the case of machine learning pipelines, we do not have an inherent distance/similarity concept that can easily determine how distant they are. For example, given two ML pipelines with the same preprocessing and different classifiers. Should they be considered closer, more distant or as distant as two pipelines with different preprocessing but the same classifier?

---

[1] bit.ly/38F0o3U.

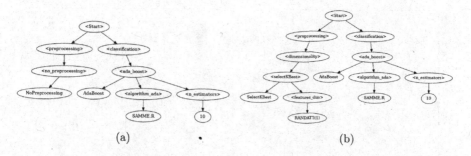

Fig. 1. Example of Machine Learning Pipelines.

Given this drawback, the authors in [12] introduced an ad-hoc technique to define the distance between AutoML pipelines represented by grammar-derivation trees, where the distance between individuals with different algorithms is greater than that of individuals that differ only in hyperparameters. [9], in turn, measured the distance between pipelines by first converting them to a binary representation using one-hot encoding and then calculating the Hamming distance between these binary representations. In addition, the authors also apply the t-SNE algorithm [21], used for dimensionality reduction, to generate a dense space and use the Euclidean distance to calculate the distance between the pipelines in the embedded space.

As the results of FLA metrics are directly influenced by the concept of distance adopted by the authors, we adopted and compared three different representation and distance methods according to previous work. We used: (i) the method proposed by [12], based on tree representations, from now on referred as *Adhoc*, (ii) the methods based on binary representation and iii) the embedded space generated by t-SNE, both proposed by [9] in the context of understanding fitness landscapes of graph neural networks. Next, we discuss each of these methods.

ad hoc distance method in tree-based representation: This method assigns constant values to represent the distance between each type of node present in the pipeline. 16 types of nodes are defined, and their distances are calculated according to constant values assigned by specialists [12]. In this case, nodes closer to the root of the tree tend to be more dissimilar than those at the leaves of the same subtree.

The final distance between two pipelines is equal to the sum of the distances of the nodes that make up each one of them. The idea of this method is to consider that the impact of changing an algorithm is more significant than changing the value of a hyperparameter and that the presence of an algorithm is greater than simply changing the algorithm. For example, the distance from the trees in Fig. 1 is 4. If we get tree (b) and replace its classifier with another one, then the distance increases to 6.

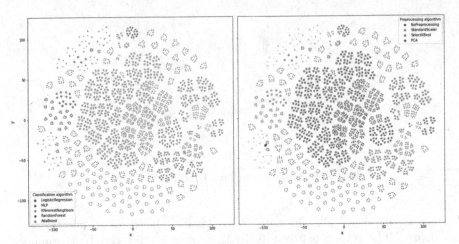

Fig. 2. Dimensionality reduction of the generated space using one-hot encoding. The one-hot encoded space has 64 dimensions, here represented as two.

Hamming distance in binary representation: The second method to define the distances between pairs of pipelines first converts the tree into a binary sequence using one-hot encoding, which is a simple way of transforming structured or categorical data into a numerical representation. The Hamming distance is then used to calculate distances between solutions.

The transformation to one-hot encoding is done as follows. Each pipeline P is represented by a binary sequence $\mathcal{S}^P$, where the presence or absence of a terminal corresponds to 1 or 0, respectively. Each terminal is assigned a specific and fixed position in the sequence, so all pipelines composed of the same terminal have 1 in the position representing that terminal. For example, if the classification algorithm X is mapped to the i-th position of the sequence $\mathcal{S}$ and a given pipeline P, then, by this algorithm, the sequence representing that pipeline has the i-th element equals to 1, i.e., $\mathcal{S}_i^P = 1$.

It is important to say that the one-hot encoding process is lexical, i.e., two different parameters that were allowed the same value in different contexts were assigned the same position in the vector. For example, the number 5 meaning trees of a Random Forest or neighbors in a KNN were mapped to the same place. Hence, instead of having a length of 96 (the number of terminals of the grammar), the one-hot encoding has a length of 64.

Euclidean distance in 2D embedded representation: The third method of distance uses the (sparse) representation generated by the one-hot encoding with high dimensionality and condenses the vector into a two-dimensional space $\mathbb{R}^2$ using the t-SNE algorithm, run with the default parameters values of its Sklearn implementation. These values are then compared using the Euclidean distance between the two representations.

Reducing the representation to a $\mathbb{R}^2$ space is interesting because it allows the visualization of the configuration space, as shown in Fig. 2. In the figure on the left, the color indicates the classification algorithm used in the pipeline. In the figure on the right, the color indicates the preprocessing algorithm used. Observe that the distribution of pipelines using a specific algorithm is not uniform and this occurs because certain algorithms have more hyperparameters than others.

3.2 Fitness Function

Having the solution space defined, we define the fitness of the solutions as the weighted average F-measure [23] (Eq. 1):

$$\text{F-measure} = \frac{2 \cdot TP}{2 \cdot TP + FP + FN} \tag{1}$$

which is the harmonic mean of precision and recall. In the equation, TP stands for the true positives, FP for the false positives, and FN for the false negatives. As some datasets have several classes, the one-vs-all strategy was employed when calculating this metric.

When evaluating the fitness, a maximum computational budget (for training and testing the pipeline) was defined, and solutions that exceeded this limit received fitness 0. This was the simple way we found to deal with the trade-off between the computational cost versus the quality of the solutions.

4 Characterization of the Fitness Landscapes

The fitness landscape of a problem depends directly on the data being analyzed. In this work, the pipelines were evaluated in 20 datasets selected from the UCI Machine Learning Repository[2] and from Kaggle[3].

Considering the search space defined in Sect. 3, we generate all the solutions and evaluated them for each of the 20 datasets, generating 20 different fitness landscapes. Table 1 presents some features of the datasets used to generate the fitness landscape. The "Code" column indicates the code used to reference each dataset, the "Instances" column indicates the number of instances, the "Features" column indicates the number of features, the "Classes" column indicates the number of classes present in the target feature. Following, the "Optimum" column indicates the fitness of the global optimum (from the space defined by the grammar) and the "#Optimum" column indicates the number of solutions that achieve the value of optimal fitness.

Figure 3 shows the boxplots of the fitness distribution of the pipelines generated for each dataset. Note that, for some datasets, the fitness of the solutions is predominantly high or low, while for others they are better distributed. Observe that this distribution does not affect FLA, but gives an insight into the difficulty of the problem.

---

[2] https://archive.ics.uci.edu/ml/index.php.
[3] https://www.kaggle.com/datasets.

Table 1. Characterization of the datasets.

Dataset	Code	Instances	Features	Classes	Optimun	#Optimum
abalone	DS01	4177	8	28	0.2842	1
bank	DS02	11162	16	2	0.8376	8
car-evaluation	DS03	1728	6	4	0.9380	8
diabetes	DS04	768	8	2	0.7900	8
dry-bean	DS05	13611	16	7	0.9309	32
fire	DS06	17442	6	2	0.9539	8
fruit	DS07	898	34	7	0.9157	1
heart	DS08	303	13	2	0.8216	96
ml-prove	DS09	6118	51	6	0.4478	21
mushrooms	DS10	8124	22	7	0.6678	10
nursery	DS11	12960	8	5	0.9937	1
pistachio-28	DS12	2148	28	2	0.9295	6
pumpkin	DS13	2500	12	2	0.8829	3
raisin	DS14	900	7	2	0.8810	1
statlog-segment	DS15	2310	19	7	0.9685	24
texture	DS16	5500	40	11	0.9980	6
vehicle	DS17	846	18	4	0.7875	7
water-potability	DS18	3276	9	2	0.6643	1
wilt	DS19	4839	5	2	0.9888	2
wine-quality-red	DS20	1599	11	6	0.6402	16

Finally, Table 2 shows the pairwise distances between the pipelines shown in Fig. 1 and a third pipeline, referred as 1bMLP – where the AdaBoost classifier is replaced by an MLP. Note that the distances are consistent among individuals, with a classifier replacement weighing more than changing a preprocess. The magnitude of these numbers and their differences, however, differ substantially.

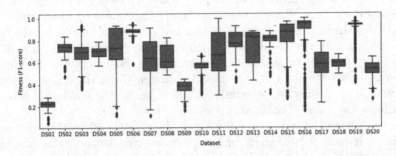

Fig. 3. Boxplot of the fitness of the pipelines in different datasets.

Table 2. Pairwise distances between pipelines using different representations. Each row refers to the pipeline represented by the figure number indicated.

Fig.	Euclidean			Hamming			Adhoc		
	1a	1b	1bMLP	1a	1b	1bMLP	1a	1b	1bMLP
1a	0.0	0.27	24.06	0.0	3.0	9.0	0.0	4.0	6.0
1b	0.27	0.0	24.14	3.0	0.0	6.0	4.0	0.0	2.0
1bMLP	24.06	24.14	0.0	9.0	6.0	0.0	6.0	2.0	0.0

5   Exploratory Fitness Landscape Analysis

According to the fitness landscapes built using the 20 datasets and three different types of representations, we calculated the Fitness Distance Correlation (FDC) [4], the dispersion metric [5] and the neutrality rate considering each of the landscapes.

Fitness Distance Correlation (FDC)) is a metric widely used in FLA and was already used in the context of AutoML [9,12,19]. The idea behind this metric is that landscapes that have a positive correlation between distance and fitness from a global optimum are proportionately easy to optimize [16]. It was originally proposed by [4], but had the limitation of depending on the knowledge of the global optimum, which is often not available. This is not the case here, as the complete space was enumerated.

Dispersion Metric (DM) measures the average distance between the top $p\%$ solutions from a set of randomly sampled points using a uniform distribution. The idea is to measure how close or dispersed the solutions with the highest fitness are in the space, and it is calculated as follows: first, it samples a fixed-length list of s_v solutions from the search space and evaluates their fitness. Then, it sorts the solutions by fitness value, and the top $s_b = s_v \times p\%$ points are selected. $disp$ is given by the average distance between these solutions.

Neutrality Rate is a metric designed to measure and identify the presence of regions of the search space with small or no variations in values of fitness, i.e., immediate neighbors with equal fitness. Flat regions in the search space can be a problem for algorithms that are gradient-driven or rely on a local search, such as Hill-Climbing. By definition, neutrality is the opposite of roughness, but both metrics are useful since part of the search space can present high roughness while other regions can present high neutrality.

Neutrality can both make the search space easier to explore [22] or get some algorithms stuck in regions of the search space with similar (or equal) fitness, preventing them from exploring areas with possibly better results [6]. Assuming a discrete representation of the solutions a_g and defining a "mutation" as a change in one of the components of a_g that leads to a neighbor solution $a_g^i \in N(a_g)$, We evaluate the neutrality of our landscape based on neutral walks (as defined by [15]), which perform a random walk and identify the number of neighbors with fitness lower than a parameter δ.

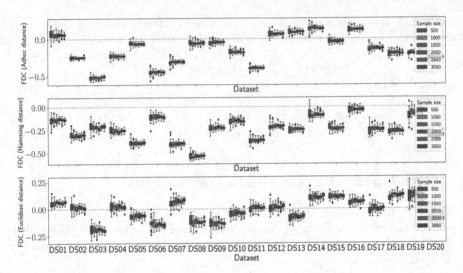

Fig. 4. FDC comparison using different distance methods and different random walk sizes

6   Results and Discussion

This section presents and discusses the results of the metrics previously discussed in the AutoML fitness landscape defined in Sect. 3. In all experiments, the metrics were calculated with 30 different samples/walks and the error lines indicate the 95% confidence interval.

6.1   Fitness Distance Correlation

The experiments were performed using random walks with lengths varying from 500 to 3,000 with steps of 500. As several datasets have more than one global optimum, the distance used was the smallest, that is, the distance from the global optimum closest to the observed solution. Figure 4 shows the boxplots grouped by dataset, where the color indicates the length of the walk.

Observe that the FDC metric tends to negative values for most datasets, but the correlation is not strong for any of the cases, as the $|FDC| \leq -0.6$ in all cases. The fact that it is negative indicates that as the distance from the solution selected in the random walk to the nearest global optimum increases, the fitness of the solution also increases. The walk length affects the variance of FDC, as observed from the largest bar in the boxplots referring to walks of length equal to 500. However, the difference between the FDC with the largest and the smallest walk length is not significant. In the case of Euclidean distance, specifically, the distribution of FDC is more uniform and 11 instances have a mean greater than 0.

Another point to be considered is that increasing the neighborhood has no effect on the result obtained by the metric, although this is a factor that directly affects the difficulty of the space, as shown in [18]. This happens because, in all cases, the distance is measured according to the same global optima. Moreover, the distance metrics used only consider the syntactic characteristics of the pipelines (they do not consider the topological structure of the search space). Therefore, the result obtained with neighborhoods of size 15, 20 and 25 result in the same values. However, these results would change if the distance was measured in relation to a local optimum, which changes when the neighborhood changes, and is often used when the space cannot be completely enumerated.

An alternative way of analyzing FDC is through a joint plot of the distance by the fitness of each solution contained in the sampled random walk, as depicted in Figs. 5a–5c. Each figure represents a different type of distance and the mean and standard deviation of the distances of each solution contained in the walk are shown in the lower left corner and the FDC in the lower right corner. The Adhoc and hamming methods concentrate the distance in a specific range while the Euclidean distance has a standard deviation of approximately half of the mean.

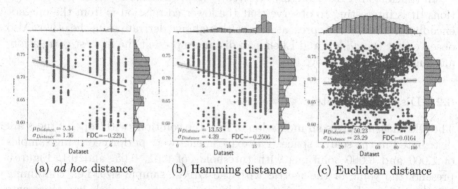

(a) *ad hoc* distance   (b) Hamming distance   (c) Euclidean distance

Fig. 5. Scatter plot of FDC calculated on DS04 (distance to global optimum).

Although in absolute terms the FDC values are different depending on the representation and distance, the results are expected to maintain the same relationship between the datasets, i.e., if the datasets are ordered according to FDC, the same order should be maintained regardless of the distance adopted.

To compare whether the order is in fact maintained, the datasets are ranked according to FDC for each distance measure used, and the rankings are compared using the Kendall coefficient τ [8] to measure the correlation between rankings. This method calculates the number of concordant/discordant pairs that are present in both rankings.

If the rankings are composed of the same N elements possibly ordered differently, then there are $\binom{N}{2}$ possible pairs. Considering that C denotes the number of concordant pairs and D the number of discordant pairs, the coefficient τ

is defined as $\tau = (C - D)/(C + D)$. If $D = 0$, then the expression reduces to $C/C = 1$, that is, if all pairs agree, then the coefficient is equal to 1. If $C=0$, then the resulting expression is $-D/D = -1$, that is, if all pairs are discordant, then $\tau = -1$. Therefore, $\tau \in [-1, +1]$, and the higher the value, the more similar the rankings.

The ideal scenario is that the correlation is positive and close to +1, indicating that the results are robust to a metric that is not strongly established, such as the distance between pipelines. However, the results show, with significance $\alpha = 0.05$, that the correlation between the rankings is low, and hence the representation and distance significantly affect the FDC results using the global optimum as a reference, as indicated below:

$$\tau(ad\ hoc,\ \text{hamming}) = 0.4421 \quad \text{p-value}=0.0983$$
$$\tau(ad\ hoc,\ \text{euclidean}) = 0.5263 \quad \text{p-value}=0.8227$$
$$\tau(\text{hamming, euclidean}) = 0.3263 \quad \text{p-value}=0.1126$$

In conclusion, FDC is highly affected by representation and distance definitions. It is interesting to observe that the lower correlation is from the one-hot encoding to the t-SNE representation, although one derivate from the other. Also observe that the *ad hoc* and Euclidean distances present the highest correlation, although it is barely above 0.5.

6.2 Dispersion Metric

The DM measure gives an indication of how the solutions with top-N fitness are distributed in the search space. The experiments were performed using samples of 1,000 and 5,000 solutions with thresholds of 0.01, 0.05, and 0.1. Figure 6 presents the value of the metric with the largest sample size, i.e., containing 5,000 solutions. Each group is formed by an instance and each bar represents the result of the metric for a different threshold.

Recall that the value of this metric represents the difference between the average of the distances between the top-N solutions, $S_{\mathcal{F}}^*$, with the other solutions $S_{\mathcal{F}}$ in the solution space. A large negative value indicates that the distance between the solutions in $S_{\mathcal{F}}^*$ are closer (concentrated) in the space than the others. However, observe that, for some datasets, the results change completely depending on the distance metric used: the metric goes from a positive to a negative value only by varying the way the distance is calculated, indicating that dispersion, as well as FDC, is highly influenced by the adopted solution representation.

Figure 6 shows the values of DM using the three different types of distances. Note that, when using the *adhoc* or Euclidean distances, the values are relatively distributed between positive and negative. However, when the hamming distance is applied (Fig. 6b), the metric value is predominantly negative. In all cases the

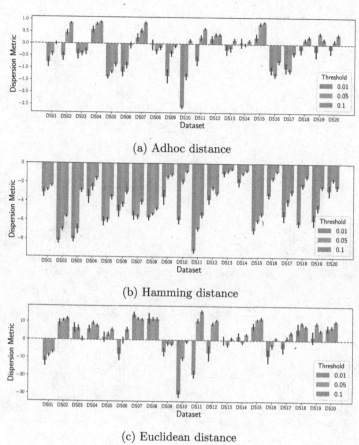

(a) Adhoc distance

(b) Hamming distance

(c) Euclidean distance

Fig. 6. Results of the dispersion metric (DM).

magnitude of the metric varies a lot: the values of the *ad hoc* distance are in the range $(-2.88, 1.27)$, while in the hamming distance, the range is $(-10.18, 0.065)$, and in the Euclidean distance, $(-41.71, 24.52)$.

DS10 presents the highest value in the ad-hoc and Euclidean methods, indicating that at this distance the solutions with the highest fitness are "dispersed" in the solution space (according to the distance metric). In the case of the hamming distance, DS11 presents the highest value for DM. This indicates that even receiving the same input data, the results vary completely due to the choice of distance.

A statistical test (ANOVA) was used to verify the difference between the means of the three distances used. For each threshold, a test was performed considering 3 samples, one for each distance calculation method. For example, for threshold 0.01, 3 samples were tested, each containing 20 values (DM for the fitness landscape of each dataset). The results show that the p-values obtained in each test were 0.06226, 1.2723×10^{-6} and 9.8619×10^{-6} for thresholds 0.01, 0.05

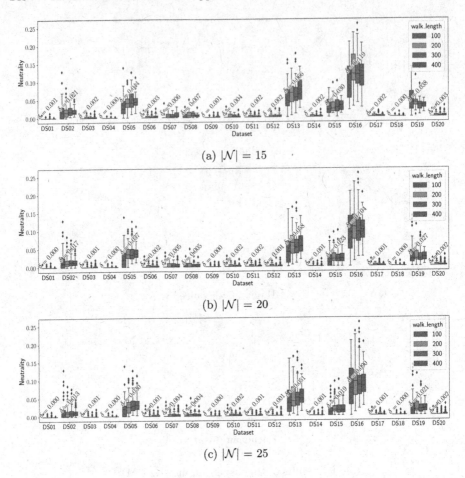

(a) $|\mathcal{N}| = 15$

(b) $|\mathcal{N}| = 20$

(c) $|\mathcal{N}| = 25$

Fig. 7. Space neutrality measured with different walk lengths.

and 0.10, respectively. Thus, considering a sensitivity of 0.05, the null hypothesis that the samples have the same mean can be rejected. Therefore, it is possible to conclude that the results obtained are statistically different due to the choice of distances and representations.

6.3   Neutrality Rate

Finally, we measured the neutrality rate of the space using random walks containing 100, 200, 300, and 400 solutions. The value of δ, the tolerance for considering whether a configuration is neutral, was defined as the standard deviation of the fitness mean of 30 random walks of size 1,000, as used in the experiments performed by [12].

The results are shown in Fig. 7. Observe that datasets DS16, DS13 and DS05 stand out for having greater neutrality than the others for all walk lengths.

However, the dataset with the highest number of repeated fitness, on average, is DS08 – where each of the 71 non-unique fitness occurs approximately 985 times. Note that the number of repeated fitness values does not directly imply space neutrality, since the solutions may not be neighbors. From the 3 datasets with highest values of neutrality, DS16 ranks 14 in terms of repeated solutions. Another factor that justifies this result is the fact that DS16 has the highest fitness variance, which affects the tolerance to consider the neutrality of the neighborhood.

Also notice that increasing the neighborhood size affects the neutrality of space, as can be seen in Figs. 7b and 7c. This is intuitive, as the more neighbors the higher the probability that one has a fitness greater than δ. In addition, note that datasets DS13, DS05, and DS02 are the ones that presented the highest neutrality according to this metric.

7 Conclusions and Future Work

This paper looked at the impact that different representations and distance metrics have on fitness landscapes of AutoML problems. We have investigated the use of three metrics, namely Fitness Distance Correlation, Dispersion Metric, and Neutrality rate, with various parameter configurations and three different representation and distance metrics for the search space.

First, we showed that the representation of the space does not necessarily change the relative distances between solutions but certainly modifies the notion of neighborhood. We also showed that traditional metrics for exploring fitness landscapes are not capable of dealing with this complex, structured AutoML space.

Finally, we believe that new metrics that account for the underlying problem being solved with AutoML may be necessary to help characterise these search spaces, as they will primarily depend on the difficulty of the dataset being investigated and not only the structure of the conditional space determined by the possible components of a machine learning pipeline.

Acknowledgements. This work was supported by FAPEMIG (through grant no. CEX-PPM-00098-17), MPMG (through the project Analytical Capabilities), CNPq (through grant no. 310833/2019-1), and CAPES.

References

1. Cleghorn, C.W., Ochoa, G.: Understanding parameter spaces using local optima networks. In: Proceedings of the Genetic and Evolutionary Computation Conference Companion, pp. 1657–1664 (2021)
2. Garciarena, U., Santana, R., Mendiburu, A.: Analysis of the complexity of the automatic pipeline generation problem. In: IEEE Congress on Evolutionary Computation, pp. 1–8 (2018)
3. Hutter, F., Kotthoff, L., Vanschoren, J. (eds.): Automated Machine Learning: Methods, Systems, Challenges. Springer, Cham (2018). http://automl.org/book

4. Jones, T., Forrest, S., et al.: Fitness distance correlation as a measure of problem difficulty for genetic algorithms. In: ICGA, vol. 95, pp. 184–192 (1995)
5. Lunacek, M., Whitley, D.: The dispersion metric and the CMA evolution strategy. In: Proceedings of the Conference on Genetic and Evolutionary Computation (2006)
6. Malan, K., Engelbrecht, A.P.: A survey of techniques for characterising fitness landscapes and some possible ways forward. Inf. Sci. **241**, 148–163 (2013)
7. Malan, K.M.: A survey of advances in landscape analysis for optimisation. Algorithms **14**(2) (2021)
8. Miller, F., Vandome, A., John, M.: Kendall Tau Rank Correlation Coefficient. VDM Publishing (2010)
9. Nunes, M., Fraga, P.M., Pappa, G.L.: Fitness landscape analysis of graph neural network architecture search spaces. In: Proceedings of the Genetic and Evolutionary Computation Conference, pp. 876–884 (2021)
10. Ochoa, G., Veerapen, N.: Neural architecture search: a visual analysis. In: Rudolph, G., Kononova, A.V., Aguirre, H., Kerschke, P., Ochoa, G., Tušar, T. (eds.) Parallel Problem Solving from Nature, pp. 603–615. Springer, Cham (2022). https://doi.org/10.1007/978-3-031-14714-2_42
11. Pedregosa, F., et al.: Scikit-learn: machine learning in python. J. Mach. Learn. Res. **12**, 2825–2830 (2011)
12. Pimenta, C.G., de Sá, A.G.C., Ochoa, G., Pappa, G.L.: Fitness landscape analysis of automated machine learning search spaces. In: Paquete, L., Zarges, C. (eds.) EvoCOP 2020. LNCS, vol. 12102, pp. 114–130. Springer, Cham (2020). https://doi.org/10.1007/978-3-030-43680-3_8
13. Pushak, Y., Hoos, H.: Algorithm configuration landscapes. In: Auger, A., Fonseca, C.M., Lourenço, N., Machado, P., Paquete, L., Whitley, D. (eds.) PPSN 2018. LNCS, vol. 11102, pp. 271–283. Springer, Cham (2018). https://doi.org/10.1007/978-3-319-99259-4_22
14. Pushak, Y., Hoos, H.: Automl loss landscapes. ACM Trans. Evol. Learn. Optim. **2**(3) (2022)
15. Reidys, C.M., Stadler, P.F.: Neutrality in fitness landscapes. Appl. Math. Comput. **117**(2–3), 321–350 (2001)
16. Richter, H.: Fitness landscapes: from evolutionary biology to evolutionary computation. In: Richter, H., Engelbrecht, A. (eds.) Recent Advances in the Theory and Application of Fitness Landscapes. ECC, vol. 6, pp. 3–31. Springer, Heidelberg (2014). https://doi.org/10.1007/978-3-642-41888-4_1
17. Rodrigues, N.M., Malan, K.M., Ochoa, G., Vanneschi, L., Silva, S.: Fitness landscape analysis of convolutional neural network architectures for image classification. Inf. Sci. **609**, 711–726 (2022)
18. Teixeira, M.C., Pappa, G.L.: Understanding AutoML search spaces with local optima networks. In: Genetic and Evolutionary Computation Conference (2022)
19. Traoré, K.R., Camero, A., Zhu, X.X.: Fitness Landscape Footprint: A Framework to Compare Neural Architecture Search Problems (2021). http://arxiv.org/abs/2111.01584
20. Treimun-Costa, G., Montero, E., Ochoa, G., Rojas-Morales, N.: Modelling parameter configuration spaces with local optima networks. In: Proceedings of the Genetic and Evolutionary Computation Conference, pp. 751–759 (2020)
21. Van Der Maaten, L., Hinton, G.: Visualizing data using t-SNE. J. Mach. Learn. Res. **9**, 2579–2625 (2008)

22. Vanneschi, L., Pirola, Y., Mauri, G., Tomassini, M., Collard, P., Verel, S.: A study of the neutrality of Boolean function landscapes in genetic programming. Theor. Comput. Sci. **425**, 34–57 (2012)
23. Witten, I.H., Frank, E.: Data Mining - Practical Machine Learning Tools and Techniques, 2nd edn. The Morgan Kaufmann Series in Data Management Systems (2005)
24. Zöller, M.A., Huber, M.F.: Benchmark and survey of automated machine learning frameworks. J. Artif. Intell. Res. **70**, 409–472 (2021)

Author Index

A
Akbay, Mehmet Anıl 16
Aranha, Claus 211

B
Blum, Christian 16, 82

C
Carlet, Claude 114
Chen, Gang 146

E
El Krari, Mehdi 1
Enderli, Cyrille 66
Escott, Kirita-Rose 146
Ettrich, Rupert 130

G
Gabonnay, Michal 34
Goudet, Olivier 66, 98
Grelier, Cyril 98
Guibadj, Rym Nesrine 1

H
Hao, Jin-Kao 66, 98
Huber, Marc 130

J
Jatschka, Thomas 50

K
Kalayci, Can Berk 16
Kaufmann, Marc 162
Kneissl, Carlo 179

L
Larcher, Maxime 162
Lavinas, Yuri 211

L
Lengler, Johannes 162
Liefooghe, Arnaud 211

M
Ma, Hui 146
Manzoni, Luca 114
Mariot, Luca 114

O
Ochoa, Gabriela 211

P
Pappa, Gisele L. 227
Picek, Stjepan 114
Pinacho-Davidson, Pedro 82
Pinsolle, Jean 66

R
Raidl, Günther R. 50, 130
Robilliard, Denis 1
Rodemann, Tobias 50
Rudová, Hana 34

S
Sassmann, Vojtěch 34
Sato, Hiroyuki 195
Sobotka, Václav 34
Sudholt, Dirk 179

T
Takadama, Keiki 195
Tanaka, Shoichiro 195
Teixeira, Matheus C. 227

W
Woodward, John 1

Z
Zou, Xun 162

L. Pérez Cáceres and T. Stützle (Eds.): EvoCOP 2023, LNCS 13987, p. 245, 2023.
https://doi.org/10.1007/978-3-031-30035-6

Printed in the United States
by Baker & Taylor Publisher Services

Printed in the United States
by Baker & Taylor Publisher Services